FOUNDATIONS OF COMPLEX SYSTEMS

Nonlinear Dynamics, Statistical Physics, Information and Prediction

FOUNDATIONS OF COMPLEX SYSTEMS

Nonlinear Dynamics, Statistical Physics, Information and Prediction

Gregoire Nicolis
University of Brussels, Belgium

Catherine Nicolis
Royal Meteorological Institute of Belgium, Belgium

World Scientific

NEW JERSEY · LONDON · SINGAPORE · BEIJING · SHANGHAI · HONG KONG · TAIPEI · CHENNAI

Published by

World Scientific Publishing Co. Pte. Ltd.
5 Toh Tuck Link, Singapore 596224
USA office: 27 Warren Street, Suite 401-402, Hackensack, NJ 07601
UK office: 57 Shelton Street, Covent Garden, London WC2H 9HE

British Library Cataloguing-in-Publication Data
A catalogue record for this book is available from the British Library.

FOUNDATIONS OF COMPLEX SYSTEMS
Nonlinear Dynamics, Statistical Physics, Information and Prediction

Copyright © 2007 by World Scientific Publishing Co. Pte. Ltd.

All rights reserved. This book, or parts thereof, may not be reproduced in any form or by any means, electronic or mechanical, including photocopying, recording or any information storage and retrieval system now known or to be invented, without written permission from the Publisher.

For photocopying of material in this volume, please pay a copying fee through the Copyright Clearance Center, Inc., 222 Rosewood Drive, Danvers, MA 01923, USA. In this case permission to photocopy is not required from the publisher.

ISBN-13 978-981-270-043-8
ISBN-10 981-270-043-9

To Helen, Stamatis and little Katy

Preface

Complexity became a major scientific field in its own right as recently as 15 years ago, and since then it has modified considerably the scientific landscape through thousands of high-impact publications as well as through the creation of specialized journals, Institutes, learned societies and University chairs devoted specifically to it. It constitutes today a paradigm for approaching a large body of phenomena of concern at the crossroads of physical, engineering, environmental, life and human sciences from a unifying point of view.

Nonlinear science and statistical physics had been addressing for some time phenomena of this kind: self-organization in nonequilibrium systems, glassy materials, pattern formation, deterministic chaos are landmarks, witnessing the success they have achieved in explaining how unexpected structures and events can be generated from the laws of nature in systems involving interacting subunits when appropriate conditions are satisfied - an issue closely related to the problematics of complexity. And yet, on the one side, for quite some time these attempts were progressing in rather disconnected ways following their own momentum and success; and on the other side, they were remaining confined to a large extent within a community of strong background in physical and mathematical science, and did not incorporate to a sufficient degree insights from the practitioner confronted with naturally occurring systems where issues eliciting the idea of complexity show up in a most pressing way. Last but not least, there was a lack of insight and of illustrative power of just what are the minimal ingredients for observing the sort of behaviors that would qualify as "complex".

A first breakthrough that contributed significantly to the birth of complexity research occurred in the late 1980's - early 1990's. It arose from the cross-fertilization of ideas and tools from nonlinear science, statistical physics and numerical simulation, the latter being a direct offspring of the increasing availability of computers. By bringing chaos and irreversibility together it showed that deterministic and probabilistic views, causality and chance, stability and evolution were different facets of a same reality when addressing certain classes of systems. It also provided insights on the relative roles of the number of elements involved in the process and the nature of the underlying dynamics. Paul Anderson's well-known aphorism, "more is different", that contributed to the awareness of the scientific community on the relevance of complexity, is here complemented in a most interesting way.

The second breakthrough presiding in the birth of complexity coincides with the increasing input of fields outside the strict realm of physical science. The intrusion of concepts that were till then not part of the vocabulary of fundamental science forced a reassessment of ideas and practices. Predictability, in connection with the increasing concern about the evolution of the atmosphere, climate and financial activities; algorithms, information, symbols, networks, optimization in connection with life sciences, theoretical informatics, computer science, engineering and management; adaptive behavior and cognitive processes in connection with brain research, ethology and social sciences are some characteristic examples.

Finally, time going on, it became clear that generic aspects of the complex behaviors observed across a wide spectrum of fields could be captured by minimal models governed by simple local rules. Some of them gave rise in their computer implementation to attractive visualizations and deep insights, from Monte Carlo simulations to cellular automata and multi-agent systems.

These developments provided the tools and paved the way to an understanding, both qualitative and quantitative, of the complex systems encountered in nature, technology and everyday experience. In parallel, natural complexity acted as a source of inspiration generating progress at the fundamental level. Spontaneously, in a very short time interval complexity became in this way a natural reference point for all sorts of communities and problems. Inevitably, in parallel with the substantial progress achieved ambiguous statements and claims were also formulated related in one way or the other to the diversity of backgrounds of the actors involved and their perceptions as to the relative roles of hard facts, mechanisms, analogies and metaphors. As a result complexity research is today both one of the most active and fastest growing fields of science and a forum for the exchange of sometimes conflicting ideas and views cutting across scientific disciplines.

In this book the foundations of complex systems are outlined. The vision conveyed is that of complexity as a part of fundamental science, in which the insights provided by its cross-fertilization with other disciplines are incorporated. What is more, we argue that by virtue of this unique blending complexity ranks among the most relevant parts of fundamental science as it addresses phenomena that unfold on our own scale, phenomena in the course of which the object and the observer are co-evolving. A unifying presentation of the concepts and tools needed to analyze, to model and to predict complex systems is laid down and links between key concepts such as emergence, irreversibility, evolution, randomness and information are established in the light of the complexity paradigm. Furthermore, the interdisciplinary dimension of complexity research is brought out through representative examples. Throughout the presentation emphasis is placed on the need for a multilevel approach to complex systems integrating deterministic and probabilistic views, structure and dynamics, microscopic, mesoscopic and macroscopic level descriptions.

The book is addressed primarily to graduate level students and to researchers in physics, mathematics and computer science, engineering, environmental and life sciences, economics and sociology. It can constitute the material of a graduate-level course and we also hope that, outside the academic community, professionals interested in interdisciplinary issues will find some interest in its reading. The choice of material, the style and the coverage of the items reflect our concern to do justice to the multiple facets of complexity. There can be no "soft" approach to complexity: observing, monitoring, analyzing, modeling, predicting and controlling complex systems can only be achieved through the time-honored approach provided by "hard" science. The novelty brought by complex systems is that in this endeavor the goals are reassessed and the ways to achieve them are reinvented in a most unexpected way as compared to classical approaches.

Chapter 1 provides an overview of the principal manifestations of complexity. Unifying concepts such as instability, sensitivity, bifurcation, emergence, self-organization, chaos, predictability, evolution and selection are sorted out in view of later developments and the need for a bottom-up approach to complexity is emphasized. In Chapter 2 the basis of a deterministic approach to the principal behaviors characteristic of the phenomenology of complex systems at different levels of description is provided, using the formalism of nonlinear dynamical systems. The fundamental mechanism underlying emergence is identified. At the same time the limitations of a universal description of complex systems within the framework of a deterministic approach are revealed and the "open future" character of their evolution is highlighted. Some prototypical ways to model complexity in physical science and beyond are also discussed, with emphasis on the role of the coupling between constituting elements. In Chapter 3 an analysis incorporating the probabilistic dimension of complex systems is carried out. It leads to some novel ways to characterize complex systems, allows one to recover universal trends in their evolution and brings out the limitations of the deterministic description. These developments provide the background for different ways to simulate complex systems and for understanding the relative roles of dynamics and structure in their behavior. The probabilistic approach to complexity is further amplified in Chapter 4 by the incorporation of the concepts of symbolic dynamics and information. A set of entropy-like quantities is introduced and their connection with their thermodynamic counterparts is discussed. The selection rules presiding the formation of complex structures are also studied in terms of these quantities and the nature of the underlying dynamics. The stage is thus set for the analysis of the algorithmic aspects of complex systems and for the comparison between algorithmic complexity as defined in theoretical computer science and natural complexity.

Building on the background provided by Chapters 1 to 4, Chapter 5 addresses "operational" aspects of complexity, such as monitoring and data analysis approaches targeted specifically to complex systems. Special emphasis is placed on the mechanisms underlying the propagation of prediction errors and the existence of a limited predictability horizon. The chapter ends with a discussion of recurrence and extreme events, two prediction-oriented topics of increasing concern. Finally, in Chapter 6 complexity is shown "in action" on a number of selected topics. The choices made in this selection out of the enormous number of possibilities reflect our general vision of complexity as part of fundamental science but also, inevitably, our personal interests and biases. We hope that this coverage illustrates adequately the relevance and range of applicability of the ideas and tools outlined in the book. The chapter ends with a section devoted to the epistemological aspects of complex systems. Having no particular background in epistemology we realize that this is a risky enterprise, but we feel that it cannot be dispensed with in a book devoted to complexity. The presentation of the topics of this final section is that of the practitioner of physical science, and contains only few elements of specialized jargon in a topic that could by itself give rise to an entire monograph.

In preparing this book we have benefitted from discussions with, comments and help in the preparation of figures by Y. Almirantis, V. Basios, A. Garcia Cantu, P. Gaspard, M. Malek Mansour, J. S. Nicolis, S. C. Nicolis, A. Provata, R. Thomas and S. Vannitsem. S. Wellens assumed the hard task of typing the first two versions of the manuscript.

Our research in the subject areas covered in this book is sponsored by The University of Brussels, the Royal Meteorological Institute of Belgium, the Science Policy Office of the Belgian Federal Government, the European Space Agency and the European Commission. Their interest and support are gratefully acknowledged.

G. Nicolis, C. Nicolis
Brussels, February 2007

Contents

Preface vii

1 The phenomenology of complex systems 1
1.1 Complexity, a new paradigm 1
1.2 Signatures of complexity 3
1.3 Onset of complexity 5
1.4 Four case studies 8
    1.4.1 Rayleigh-Bénard convection 8
    1.4.2 Atmospheric and climatic variability 11
    1.4.3 Collective problem solving: food recruitment in ants . . 15
    1.4.4 Human systems 19
1.5 Summing up 23

2 Deterministic view 25
2.1 Dynamical systems, phase space, stability 25
    2.1.1 Conservative systems 27
    2.1.2 Dissipative systems 27
2.2 Levels of description 34
    2.2.1 The microscopic level 34
    2.2.2 The macroscopic level 36
    2.2.3 Thermodynamic formulation 38
2.3 Bifurcations, normal forms, emergence 41
2.4 Universality, structural stability 46
2.5 Deterministic chaos 49
2.6 Aspects of coupling-induced complexity 53
2.7 Modeling complexity beyond physical science 59

3 The probabilistic dimension of complex systems 64
3.1 Need for a probabilistic approach 64
3.2 Probability distributions and their evolution laws 65
3.3 The retrieval of universality 72

3.4		The transition to complexity in probability space	77
3.5		The limits of validity of the macroscopic description	82
	3.5.1	Closing the moment equations in the mesoscopic description	82
	3.5.2	Transitions between states	84
	3.5.3	Average values versus fluctuations in deterministic chaos	88
3.6		Simulating complex systems	90
	3.6.1	Monte Carlo simulation	91
	3.6.2	Microscopic simulations	92
	3.6.3	Cellular automata	94
	3.6.4	Agents, players and games	95
3.7		Disorder-generated complexity	96

4 Information, entropy and selection   101

4.1		Complexity and information	101
4.2		The information entropy of a history	104
4.3		Scaling rules and selection	106
4.4		Time-dependent properties of information. Information entropy and thermodynamic entropy	115
4.5		Dynamical and statistical properties of time histories. Large deviations, fluctuation theorems	117
4.6		Further information measures. Dimensions and Lyapunov exponents revisited	120
4.7		Physical complexity, algorithmic complexity, and computation	124
4.8		Summing up: towards a thermodynamics of complex systems	128

5 Communicating with a complex system: monitoring, analysis and prediction   131

5.1		Nature of the problem	131
5.2		Classical approaches and their limitations	131
	5.2.1	Exploratory data analysis	132
	5.2.2	Time series analysis and statistical forecasting	135
	5.2.3	Sampling in time and in space	138
5.3		Nonlinear data analysis	139
	5.3.1	Dynamical reconstruction	139
	5.3.2	Symbolic dynamics from time series	143
	5.3.3	Nonlinear prediction	148
5.4		The monitoring of complex fields	151

	5.4.1	Optimizing an observational network 153
	5.4.2	Data assimilation 157
5.5	The predictability horizon and the limits of modeling 159
	5.5.1	The dynamics of growth of initial errors 160
	5.5.2	The dynamics of model errors 164
	5.5.3	Can prediction errors be controlled? 170
5.6	Recurrence as a predictor 171
	5.6.1	Formulation 172
	5.6.2	Recurrence time statistics and dynamical complexity 176
5.7	Extreme events 180
	5.7.1	Formulation 180
	5.7.2	Statistical theory of extremes 182
	5.7.3	Signatures of a deterministic dynamics in extreme events 185
	5.7.4	Statistical and dynamical aspects of the Hurst phenomenon 191

6 Selected topics 195

6.1	The arrow of time 195
	6.1.1	The Maxwell-Boltzmann revolution, kinetic theory, Boltzmann's equation 196
	6.1.2	First resolution of the paradoxes: Markov processes, master equation 200
	6.1.3	Generalized kinetic theories 202
	6.1.4	Microscopic chaos and nonequilibrium statistical mechanics 204
6.2	Thriving on fluctuations: the challenge of being small 208
	6.2.1	Fluctuation dynamics in nonequilibrium steady states revisited 210
	6.2.2	The peculiar energetics of irreversible paths joining equilibrium states 211
	6.2.3	Transport in a fluctuating environment far from equilibrium 214
6.3	Atmospheric dynamics 217
	6.3.1	Low order models 218
	6.3.2	More detailed models 222
	6.3.3	Data analysis 223
	6.3.4	Modeling and predicting with probabilities 224
6.4	Climate dynamics 226
	6.4.1	Low order climate models . . . . , ... 227

	6.4.2	Predictability of meteorological versus climatic fields	. 230
	6.4.3	Climatic change	233
6.5	Networks		235
	6.5.1	Geometric and statistical properties of networks	236
	6.5.2	Dynamical origin of networks	239
	6.5.3	Dynamics on networks	244
6.6	Perspectives on biological complexity		247
	6.6.1	Nonlinear dynamics and self-organization at the biochemical, cellular and organismic level	249
	6.6.2	Biological superstructures	251
	6.6.3	Biological networks	253
	6.6.4	Complexity and the genome organization	260
	6.6.5	Molecular evolution	263
6.7	Equilibrium versus nonequilibrium in complexity and self-organization		267
	6.7.1	Nucleation	268
	6.7.2	Stabilization of nanoscale patterns	272
	6.7.3	Supramolecular chemistry	274
6.8	Epistemological insights from complex systems		276
	6.8.1	Complexity, causality and chance	277
	6.8.2	Complexity and historicity	279
	6.8.3	Complexity and reductionism	283
	6.8.4	Facts, analogies and metaphors	285

Color plates   **287**

Suggestions for further reading   **291**

Index   **321**

> Τό πᾶν ἀλλ' ἔστι τι τό ὅλον παρά
> τά μόρια
> The whole is more than the sum
> of its parts
>
> Aristotle Metaphysica 1045a

Chapter 1

The phenomenology of complex systems

1.1 Complexity, a new paradigm

Complexity is part of our ordinary vocabulary. It has been used in everyday life and in quite different contexts for a long time and suddenly, as recently as 15 years ago it became a major field of interdisciplinary research that has since then modified considerably the scientific landscape. What is in the general idea of complexity that was missing in our collective knowledge -one might even say, in our collective consciousness- which, once recognized, conferred to it its present prominent status? What makes us designate certain systems as "complex" distinguishing them from others that we would not hesitate to call "simple", and to what extent could such a distinction be the starting point of a new approach to a large body of phenomena at the crossroads of physical, engineering, environmental, life and human sciences?

For the public and for the vast majority of scientists themselves science is usually viewed as an algorithm for predicting, with a theoretically unlimited precision, the future course of natural objects on the basis of their present state. Isaac Newton, founder of modern physics, showed more than three centuries ago how with the help of a few theoretical concepts like the law of universal gravitation, whose statement can be condensed in a few lines, one can generate data sets as long as desired allowing one to interpret the essence of the motion of celestial bodies and predict accurately, among others, an eclipse of the sun or of the moon thousands of years in advance. The impact of this historical achievement was such that, since then, scientific thinking has been dominated by the *Newtonian paradigm* whereby the world is reducible to a few fundamental elements animated by a regular, reproducible and hence predictable behavior: a world that could in this sense be qualified as fundamentally simple.

During the three-century reign of the Newtonian paradigm science reached a unique status thanks mainly to its successes in the exploration of the very small and the very large: the atomic, nuclear and subnuclear constitution of matter on the one side; and cosmology on the other. On the other hand man's intuition and everyday experience are essentially concerned with the intermediate range of phenomena involving objects constituted by a large number of interacting subunits and unfolding on his own, macroscopic, space and time scales. Here one cannot avoid the feeling that in addition to regular and reproducible phenomena there exist other that are, manifestly, much less so. It is perfectly possible as we just recalled to predict an eclipse of the sun or of the moon thousands of years in advance but we are incapable of predicting the weather over the region we are concerned more than a few days in advance or the electrical activity in the cortex of a subject a few minutes after he started performing a mental task, to say nothing about next day's Dow Jones index or the state of the planet earth 50 years from now. Yet the movement of the atmosphere and the oceans that governs the weather and the climate, the biochemical reactions and the transport phenomena that govern the functioning of the human body and underlie, after all, human behavior itself, obey to the same dispassionate laws of nature as planetary motion.

It is a measure of the fascination that the Newtonian paradigm exerted on scientific thought that despite such indisputable facts, which elicit to the observer the idea of "complexity", the conviction prevailed until recently that the irregularity and unpredictability of the vast majority of phenomena on our scale are not authentic: they are to be regarded as temporary drawbacks reflecting incomplete information on the system at hand, in connection with the presence of a large number of variables and parameters that the observer is in the practical impossibility to manage and that mask some fundamental underlying regularities.

If evidence on complexity were limited to the intricate, large scale systems of the kind mentioned above one would have no way to refute such an assertion and fundamental science would thus have nothing to say on complexity. But over the years evidence has accumulated that quite ordinary systems that one would tend to qualify as "simple", obeying to laws known to their least detail, in the laboratory, under strictly controlled conditions, generate unexpected behaviors similar to the phenomenology of complexity as we encounter it in nature and in everyday experience: Complexity is not a mere metaphor or a nice way to put certain intriguing things, it is a phenomenon that is deeply rooted into the laws of nature, where systems involving large numbers of interacting subunits are ubiquitous.

This realization opens the way to a systematic search of the physical and mathematical laws governing complex systems. The enterprise was crowned with success thanks to a multilevel approach that led to the development of highly original methodologies and to the unexpected cross-fertilizations and blendings of ideas and tools from nonlinear science, statistical mechanics and thermodynamics, probability theory and numerical simulation. Thanks to the progress accomplished complexity is emerging as the new, post-Newtonian paradigm for a fresh look at problems of current concern. On the one side one is now in the position to gain new understanding, both qualitative and quantitative, of the complex systems encountered in nature and in everyday experience based on advanced modeling, analysis and monitoring strategies. Conversely, by raising issues and by introducing concepts beyond the traditional realm of physical science, natural complexity acts as a source of inspiration for further progress at the fundamental level. It is this sort of interplay that confers to research in complexity its unique, highly interdisciplinary character.

The objective of this chapter is to compile some representative facts illustrating the phenomenology associated with complex systems. The subsequent chapters will be devoted to the concepts and methods underlying the paradigm shift brought by complexity and to showing their applicability on selected case studies.

1.2 Signatures of complexity

The basic thesis of this book is that a system perceived as complex induces a characteristic phenomenology the principal signature of which is *multiplicity*. Contrary to elementary physical phenomena like the free fall of an object under the effect of gravity where a well-defined, single action follows an initial cause at any time, several outcomes appear to be possible. As a result the system is endowed with the capacity to choose between them, and hence to explore and to adapt or, more generally, to evolve. This process can be manifested in the form of two different expressions.

- The *emergence*, within a system composed of many units, of global traits encompassing the system as a whole that can in no way be reduced to the properties of the constituent parts and can on these grounds be qualified as "unexpected". By its non-reductionist character emergence has to do with the creation and maintenance of hierarchical structures in which the disorder and randomness that inevitably exist at the local level are controlled, resulting in states of order and long range coherence. We refer to this process as *self-organization*. A classical example of this behavior is provided by the communication and control networks in living matter, from the subcellular to the organismic level.

- The intertwining, within the same phenomenon, of large scale regularities and of elements of "surprise" in the form of seemingly erratic evolutionary events. Through this coexistence of order and disorder the observer is bound to conclude that the process gets at times out of control, and this in turn raises the question of the very possibility of its long-term prediction. Classical examples are provided by the all-familiar difficulty to issue satisfactory weather forecasts beyond a horizon of a few days as well as by the even more dramatic extreme geological or environmental phenomena such as earthquakes or floods.

If the effects generated by some underlying causes were related to these causes by a simple proportionality -more technically, by *linear* relationships- there would be no place for multiplicity. Nonlinearity is thus a necessary condition for complexity, and in this respect nonlinear science provides a natural setting for a systematic description of the above properties and for sorting out generic evolutionary scenarios. As we see later nonlinearity is ubiquitous in nature on all levels of observation. In macroscopic scale phenomena it is intimately related to the presence of feedbacks, whereby the occurrence of a process affects (positively or negatively) the way it (or some other coexisting process) will further develop in time. Feedbacks are responsible for the onset of cooperativity, as illustrated in the examples of Sec. 1.4.

In the context of our study a most important question to address concerns the transitions between states, since the question of complexity would simply not arise in a system that remains trapped in a single state for ever. To understand how such transitions can happen one introduces the concept of *control parameter*, describing the different ways a system is coupled to its environment and affected by it. A simple example is provided by a thermostated cell containing chemically active species where, depending on the environmental temperature, the chemical reactions will occur at different rates. Another interesting class of control parameters are those associated to a *constraint* keeping the system away of a state of equilibrium of some sort. The most clearcut situation is that of the state of thermodynamic equilibrium which, in the absence of phase transitions, is known to be unique and lack any form of dynamical activity on a large scale. One may then choose this state as a reference, switch on constraints driving the system out of equilibrium for instance in the form of temperature or concentration differences across the interface between the system and the external world, and see to what extent the new states generated as a response to the constraint could exhibit qualitatively new properties that are part of the phenomenology of complexity. These questions, which are at the heart of complexity theory, are discussed in the next section.

1.3 Onset of complexity

The principal conclusion of the studies of the response of a system to changes of a control parameter is that the onset of complexity is not a smooth process. Quite to the contrary, it is manifested by a cascade of transition phenomena of an explosive nature to which is associated the universal model of *bifurcation* and the related concepts of *instability* and *chaos*. These catastrophic events are not foreseen in the fundamental laws of physics in which the dependence on the parameters is perfectly smooth. To use a colloquial term, one might say that they come as a "surprise".

Figure 1.1 provides a qualitative representation of the foregoing. It depicts a typical evolution scenario in which, for each given value of a control parameter λ, one records a certain characteristic property of the system as provided, for instance, by the value of one of the variables X (temperature, chemical concentration, population density, etc.) at a given point. For values of λ less than a certain limit λ_c only one state can be realized. This state possesses in addition to uniqueness the property of *stability*, in the sense that the system is capable of damping or at least of keeping under control the influence of the external perturbations inflicted by the environment or of the internal fluctuations generated continuously by the locally prevailing disorder, two actions to which a natural system is inevitably subjected. Clearly, complexity has no place and no meaning under these conditions.

The situation changes radically beyond the critical value λ_c. One sees that if continued, the unique state of the above picture would become unstable: under the influence of external perturbations or of internal fluctuations the system responds now as an amplifier, leaves the initial "reference" state and is driven to one or as a rule to several new behaviors that merge to the previous state for $\lambda = \lambda_c$ but are differentiated from it for λ larger than λ_c. This is the phenomenon of bifurcation: a phenomenon that becomes possible thanks to the nonlinearity of the underlying evolution laws allowing for the existence of multiple solutions (see Chapter 2 for quantitative details). To understand its necessarily catastrophic character as anticipated earlier in this section it is important to account for the following two important elements.

(a) An experimental measurement -the process through which we com-

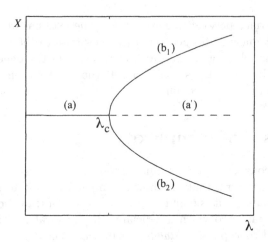

Fig. 1.1. A bifurcation diagram, describing the way a variable X characterizing the state of a system is affected by the variations of a control parameter λ. Bifurcation takes place at a critical value λ_c beyond which the original unique state (a) loses its stability, giving rise to two new branches of solutions (b_1) and (b_2).

municate with a system- is necessarily subjected to finite precision. The observation of a system for a given value of control parameter entails that instead of the isolated point of the λ axis in Fig. 1.1 one deals in reality with an "uncertainty ball" extending around this axis. The system of interest lies somewhere inside this ball but we are unable to specify its exact position, since for the observer all of its points represent one and the same state.

(b) Around and beyond the criticality λ_c we witness a selection between the states available that will determine the particular state to which the system will be directed (the two full lines surrounding the intermediate dotted one -the unstable branch in Fig. 1.1- provide an example). Under the conditions of Fig. 1.1 there is no element allowing the observer to determine beforehand this state. Chance and fluctuations will be the ones to decide. The system makes a series of attempts and eventually a particular fluctuation takes over. By stabilizing this choice it becomes a historical object, since its subsequent evolution will be conditioned by this critical choice. For the observer, this pronounced *sensitivity to the parameters* will signal its inability to predict the system's evolution beyond λ_c since systems within the uncertainty ball, to him identical in any respect, are differentiated and end up in states whose distance is much larger than the limits of resolution of the experimental measurement.

The Phenomenology of Complex Systems

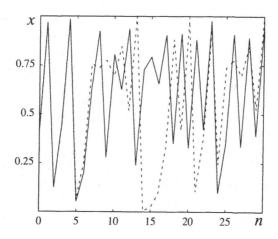

Fig. 1.2. Illustration of the phenomenon of sensitivity to the initial conditions in a model system giving rise to deterministic chaos. Full and dashed lines denote the trajectories (the set of successive values of the state variable X) emanating from two initial conditions separated by a small difference $\epsilon = 10^{-3}$.

We now have the basis of a mechanism of generation of complexity. In reality this mechanism is the first step of a cascade of successive bifurcations through which the multiplicity of behaviors may increase dramatically, culminating in many cases in a state in which the system properties change in time (and frequently in space as well) in a seemingly erratic fashion, not any longer because of external disturbances or random fluctuations as before but, rather, as a result of deterministic laws of purely intrinsic origin. The full line of Fig. 1.2 depicts a *time series* -a succession of values of a relevant variable in time- corresponding to this state of *deterministic chaos*. Its comparison with the dotted line reveals what is undoubtedly the most spectacular property of deterministic chaos, the *sensitivity to the initial conditions*: two systems whose initial states are separated by a small distance, smaller than the precision of even the most advanced method of experimental measurement, systems that will therefore be regarded by the observer as indistinguishable (see also point (a) above) will subsequently diverge in such a way that the distance between their instantaneous states (averaged over many possible initial states, see Chapters 2 and 3) will increase exponentially. As soon as this distance will exceed the experimental resolution the systems will cease to be indistinguishable for the observer. As a result, it will be impossible to predict their future evolution beyond this temporal horizon.

We here have a second imperative reason forcing us to raise the question of predictability of the phenomena underlying the behavior of complex systems.

All elements at our disposal from the research in nonlinear science and chaos theory lead to the conclusion that one cannot anticipate the full list of the number or the type of the evolutionary scenarios that may lead a system to complex behavior. In addition to their limited predictability complex systems are therefore confronting us with the fact that we seem to be stuck with a mode of description of a limited universality. How to reconcile this with the requirement that the very mission of science is to provide a universal description of phenomena and to predict their course? The beauty of complex systems lies to a great extent in that despite the above limitations this mission can be fulfilled, but that its realization necessitates a radical reconsideration of the concepts of universality and prediction. We defer a fuller discussion of this important issue to Chapters 2 and 3.

1.4 Four case studies

1.4.1 Rayleigh-Bénard convection

Consider a shallow layer of a fluid limited by two horizontal plates brought to identical temperatures. As prescribed by the second law of thermodynamics, left to itself the fluid will tend rapidly to a state where all its parts along the horizontal are macroscopically identical and where there is neither bulk motion nor internal differentiation of temperatures: $T = T_1 = T_2$, T_2 and T_1 being respectively the temperatures of the lower and upper plate. This is the state we referred to in Sec. 1.2 as the state of thermodynamic equilibrium.

Imagine now that the fluid is heated from below. By communicating to it in this way energy in the form of heat one removes it from the state of equilibrium, since the system is now submitted to a constraint $\Delta T = T_2 - T_1 > 0$, playing in this context the role of the control parameter introduced in Sec. 1.2. As long at ΔT remains small the flux of energy traversing the system will merely switch on a process of heat conduction, in which temperature varies essentially linearly between the hot (lower) zone and the cold (upper) one. This state is maintained thanks to a certain amount of energy that remains trapped within the system -one speaks of *dissipation*- but one can in no way speak here of complexity and emergence, since the state is unique and the differentiation observed is dictated entirely by the way the constraint has been applied: the behavior is as "simple" as the one in the state of equilibrium.

If one removes now the system progressively from equilibrium, by increas-

The Phenomenology of Complex Systems

Fig. 1.3. Rayleigh-Bénard convection cells appearing in a liquid maintained between a horizontal lower hot plate and an upper cold one, below a critical value of the temperature difference ΔT (see Color Plates).

ing ΔT, one suddenly observes, for a critical value ΔT_c, the onset of bulk motion in the layer. This motion is far from sharing the randomness of the motion of the individual molecules: the fluid becomes structured and displays a succession of cells along a direction transversal to that of the constraint, as seen in Fig. 1.3. This is the regime of thermal, or Rayleigh-Bénard convection. Now one is entitled to speak of complexity and emergence, since the spatial differentiation along a direction free from any constraint is the result of processes of internal origin specific to the system, maintained by the flow of energy communicated by the external world and hence by the dissipation. We have thus witnessed a particular manifestation of emergence, in the form of the birth of a *dissipative structure*. In a way, one is brought from a static, geometric view of space, to one where space is modeled by the dynamical processes switched on within the system. One can show that the state of rest is stable below the threshold ΔT_c but loses its stability above it while still remaining a solution -in the mathematical sense of the term- of the evolution laws of the fluid. As for the state of thermal convection, it simply does not exist below ΔT_c and inherits above it the stability of the state of rest. For $\Delta T = \Delta T_c$ there is degeneracy in the sense that the two states merge. We here have a concrete illustration of the generic phenomenon of bifurcation introduced in Sec. 1.3, see Fig. 1.1. Similar phenomena are observed in a wide range of laboratory scale systems, from fluid mechanics to chemical kinetics, optics, electronics or materials science. In each case one encounters essentially the same phenomenology. The fact that this is taking place under perfectly well controlled conditions allows one to sort out common features and set up a quantitative theory, as we see in detail in the subsequent chapters.

A remarkable property of the state of thermal convection is to possess a characteristic space scale -the horizontal extent of a cell (Fig. 1.3) related, in turn, to the depth of the layer. The appearance of such a scale reflects the fact that the states generated by the bifurcation display *broken symmetries*. The laws of fluid dynamics describing a fluid heated from below and contained between two plates that extend indefinitely in the horizontal direction remain invariant -or more plainly look identical- for all observers displaced to one another along this direction (translational invariance). This invariance property is shared by the state realized by the fluid below the threshold ΔT_c but breaks down above it, since a state composed of a succession of Bénard cells displays an intrinsic differentiation between its different parts that makes it less symmetrical than the laws that generated it. A differentiation of this sort may become in many cases one of the prerequisites for further complexification, in the sense that processes that would be impossible in an undifferentiated medium may be switched on. In actual fact this is exactly what is happening in the Rayleigh-Bénard and related problems. In addition to the first bifurcation described above, as the constraint increases beyond ΔT_c the system undergoes a whole series of successive transitions. Several scenarios have been discovered. If the horizontal extent of the cell is much larger than the depth the successive transition thresholds are squeezed in a small vicinity of ΔT_c. The convection cells are first maintained globally but are subsequently becoming fuzzy and eventually a regime of *turbulence* sets in, characterized by an erratic-looking variability of the fluid properties in space (and indeed in time as well). In this regime of extreme spatio-temporal chaos the motion is ordered only on a local level. The regime dominated by a characteristic space scale has now been succeeded by a *scale-free* state in which there is a whole spectrum of coexisting spatial modes, each associated to a different space scale. Similar phenomena arise in the time domain, where the first bifurcation may lead in certain types of systems to a strictly periodic clock-like state which may subsequently lose its coherence and evolve to a regime of deterministic chaos in which the initial periodicity is now part of a continuous spectrum of coexisting time scales.

As we see throughout this book states possessing a characteristic scale and scale-free states are described, respectively, by exponential laws and by power laws. There is no reason to restrict the phenomenology of complexity to the class of scale free states as certain authors suggest since, for one thing, coherence in living matter is often reflected by the total or partial synchronization of the activities of the individual cells to a dominant temporal or spatial mode.

In concluding this subsection it is appropriate to stress that configurations of matter as unexpected a priori as the Bénard cells, involving a number of molecules (each in disordered motion !) of the order of the Avogadro number $N \approx 10^{23}$ are born spontaneously, inevitably, at a modest energetic and informational cost, provided that certain conditions related to the nature of the system and the way it is embedded to its environment are fulfilled. Stated differently the overall organization is not ensured by a centralized planification and control but, rather, by the "actors" (here the individual fluid parcels) present. We refer to this process as the *bottom-up* mechanism.

1.4.2 Atmospheric and climatic variability

Our natural environment plays a central role in this book, not only on the grounds of its importance in man's everyday activities but also because it qualifies in any respect as what one intuitively means by complex system and forces upon the observer the need to cope with the problem of prediction. Contrary to the laboratory scale systems considered in the previous subsection we have no way to realize at will the successive transitions underlying its evolution to complexity. The best one can expect is that a monitoring in the perspective of the complex systems approach followed by appropriate analysis and modeling techniques, will allow one to constitute the salient features of the environment viewed as a dynamical system and to arrive at a quantitative characterization of the principal quantities of interest.

To an observer caught in the middle of a hurricane, a flood or a long drought the atmosphere appears as an irrational medium. Yet the atmospheric and climatic variables are far from being distributed randomly. Our environment is structured in both space and time, as witnessed by the stratification of the atmospheric layers, the existence of global circulation patterns such as the planetary waves, and the periodicities arising from the daily or the annual cycle. But in spite of this global order one observes a pronounced superimposed variability, reflected by marked deviations from perfect or even approximate regularity.

An example of such a variability is provided by the daily evolution of air temperature at a particular location (Fig. 1.4). One observes small scale irregular fluctuations that are never reproduced in an identical fashion, superimposed on the large scale regular seasonal cycle of solar radiation. A second illustration of variability pertains to the much larger scale of global climate. All elements at our disposal show indeed that the earth's climate has undergone spectacular changes in the past, like the succession of glacial-

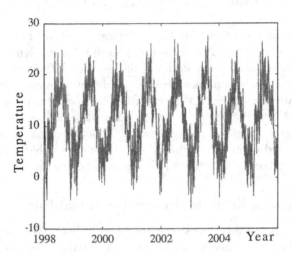

Fig. 1.4. Mean daily temperature at Uccle (Brussels) between January 1st, 1998 and December 31, 2006.

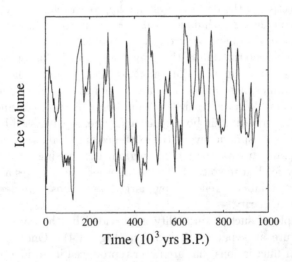

Fig. 1.5. Evolution of the global ice volume on earth during the last million years as inferred from oxygen isotope data.

interglacial periods. Figure 1.5 represents the variation of the volume of continental ice over the last million years as inferred from the evolution of the composition of marine sediments in oxygen 16 and 18 isotopes. Again, one is struck by the intermittent character of the evolution, as witnessed by a marked aperiodic component masking to a great extent an average time scale of 100 000 years that is sometimes qualified as the Quaternary glaciation "cycle". An unexpected corollary is that the earth's climate can switch between quite different modes over a short time in the geological scale, of the order of a few thousand years.

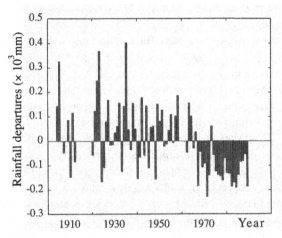

Fig. 1.6. Yearly rainfall departures from the long time average value at Kaédi (Mauritania) between 1904 and 1987.

Figure 1.6 depicts another example of climatic variability and regime switching, on a scale that is intermediate between those in Figs 1.4 and 1.5. It has to do with the time variation of the precipitation in western Sahel, and signals the onset of a regime of drought in this region, a phenomenon known to occur in several other areas of the globe. Again, one is struck by the irregular character of the process. The new element as compared to Figs 1.4 and 1.5 is that in the language of statistics the signal is no longer stationary: rather than succeeding each other without exhibiting a systematic trend, the states are here undergoing an abrupt transition between a regime of a quasi-normal and a weak rainfall that one can locate using traditional statistical analysis around the mid-1960's. It is likely that observations over a much longer time scale will reestablish the stationarity of the process, in the sense that the state of drought will sooner or later be succeeded by a quasi-normal state which will subsequently switch again to a state of drought, and so forth.

A fundamental consequence of the aperiodicity of the atmospheric and climate dynamics is the well-known difficulty to make reliable predictions. Contrary to simple periodic or multiperiodic phenomena for which a long term prediction is possible, predictions in meteorology are limited in time. The most plausible (and currently admitted) explanation is based on the realization that a small uncertainty in the initial conditions used in a prediction scheme (usually referred as "error") seems to be amplified in the course of the evolution. Such uncertainties are inherent in the process of experimental measurement, as pointed out already in Sec. 1.3. A great deal of effort is devoted in atmospheric sciences in the development of *data assimilation* techniques aiming to reduce them as much as possible (cf. also Sec. 5.4), but it is part of the laws of nature that they will never be fully eliminated. This brings us to the picture drawn in connection with Fig. 1.2, suggesting that the atmosphere displays sensitivity to the initial conditions because it is in a state of deterministic chaos. This conjecture seems to be compatible both with the analysis of the data available and with the modeling of atmospheric dynamics. This aspect is discussed more amply in Chapters 5 and 6, but one may already notice at this stage that much like experiment, modeling is also limited in practice by a finite resolution (of the order of several kilometers) and the concomitant omission of "subgrid" processes like e.g. local turbulence. Furthermore, many of the parameters are not known to a great precision. In addition to initial errors prediction must thus cope with *model errors*, reflecting the fact that a model is only an approximate representation of nature. This raises the problem of sensitivity to the parameters and brings us to the picture drawn in connection with Fig. 1.1. If the dynamics were simple like in the part of Fig. 1.1 left to λ_c neither of these errors would matter. But this is manifestly not the case. Initial and model errors can thus be regarded as probes revealing the fundamental instability and complexity underlying the atmosphere.

In all the preceding examples it was understood that the characteristic parameters of the atmosphere remained fixed. Over the last years there has been growing interest in the response of the weather and climate to changing parameter values - for instance, as a result of anthropogenic effects. In the representation of Fig. 1.1, the question would then be, whether the underlying dynamical system would undergo transitions to new regimes and if so, what would be the nature of the most plausible transition scenarios. This raises a whole new series of problems, some of which will be taken up in the sequel.

As pointed out earlier in this subsection, in certain environmental phenomena the variability is so considerable that no underlying regularity seems to be present. This property, especially pronounced in hydrology and in particular in the regime of river discharges, entails that the average and other quantifiers featured in traditional statistics are irrelevant. An ingenious way to handle such records, suggested some time ago by Harold Hurst, is to monitor the way the distance R between the largest and smallest value in a certain time window τ -usually referred to as *the range*- varies with τ. Actually, to deal with a dimensionless quantity one usually reduces R by the standard deviation C around the mean measured over the same interval. A most surprising result is that in a wide spectrum of environmental records R/C varies with τ as a power law of the form τ^H, where the *Hurst exponent* H turns out to be close to 0.70. To put this in perspective, for records generated by statistically independent processes with finite standard deviation, H is bound to be $1/2$ and for records where the variability is organized around a characteristic time scale there would simply not be a power law at all. Environmental dynamics provides therefore yet another example of the coexistence of phenomena possessing a characteristic scale and of scale free ones.

An interesting way to differentiate between these processes is to see how the law is changing upon a transformation of the variable (here the window τ). For an exponential law, switching from τ to $\lambda\tau$ (which can be interpreted as a change of scale in measuring τ) maintains the exponential form but changes the exponent multiplying τ, which provides the characteristic scale of the process, by a factor of λ. But in a power law the same transformation keeps the exponent H invariant, producing merely a multiplicative factor. We express this by qualifying this law as *scale invariant*. The distinction breaks down for nonlinear transformations, for which a power law can become exponential and vice versa.

As we see later deterministic chaos can be associated with variabilities of either of the two kinds, depending on the mechanisms presiding in its generation.

1.4.3 Collective problem solving: food recruitment in ants

In the preceding examples the elements constituting the system of interest were the traditional ones considered in physical sciences: molecules, volume elements in a fluid or in a chemical reagent, and so forth. In this subsection we are interested in situations where the actors involved are living organisms. We will see that despite this radical change, the principal manifestations of complexity will be surprisingly close to those identified earlier. Our discussion will focus on social insects, in particular, the process of food searching in ants.

Ants, like bees, termites and other social insects represent an enormous ecological success in biological evolution. They are known to be able to accomplish successfully number of collective activities such as nest construction, recruitment, defense etc. Until recently the view prevailed that in such highly non-trivial tasks individual insects behave as small, reliable automatons executing a well established genetic program. Today this picture is fading and replaced by one in which adaptability of individual behavior, collective interactions and environmental stimuli play an important role. These elements are at the origin of a two-scale process. One at the level of the individual, characterized by a pronounced probabilistic behavior, and another at the level of the society as a whole, where for many species despite the inefficiency and unpredictability of the individuals, coherent patterns develop at the scale of the entire colony.

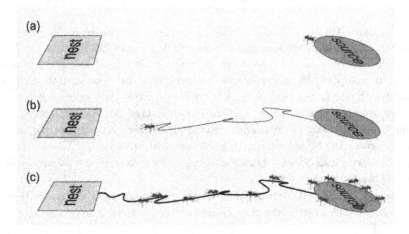

Fig. 1.7. Schematic representation of recruitment: (a) discovery of the food source by an individual; (b) return to the nest with pheromone laying; (c) the pheromone trail stimulates additional individuals to visit the source, which contribute to its reinforcement by further pheromone laying.

Let us see how these two elements conspire in the process of food searching by ants. Consider first the case where a single food source (for instance a saccharose solution) is placed close to the nest, as in Fig. 1.7 (here and in the sequel laboratory experiments emulating naturally occurring situations while allowing at the same time for detailed quantitative analyses are instru-

mental). A "scout" discovers the source in the course of a random walk. After feeding on the source it returns to the nest and deposits along the way a chemical signal known as *trail pheromone*, whose quantity is correlated to the sugar concentration in the source. Subsequently a process of *recruitment* begins in which two types of phenomena come into play:
- a first mechanism in which the scout-recruiter and/or the trail stimulate individuals that were till then inactive to go out of the nest;
- and a second one where the trail guides the individuals so recruited to the food source, entailing that as recruited individuals will sooner or later become recruiters in their turn the process will be gradually amplified and a substantial traffic will be established along the trail.

Consider now the more realistic situation where the colony disposes of several food sources. A minimal configuration allowing one to study how it then copes with the problem of choice to which it is confronted is depicted in Fig. 1.8a: two equivalent paths leading from the nest to two simultaneously present identical food sources. In a sufficiently numerous colony after a short period of equal exploitation a bifurcation, in the precise sense of Fig. 1.1, is then observed marking a preferential exploitation of one of the sources relative to the other, to its exhaustion (Fig. 1.8b). Thereafter the second source is fully colonized and its exploitation is intensified. When the colony is offered two sources with different sugar concentrations and the richest source is discovered before or at the same time as the poorer one, it is most heavily exploited. But when it is discovered after the poorer one, it is only weakly exploited. This establishes the primordial importance of the long-range cooperativity induced by the presence of the trail.

It is tempting to conjecture that far from being a curiosity the above phenomenon, which shares with the Rayleigh-Bénard instability the property of spontaneous emergence of an a priori highly unexpected behavioral pattern, is prototypical of a large class of systems, including socio-economic phenomena in human populations (see also Sec. 1.4.4 below). The key point lies in the realization that nature offers a bottom-up mechanism of organization that has no recourse to a central or hierarchical command process as in traditional modes of organization. This mechanism leads to collective decisions and to problem solving on the basis of (a) the local information available to each "agent"; and (b) its implementation on global level without the intervention of an information-clearing center. It opens the way to a host of applications in the organization of distributed systems of interacting agents as seen, for example, in communication networks, computer networks and networks of mobile robots or static sensory devices. Such analogy-driven considerations can stimulate new ideas in a completely different context by serving as archetypes. They are important elements in the process of model

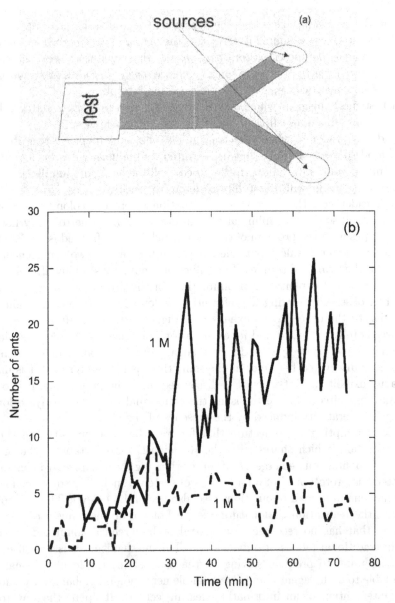

Fig. 1.8. (a) A typical experimental set up for the study of the process of choice between two options. (b) Time evolution of the number of individuals (here ants of the species *Lasius niger*) exploiting two equivalent (here 1 molar saccharose rich) food sources offered simultaneously, in an experimental set up of the type depicted in Fig. 1.8(a).

building -an essential part of the research in complex systems- in situations in which the evolution laws of the variables involved may not be known to any comparable degree of detail as in physical systems.

1.4.4 Human systems

We now turn to a class of complexity related problems in which the actors involved are human beings. Here the new element that comes into play is the presence of such concepts as strategy, imitation, anticipation, risk assessment, information, history, quite remote at first sight from the traditional vocabulary of physical science. The expectation would be that thanks to the rationality underlying these elements, the variability and unpredictability should be considerably reduced. The data at our disposal show that this is far from being the case. Human systems provide, in fact, one of the most authentic prototypes of complexity. They also constitute a source of inspiration for raising number of new issues, stimulating in turn fundamental research in the area of complex systems.

A first class of instances pertains to cooperativity (imitation) driven sociocultural phenomena. They usually lead to bifurcations very similar to those considered in the previous subsection in which the variability inherent in the dynamics of the individuals is eventually controlled to yield an emergent pattern arising through a sharp transition in the form of a bifurcation. The propagation of rumors or of opinions is the most classical example in this area, but in recent years some further unexpected possibilities have been suggested, such as the genesis of a phonological system in a human society. Ordinarily, the inherent capacity of humans to emit and recognize sounds and to attribute them to objects is advanced as the most plausible mechanism of this process. On the other hand, consider a population of N individuals capable to emit M sounds to designate a given object. When two individuals pronouncing sounds i and j meet, each one of them can convince, with certain probabilities, the other that his sound is more appropriate to designate the object. This switches on a cooperativity in the process of competition between the options available very similar to that between the two trails in Fig. 1.8a, leading to the choice of one of them by the overwhelming part of the population (being understood that N is large enough). This scenario opens interesting perspectives, which need to be implemented by linguistic analyses and real-time experiments.

Competition between different options is also expected to underlie the origin of a variety of spatial patterns and organizational modes observed in human systems. An example is provided by the formation and the evolution of urban structures, as certain areas specialize in specific economic activities and as residential differentiation produces neighborhoods differing in their living conditions and access to jobs and services. In many cases this occurs as a spontaneous process of endogenous origin. In addition to this evolutionary scenario central planning may be present as well and provide a bias in the individual decision making. It is, however, most unlikely that under present conditions it will supersede the bottom-up mechanism operating in complex systems: the chance of a modern Deinokrates or a modern Constantine the Great designing from scratch an Alexandria or a Constantinople-like structure are nowadays practically nil.

It is, perhaps, in the domain of economic and financial activities that the specificity of the human system finds its most characteristic expression. In addition to steps involving self-organization and emergence through bifurcation one witnesses here the entrance in force of the second fingerprint of complexity, namely, the intertwining of order and disorder. This raises in turn the problem of prediction in a most acute manner. The economics of the stock market provides a striking example. On October 19, 1987 the Dow Jones index of New York stock exchange dropped by 22.6%. This drop, the highest registered ever in a single day, was preceded by three other substantial ones on October 14, 15, 16. Impressive as they are, such violent phenomena are far from being unique: financial history is full of stock market crises such as the famous October 1929 one in which on two successive days the values were depreciated cumulatively by 23.1%.

The first reaction that comes to mind when witnessing these events is that of irrationality yet, much like in our discussion of subsection 1.4.2, the evidence supports on the contrary the idea of perfectly rational attitudes being at work. Ideally, in a market a price should be established by estimating the capacity of a company to make benefits which depends in turn on readily available objective data such as its technological potential, its developmental strategy, its current economic health and the quality of its staff. In reality, observing the market one realizes that for a given investor these objective criteria are in many instances superseded by observing the evolution of the index in the past and, especially, by watching closely the attitude of the other investors at the very moment of action. This may lead to strong cooperative effects in which a price results in from an attitude adopted at a certain time, and is subsequently affecting (e.g. reinforcing) this very attitude (which was perhaps initially randomly generated). As a matter of fact this largely endogenous mechanism seems to be operating not only during major crises but also under "normal" conditions, as illustrated by Fig. 1.9 in which the "real" (full line) versus the "objective" (dashed line) value of a certain product in the New York stock exchange is depicted for a period of about 50 years. It may result in paradoxical effects such as the increase of

The Phenomenology of Complex Systems 21

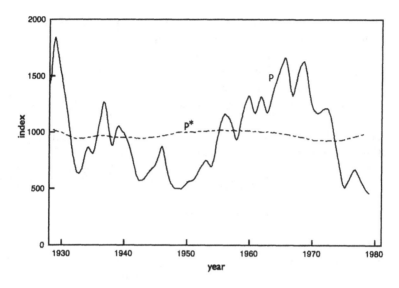

Fig. 1.9. Dow Jones industrial average p and a posteriori estimated rational price p^* of the New York stock market during the period 1928 to 1979. Raw data have been detrended by dividing by the systematic growth factor.

a certain value merely because the investors anticipate at a certain moment that this is indeed going to happen, though it has not happened yet! In this logic the product that is supposed to guarantee this high value might even be inferior to others, less well quoted ones. That such a priori unexpected events actually occur with appreciable probability is reminiscent of the comments made in subsections 1.4.1 and 1.4.3 in connection with the emergence of Rayleigh-Bénard cells and pheromone trails. It suggests that key manifestations of economic activities are the result of constraints acting on the system and activating intrinsic nonlinearities, as a result of which the concept of economic equilibrium often becomes irrelevant. Of equal importance is also the variability of the individual agents, reflected by the presence of different goals and strategies amongst them (cf. also Sec. 3.7).

It is important to realize that the speculative character of the process underlying Fig. 1.9 coexists with regular trends reflected by the generally admitted existence of economic cycles. While the latter are manifested on a rather long time scale, the behavior on a wide range covering short to intermediate scales seems rather to share the features of a scale free process. Again the situation looks similar in this respect to that encountered in the previous subsections. An analysis of the range of variability normalized by its standard deviation confirms this, with Hurst exponents H close to 0.5 for products most easily subject to speculation, and higher for products that are less negotiable. As mentioned in connection with subsection 1.4.2 this implies that the corresponding processes are, respectively, uncorrelated and subjected to long range correlations.

An alternative view of financial fluctuations is provided by the construction of their histograms from the available data. Let P_t be the present price of a given stock. The stock price return r_t is defined as the change of the logarithm of the stock price in a given time interval Δt, $r_t = \ln P_t - \ln P_{t-\Delta t}$. The probability that a return is (in absolute value) larger than x is found empirically to be a power law of the form

$$P(|r_t| > x) \approx x^{-\gamma_t} \qquad (1.1)$$

with $\gamma_t \approx 3$. This law which belongs to the family of probability distributions known as Pareto distributions holds for about 80 stocks with Δt ranging from one minute to one month, for different time periods and for different sizes of stocks. It may thus be qualified as "universal" in this precise sense. The scale invariant (in Δt and in size) behavior that it predicts in the above range suggests that large deviations can occur with appreciable probability, much more appreciable from what would be predicted by an exponential or a Gaussian distribution. As a matter of fact such dramatic events as the 1929 and 1987 market crashes conform to this law. Surprisingly, Pareto's law seems also to describe the distribution of incomes of individuals in a country, with an exponent that is now close to 1.5.

In an at first sight quite different context, power laws concomitant to self-similarity and scale free behavior are also present whenever one attempts to rank objects according to a certain criterion and counts how the frequency of their occurrence depends on the rank. For instance, if the cities of a given country are ranked by the integers 1, 2, 3,... according to the decreasing order of population size, then according to an empirical discovery by George Zipf the fraction of people living in the nth city varies roughly as

$$P(n) \approx n^{-1} \qquad (1.2)$$

Zipf has found a similar law for the frequency of appearance of words in the English prose, where $P(n)$ represents now the relative frequency of the nth most frequent word ("the", "of", "and" and "to" being the four successively more used words in a ranking that extends to 10 000 or so).

Eq. (1.2) is parameter free, and on these grounds one might be tempted to infer that it applies universally to all populations and to all languages. Benoît Mandelbrot has shown that this is not the case and proposed a two-parameter extension of Zipf's law accounting for the differences between subjects and languages, in the form

$$P(n) \approx (n + n_0)^{-B} \qquad (1.3)$$

where n_0 plays the role of a cutoff.

1.5 Summing up

The fundamental laws of nature governing the structure of the building blocks of matter and their interactions are deterministic: a system whose state is initially fully specified will follow a unique course. Yet throughout this chapter we have been stressing multiplicity as the principal manifestation of complexity; and have found it natural -and necessary- to switch continuously on many occasions between the deterministic description of phenomena and a probabilistic view.

Far from reflecting the danger of being caught in a contradiction already at the very start of this book this opposition actually signals what is going to become the leitmotiv of the chapters to come, namely, that when the fundamental laws of nature are implemented on complex systems the deterministic and the probabilistic dimensions become two facets of the same reality: because of the limited predictability of complex systems in the sense of the traditional description of phenomena one is forced to adopt an alternative view, and the probabilistic description offers precisely the possibility to sort out regularities of a new kind; but on the other side, far from being applied in a heuristic manner in which observations are forced to fit certain a priori laws imported from traditional statistics, the probabilistic description one is dealing with here is intrinsic in the sense that it is generated by the underlying dynamics. Depending on the scale of the phenomenon, a complex system may have to develop mechanisms for controlling randomness in order to sustain a global behavioral pattern thereby behaving deterministically or, on the contrary, to thrive on randomness in order to acquire transiently the variability and flexibility needed for its evolution between two such configurations.

Similarly to the determinism versus randomness, the structure versus dynamics dualism is also fading as our understanding of complex systems is improving. Complex systems shape in many respects the geometry of the space in which they are embedded, through the dynamical processes that they generate. This intertwining can occur on the laboratory time scale as in the Rayleigh-Bénard cells and the pheromone trails (1.4.1, 1.4.3); or on the much longer scale of geological or biological evolution, as in e.g. the composition of the earth's atmosphere or the structure of biomolecules.

Complexity is the conjunction of several properties and, because of this, no single formal definition doing justice to its multiple facets and manifestations can be proposed at this stage. In the subsequent chapters a multilevel approach capable of accounting for these diverse, yet tightly intertwined elements will be developed. The question of complexity definition(s) will be taken up again in the end of Chapter 4.

Chapter 2

Deterministic view

2.1 Dynamical systems, phase space, stability

Complexity finds its natural expression in the language of the theory of dynamical systems. Our starting point is to observe that the knowledge of the instantaneous state of a system is tantamount to the determination of a certain set of *variables* as a function of time: $x_1(t), ..., x_n(t)$. The time dependence of these variables will depend on the structure of the evolution laws and, as stressed in Sec. 1.2, on the set of *control parameters* $\lambda_1, ..., \lambda_m$ through which the system communicates with the environment. We qualify this dependence as *deterministic* if it is of the form

$$\mathbf{x}_t = \mathbf{F}^t(\mathbf{x}_0, \lambda) \tag{2.1}$$

Here $\mathbf{x}_t$ is the state at time t ; $\mathbf{x}_0$ is the initial state, and $\mathbf{F}^t$ is a smooth function such that for each given $\mathbf{x}_0$ there exists only one $\mathbf{x}_t$. For compactness we represented the state as a vector whose components are $x_1(t), ..., x_n(t)$. $\mathbf{F}^t$ is likewise a vector whose components $F_1(x_1(0), ... x_n(0); t, \lambda), ..., F_n(x_1(0), ... x_n(0); t, \lambda)$ describe the time variation of the individual $x's$.

In many situations of interest the time t is a continuous (independent) variable. There exists then, an *operator* $\mathbf{f}$ determining the rate of change of x_t in time :

Rate of change of $\mathbf{x}_t$ in time = function of the $\mathbf{x}_t$ and λ
or, more quantitatively

$$\frac{\partial \mathbf{x}}{\partial t} = \mathbf{f}(\mathbf{x}, \lambda) \tag{2.2}$$

As stressed in Secs 1.2 and 1.3 in a complex system $\mathbf{f}$ depends on $\mathbf{x}$ in a nonlinear fashion, a feature that reflects, in particular, the presence of cooperativity between its constituent elements.

An important class of complex systems are those in which the variables $\mathbf{x}_t$ depend only on time. This is not a trivial statement since in principle the properties of a system are expected to depend on space as well, in which case the $\mathbf{x}_t$'s define an infinite set (actually a continuum) of variables constituted by their instantaneous values at each space point. Discounting this possibility for the time being (cf. Sec. 2.2.2 for a full discussion), a very useful geometric representation of the relations (2.1)-(2.2) is provided then by their embedding onto the *phase space*. The phase space, which we denote by Γ, is an abstract space spanned by coordinates which are the variables $x_1, ..., x_n$ themselves. An instantaneous state corresponds in this representation to a point P_t and a time evolution between the initial state and that at time t to a curve γ, the *phase trajectory* (Fig. 2.1). In a deterministic system (eq. (2.1)) the phase trajectories emanating from different points will never intersect for any finite time t, and will possess at any of their points a unique, well-defined tangent.

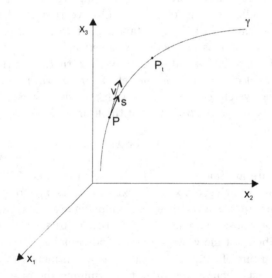

Fig. 2.1. Phase space trajectory γ of a dynamical system embedded in a three-dimensional phase space Γ spanned by the variables x_1, x_2 and x_3.

The set of the evolutionary processes governed by a given law $\mathbf{f}$ will be provided by the set of the allowed phase trajectories, to which we refer as *phase portrait*. There are two qualitatively different topologies describing

Deterministic View

these processes which define the two basic classes of dynamical systems encountered in theory and in practice, the conservative and the dissipative systems.

In the discussion above it was understood that the control parameters λ are time independent and that the system is not subjected to time-dependent external forcings. Such *autonomous* dynamical systems constitute the core of nonlinear dynamics. They serve as a reference for identifying the different types of complex behaviors and for developing the appropriate methodologies. Accordingly, in this chapter we will focus entirely on this class of systems. Non-autonomous systems, subjected to random perturbations of intrinsic or environmental origin will be considered in Chapters 3, 4 and onwards. The case of time-dependent control parameters will be briefly discussed in Sec. 6.4.3.

2.1.1 Conservative systems

Consider a continuum of initial states, enclosed within a certain phase space region $\Delta\Gamma_0$. As the evolution is switched on, each of these states will be the point from which will emanate a phase trajectory. We collect the points reached on these trajectories at time t and focus on the region $\Delta\Gamma_t$ that they constitute. We define a *conservative system* by the property that $\Delta\Gamma_t$ will keep the same volume as $\Delta\Gamma_0$ in the course of the evolution, $|\Delta\Gamma_t| = |\Delta\Gamma_0|$ although it may end up having a quite different shape and location in Γ compared to $\Delta\Gamma_0$. It can be shown that this property entails that the phase trajectories are located on phase space regions which constitute a continuum, the particular region enclosing a given trajectory being specified uniquely by the initial conditions imposed on $x_1, ..., x_n$. We refer to these regions as *invariant sets*.

A simple example of conservative dynamical system is the frictionless pendulum. The corresponding phase space is two-dimensional and is spanned by the particle's position and instantaneous velocity. Each trajectory with the exception of the equilibrium state on the downward vertical is an ellipse, and there is a continuum of such ellipses depending on the total energy (a combination of position and velocity variables) initially conferred to the system.

2.1.2 Dissipative systems

Dissipative systems are defined by the property that the dynamics leads to eventual contraction of the volume of an initial phase space region. As a result the invariant sets containing the trajectories once the transients have died out are now isolated objects in the phase space and their dimension is strictly less than the dimension n of the full phase space. The most important invariant sets for the applications are the *attractors*, to which tend all the trajectories emanating from a region around the attractor time going on (Fig. 2.2). The set of the trajectories converging to a given attractor is its *attraction basin*. Attraction basins are separated by non-attracting invariant sets which may have a quite intricate topology.

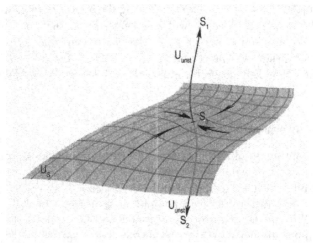

Fig. 2.2. Attraction basins in a 3-dimensional phase space separated by an unstable fixed point possessing a 2-dimensional stable manifold and a one-dimensional unstable one.

The simplest example of dissipative system is a one-variable system, for which the attractors are necessarily isolated points. Once on such a point the system will no longer evolve. Point attractors, also referred as *fixed points*, are therefore models of steady-state solutions of the evolution equations.

A very important property providing a further characterization of the solutions of eqs (2.1)-(2.2) and of the geometry of the phase space portrait is *stability*, to which we referred already in qualitative terms in Sec. 1.3. Let γ_s be a "reference" phase trajectory describing a particular long-time behavior of the system at hand. This trajectory lies necessarily on an invariant set like an attractor, or may itself constitute the attractor if it reduces to e.g. a fixed point. Under the influence of the perturbations to which all real world systems are inevitably subjected (see discussion in Secs 1.2 and 1.3) the trajectory that will in fact be realized will be a displaced one, γ whose instantaneous displacement from γ_s we denote by $\delta \mathbf{x}_t$ (Fig. 2.3). The question is, then, whether the system will be able to control the perturbations or,

Deterministic View

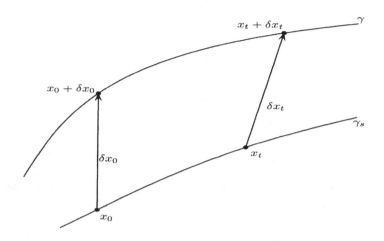

Fig. 2.3. Evolution of two states on the reference trajectory γ_s and on a perturbed one γ separated initially by a perturbation δx_0, leading to a separation δx_t at time t.

on the contrary, it will be removed from γ_s as a result of their action. These questions can be formulated more precisely by comparing the initial distance $|\delta \mathbf{x}_0|$ between γ and γ_s (where the bars indicate the length (measure) of the vector $\delta \mathbf{x}_0$) and the instantaneous one $|\delta \mathbf{x}_t|$ in the limit of long times. The following situations may then arise:

(i) For each prescribed "level of tolerance", ϵ for the magnitude of $|\delta \mathbf{x}_t|$, it is impossible to find an initial vicinity of γ_s in which $|\delta \mathbf{x}_0|$ is less than a certain δ, such that $|\delta \mathbf{x}_t|$ remains less than ϵ for all times. The reference trajectory γ_s will then be qualified as *unstable*.

(ii) Such a vicinity can be found, in which case γ_s will be qualified as *stable*.

(iii) γ_s is stable and, in addition, the system damps eventually the perturbations thereby returning to the reference state. γ_s will then be qualified as *asymptotically stable*.

Typically, these different forms of stability are not manifested uniformly in phase space: there are certain directions around the initial state $\mathbf{x}_0$ on the reference trajectory along which there will be expansion, others along which there will be contraction, still other ones along which distances neither explode nor damp but simply remain in a vicinity of their initial values. This classification becomes more transparent in the limit where $|\delta \mathbf{x}_0|$ is taken to be small. There is a powerful theorem asserting that instability or asymptotic

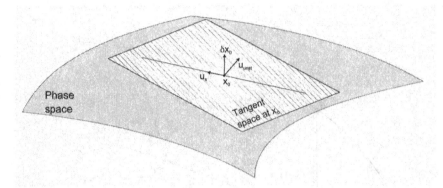

Fig. 2.4. Decomposition of an initial perturbation along the stable and unstable manifolds $\mathbf{u}_s$ and $\mathbf{u}_{unst}$ of the reference trajectory γ_s.

stability in this limit of *linear stability analysis* guarantee that the same properties hold true in the general case as well.

Figure 2.4 depicts a schematic representation of the situation. A generic small perturbation $\delta\mathbf{x}_0$ possesses non-vanishing projections on directions $\mathbf{u}_s$ and $\mathbf{u}_{unst}$ along which there are, respectively, stabilizing and non-stabilizing trends. One of the $\mathbf{u}_s$'s lies necessarily along the local tangent of γ_s on $\mathbf{x}_0$, the other $\mathbf{u}_s$ and $\mathbf{u}_{unst}$'s being transversal to γ_s. The hypersurface they define is referred as the *tangent space* of γ_s, and is the union of the *stable* and *unstable manifolds* associated to γ_s.

Analytically, upon expanding $\mathbf{F}^t$ in (2.1) around $\mathbf{x}_0$ and neglecting terms beyond the linear ones in $|\delta\mathbf{x}_0|$ one has

$$\begin{aligned}\delta\mathbf{x}_t &= \frac{\partial \mathbf{F}^t(\mathbf{x}_0, \lambda)}{\partial \mathbf{x}_0} \cdot \delta\mathbf{x}_0 \\ &= \mathbf{M}(t, \mathbf{x}_0) \cdot \delta\mathbf{x}_0\end{aligned} \qquad (2.3)$$

Here $\mathbf{M}$ has the structure of an $n \times n$ matrix and is referred as the *fundamental matrix*. An analysis of this equation shows that in the limit of long times $|\delta\mathbf{x}_t|$ increases exponentially along the $\mathbf{u}_{unst}$'s, and decreases exponentially or follows a power law in t along the $\mathbf{u}_s$'s. To express the privileged status of this exponential dependence it is natural to consider the logarithm of $|\delta\mathbf{x}_t|/|\delta\mathbf{x}_0|$ divided by the time t,

$$\sigma(\mathbf{x}_0) = \frac{1}{t} \ln \frac{|\delta\mathbf{x}_t|}{|\delta\mathbf{x}_0|} \qquad (2.4)$$

in the double limit where $|\delta\mathbf{x}_0|$ tends to zero and t tends to infinity. A more detailed description consists in considering perturbations along the $\mathbf{u}_j$'s and evaluating the quantities $\sigma_j(\mathbf{x}_0)$ corresponding to them. We refer to these

Deterministic View   31

quantities as the *Lyapunov exponents* . They can be ordered in size, $\sigma_1 \geq \cdots \geq \sigma_n$ and for a generic initial perturbation $\sigma(\mathbf{x}_0)$ in (2.4) coincides with σ_1. It can be shown that the σ_j's are intrinsic properties of the dynamical system at hand, in the sense that they are independent of the way one measures distances in phase space.

Clearly, if at least one of the σ_j's is positive $|\delta \mathbf{x}_t|$ will increase exponentially in time: the dynamics will be unstable and will display sensitivity to the initial conditions - a property which, as anticipated in Sec. 1.3, is one of the principal signatures of deterministic chaos. If on the contrary there is no positive Lyapunov exponent σ_j, the dynamics is stable. Technically speaking the σ_j's can be related to the properties (more precisely, the *eigenvalues*) of an $n \times n$ matrix constructed by multiplying the transposed of $\mathbf{M}$ (the matrix obtained from $\mathbf{M}$ by interchanging rows and columns) with $\mathbf{M}$ itself.

Stability allows one to formulate quantitatively the onset of complexity within a system. Specifically, suppose that the reference state γ_s is taken to represent a simple behavior like e.g. a unique fixed point. As long as this state is stable the system will be trapped in this simple world. Conversely, the onset of complex behaviors will be signaled by the failure of the stability of the corresponding steady-state solution, whereupon the system will be bound to evolve toward the coexistence of several fixed point solutions or toward invariant sets of a new type. The most elementary class of such sets are closed curves in phase space, to which one refers as *limit cycles*. Once on such a set the system will evolve in such a way that it will visit regularly, again and again, the same states. This is the signature of periodic behavior in time. Multiperiodic behavior is then associated with sets whose section along different directions is a closed curve, known as *tori*.

Figure 2.4 allows us to guess the type of invariant sets that can be associated with chaotic behavior. The existence of positive Lyapunov exponents implies that during the most part of the evolution there will be phase space directions along which distances between two initially nearby trajectories (and phase space volumes containing their initial states as well) will expand. In the long run it would seem that this is bound to lead to explosion, which is not allowed on physical grounds since it will imply that one of the system's properties like e.g. its energy becomes infinite. Unless if, at certain stages, the trajectories fold, being thereby reinjected in a finite portion of phase space. This process must clearly be repeated indefinitely. It should also be such that the trajectories, which are now crowded within a finite region, are not intersecting since as we saw earlier this would contradict the deterministic nature of the process. It turns out that this is indeed an allowable -and generic- possibility as long as the number of variables present is at least three.

This process of successive stretchings and foldings creates intricate struc-

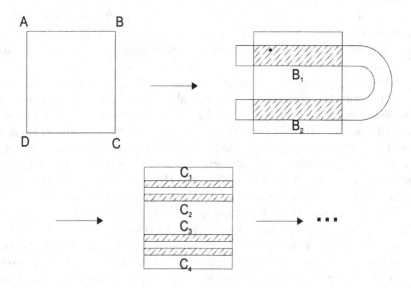

Fig. 2.5. Illustration of the formation of a fractal manifold in phase space. A square ABCD is stretched in the horizontal direction, contracted in the vertical one and folded to form a horseshoe such that the two disconnected horizontal bands B_1, B_2 are confined within the initial square. Repeating the process leads to a more tightly folded object generating the four disconnected bands C_1 to C_4 confined within the square. These bands, which arise from the fragmentation of B_1 and B_2, along the original ones constitute the first steps of an iterative process leading to a Cantor set.

tures referred as fractals that can be nicely visualized in the model of Smale's *horseshoe*. Consider a square R (Fig. 2.5). The square is contracted in the vertical direction and expanded horizontally to form a bar that is then bent into a horseshoe shape such that the arc portions and the ends of the horseshoe project outside the area of the initial square. In this way only the two bands of the straight legs are contained in R. We denote this projection by TR, T being the above described transformation performed on R. Repeating the process leads to an object $T^2 R$, producing four disconnected bands contained in R. By continuing the iterations we see that the points that will be contained in R are situated in 2^N increasingly thin bands. Let us consider the intersection between these bands and a line cutting across them. In the limit of N going to infinity (infinite fragmentation) the resulting set will be

Deterministic View 33

constituted of an infinity of points of total length zero, which can be shown to be non-denumerable. It is referred as *Cantor set*. It is through this type of sets that deterministic chaos can be embedded in phase space.

One may characterize a fractal object like a Cantor set further by choosing cells of size ϵ and counting the minimum number of cells N_ϵ necessary to cover it. Dividing the logarithm of N_ϵ by the logarithm of $1/\epsilon$ and taking the limit of very small ϵ we obtain a quantity D_0 intrinsic to the object under consideration,

$$D_0 = \lim_{\epsilon \to 0} \frac{\ln N_\epsilon}{\ln(1/\epsilon)} \qquad (2.5a)$$

Applied to the conventional objects of Euclidean geometry this expression yields $D_0 = 0$ for a set of isolated points, $D_0 = 1$ for a line, $D_0 = 2$ for a surface, and so on. For a Cantor set one would be inclined to assert that the answer should be $D_0 = 0$. In reality, since as we saw earlier ϵ decreases as a power of 3 (in the simplest model of equidistant bars) and N_ϵ increases as a power of 2, one obtains a finite value $D_0 = \ln 2/\ln 3 \approx 0.63$. We are therefore entitled to consider (2.5a) as a *generalized dimension*. The concept can be extended to a variety of sets obtained by a process of successive stretchings and foldings also referred as *strange sets*. It is of central importance not only in chaos theory but also as an indicator of a wide class of natural objects embedded in ordinary space (rather than the phase space) which are highly irregular and fragmented, from clouds and coastlines to the respiratory system of vertebrates.

Coming back to phase space, an interesting class of strange attracting sets are the *strange non-chaotic attractors* in which the maximum (mean) Lyapunov exponent is zero, in spite of the fact that their geometric structure is fractal. It can be shown that these objects are generic, in the sense that they do not only arise at specific values of the control parameters. Finally, strange non-attracting sets constitute yet another generic class of fractal objects separating different attraction basins in certain classes of dynamical systems. Systems possessing such sets present the interesting property of *final state sensitivity*, discovered by James Yorke, Edward Ott and Kelso Gregobi, whereby the attractors to which a trajectory will evolve depend upon minute changes of the initial conditions.

It is instructive to write (2.5a) prior to the limit $\epsilon \to 0$,

$$N_\epsilon \approx \epsilon^{-D_0} \qquad (2.5b)$$

We obtain a power law reminiscent of those considered in Secs 1.4.2 and 1.4.4. This law is indicative of *scaling behavior*, in the sense that upon a change of scale a part of a fractal object, however small, will display the structure of the initial, full object. Scaling behavior is therefore exhibited not only in the somewhat abstract form of Hurst, Pareto or Zipf type laws, in which no reference whatsoever to "shape" is made, but is also ubiquitous in a wide variety of objects embedded in physical or in the phase space. To mark this difference one sometimes refers to these laws as *fractal laws*.

In the examples of the horseshoe and of the Cantor set scaling behavior is referred to a single scale, ϵ. In many natural objects and in the overwhelming majority of strange phase space sets generated by deterministic chaos one deals with *multifractal* structures, in which variations take place along at least two independent scales. These structures are interesting in their own right, but a detailed study is outside the scope of this book.

2.2 Levels of description

Our direct perception of the natural environment takes place on a macroscopic level, where distances and events are on the scale of the meter and the second. On the other hand, any natural object is constituted of elementary entities of microscopic dimensions that evolve on scales differing by several orders of magnitude from the macroscopic ones. This separation prompts us to approach complex systems through a hierarchy of spatio-temporal scales and levels of description.

2.2.1 The microscopic level

This is the level on which the fundamental interactions between the elementary constituents of matter (Newton's law of universal gravitation, Coulomb's law of electrostatics, etc.) are formulated. In systems described by classical physics the variables $x_1, \cdots, x_n$ in eqs (2.1)-(2.2) refer to the coordinates, $\mathbf{q}_i$ and the momenta, $\mathbf{p}_i$ of the constituting particles. In a macroscopic system n runs up to a value of the order of the Avogadro number, $N \approx 6 \times 10^{23}$.

The time dependence of these variables is given by the fundamental laws of classical mechanics (Newton's or Hamilton's equations). These laws are expressed in the form of a set of ordinary differential equations which are nonlinear in most situations of interest, owing to the nonlinear dependence of the fundamental interactions on the particles' positions. As it turns out the dynamical system defined by these equations is conservative, in the sense of Sec. 2.1. Furthermore, the structure of the equations remains invariant under time reversal. The conservative character and the absence of arrow of time entail that at this level of description there can be no attractors, nor strong stability properties in the form of asymptotic stability.

Deterministic View

When a quantum description is adopted, the set $\{\mathbf{q}_i, \mathbf{p}_i\}$ is replaced by that of the $\mathbf{q}_i$ or the $\mathbf{p}_i$, owing to the uncertainty principle. These variables determine at each time the value of a new object, the wave function, from which all properties associated to the instantaneous state of the system can be deduced. This object evolves in time according to the Schrödinger equation, a linear equation in which the nonlinearity of the fundamental interactions is manifested through the appearance of nonlinear coefficients.

An important feature underlying the microscopic level description is *universality*: the basic structure of the evolution equations is given once for all, the specificity of the system at hand being reflected in the explicit form of a function known as Hamiltonian, which in a classical isolated system is just the total energy. This universality finds an elegant expression in the formulation of *variational principles*. The idea (in a classical system) is that, whatever the specifics might be, among all possible paths that may lead from an initial to a final state, the one that will actually be followed under the action of the evolution laws extremizes a certain quantity. A celebrated variational principle of this sort is the *principle of least action*: the evolution between an initial and a final state in which the values of the position coordinates are prescribed minimizes the action W - the integral over time of the product of the momenta multiplied by the time derivatives of the positions, summed over all particles. Variational principles also capture the most essential features of quantum mechanical systems.

Natural laws of the kind mentioned above can be looked at as minimal algorithms, defining the set of rules under which events will unfold in time. As stressed in Sec. 1.1 such rules can be condensed in a few lines but the events that they generate will extend on an in principle unlimited horizon. The situation is somewhat similar to that of a linguistic system where, under a given set of grammatical rules, one can generate texts of unlimited length. A great deal of attention has been devoted to the characterization of the type of dynamics associated with this description. Under the influence of the Newtonian paradigm, it was believed until the middle of the 20th century that microscopic level has to do with *integrable systems* which, under suitable transformation of variables, are mapped to uncoupled units each of which undergoes a time periodic motion. Since then it has been demonstrated that integrable systems constitute the exception rather than the rule. Microscopic level dynamics in most systems obeying to classical physics belongs to a quite different class, characterized by the prevalence of deterministic chaos. Actually in systems composed of many particles one deals with a particularly pronounced form of chaos where strong variability is generated owing to the defocusing action of interparticle collisions, which tend to amplify minute differences in the initial conditions, see also Sec. 6.1.

In complex systems research one often deals with situations in which the constituent units are not molecules or other elementary constituents of matter amenable to exhaustive experimentation and obeying to the by now well established laws and principles summarized in this subsection. One may deal, instead, with individuals in an animal ecosystem, financial agents trading in the stock market or cars in a traffic flow on a highway. The rules governing the behavior of these individuals are not known to any comparable degree of detail. True, they are all constituted of molecules satisfying Hamilton's or Schrödinger's equations, or the principle of least action. But one could hardly dispute the claim that e.g. tomorrow's Dow Jones index will not spring out of these laws and principles in any direct way. Still, observation shows that despite such differences individual level behavior (the analog of the "microscopic" description) is here also characterized by a very pronounced variability. We are thus led to the conclusion that to sort out systematic trends it is necessary to complement the microscopic/individual level description by some new considerations. This will be the subject of the next subsection.

2.2.2 The macroscopic level

The starting point is to introduce a new set of variables which are suitable combinations of the microscopic ones and describe *collective properties* such as the temperature and the bulk velocity of a fluid, the concentrations of chemically active species in a chemical reagent or, to take up an example beyond the strict realm of physical sciences, the population density in a multi-agent system. These variables to which we refer as *macroscopic variables* depend, by construction, not only on time but on the spatial location as well. By definition, their evolution is induced by that of the microscopic variables. Microscopic effects with their characteristic strong variability stressed in the previous subsection are therefore coming into play. Interestingly, under certain conditions it becomes possible to dissociate them from the effects involving solely the macroscopic variables, and to account for them through the values of a set of appropriate control parameters (in the sense of Sec. 1.2). In particular, such a decoupling holds true for certain combinations of the microscopic variables which in the context of physical sciences are known as *hydrodynamic modes*. The resulting macroscopic variables, whose number is much less than the number of the constituent entities, obey then to a closed set of evolution equations involving no extra variables beyond those belonging to the set. The latter have the general structure of eqs (2.2), in which the right hand side now depends not only on the variables themselves but also on their space derivatives. They are thus partial differential equa-

Deterministic View

tions rather than ordinary ones as encountered in the previous subsection and need to be supplemented with appropriate *boundary conditions*, involving combinations of the relevant variables and their space derivatives on the surface separating the system from the external world. The specific form of these relationships depends crucially on whether the system is isolated or exchanges energy, matter and perhaps some other kinds of properties with the environment.

Evolution equations of the macroscopic observables can be viewed as *balance equations*, describing how the time variation of the relevant variable arises as a balance between source (or inflow) and sink (or outflow) terms. Examples of such a description, to which we refer as the *macroscopic description*, are the Navier-Stokes equations of fluid mechanics and the equations of chemical kinetics. Nonlinearities are here also ubiquitous. In fluid mechanics they arise from the fact that the different fluid properties like e.g. the concentration c of a certain substance, are transported in space by the fluid velocity **v**, which is one of the variables of the macroscopic description. This gives rise in the right hand side of (2.2) to the product of **v** and the spatial derivative of c. In chemistry transport phenomena like diffusion are on the contrary described by linear laws to a very good approximation. Nonlinearities arise here from localized processes and express the fact that reactions other than unimolecular ones result from collisions between molecules of the reactive species, thereby giving rise to terms containing products of concentrations. The combined effect of reactions and diffusion gives rise to the following more explicit form of eqs (2.2), known as reaction-diffusion equations: rate of change of c_i in time = sum over reactions α of terms containing products of concentrations + diffusion coefficient of i multiplying the sum of second order spatial derivatives of c_i or, more formally,

$$\frac{\partial c_i}{\partial t} = \sum_\alpha k_\alpha \nu_{i\alpha} c_1^{m_{1\alpha}} \cdots c_n^{m_{n\alpha}} + D_i \nabla^2 c_i \qquad (2.6)$$

Here k_α is the rate constant of reaction α, measuring the fraction of particle encounters that are actually resulting in a chemical transformation; $\nu_{i\alpha}$ is an integer number known as stoichiometric coefficient, measuring the number of particles (or moles) of i synthesized or consumed in reaction α; $m_{1\alpha}, \cdots$ are positive integers or zeros; and ∇^2 is the Laplace operator.

An essential feature of equations like (2.6) describing macroscopic level dynamics is that they define a dissipative dynamical system. Such systems possess attractors and are endowed with asymptotic stability, as we saw in Sec. 2.1. These properties reflect the presence of the *arrow of time* -the irreversibility of macroscopic level phenomena, a well-known fact familiar from everyday experience. They thus seem at first sight to contradict the laws operating at the microscopic level even though, as we stressed above, they are supposed to be induced by the microscopic level dynamics. This problem is addressed in Chapter 6, where it is shown that truncations leading to a closed description of the evolution of macroscopic variables are under certain conditions an exact, albeit coarse-grained expression of the full microscopic description.

Notice that the dynamical system defined by equations like (2.6) lives in an infinitely-dimensional phase space, as anticipated already in Sec. 2.1. Fortunately, in many problems of interest a discretized form containing a finite (possibly quite large) number of variables turns out to provide all the relevant information. This allows one to use the phase space description, a very useful tool for visualizing certain aspects of the evolution of complex systems. Examples of this reduction process will be seen in Secs 2.6 and 2.7.

Whatever the details of the solutions of the macroscopic evolution equations might be, one expects that the variability inherent in the microscopic description will be drastically reduced. In a sense molecules are here substituted by macroscopic objects like volume elements. Within these volume elements each molecule is perceiving an effective interaction created by all molecules present, including itself. Such a *mean-field* description is obviously smoothing out fine details of microscopic level processes, retaining only large scale features. This view is also useful when modeling complex systems beyond physical sciences, like e.g. those involved in biological and socio-economic phenomena. In particular one may expect that replacing individuals in a multi-agent system by population densities will result in a closed description similar to eq. (2.6) that can yield some salient features of the system, otherwise masked by individual variability. An example of this description is provided by the equations of population dynamics. Contrary to (2.6) and the like, these equations are to be seen as *models*, established empirically or proposed ad hoc to be verified at a later stage by experiment. Once again nonlinearities are ubiquitous, reflecting the presence of cooperative interactions between the constituent entities such as activatory or inhibitory processes (imitation, competition, etc.). This point is taken up further in Sec. 2.7.

2.2.3 Thermodynamic formulation

The advantage of disposing of a description like the macroscopic description, limited to a small set of key variables, is gained at the considerable expense that the universality of the microscopic description is here lost. To begin with the structure of the equations seems to be system or model-dependent,

Deterministic View

contrary to that of the microscopic description. Furthermore, there seems to exist no general trend in the evolution embodied in a property like a variational principle. Macroscopic level dynamics is in this sense "opportunistic", as it does not strive to optimize any well-defined property. There is, however, an exception to this rule. It refers to physico-chemical systems in the state of thermodynamic equilibrium, or evolving in its immediate vicinity, for which a limited universality holds in the following sense.

(i) There exist certain functions depending in a unique way on the state of the system, which take their extremal values in thermodynamic equilibrium. For instance, in an isolated system one can define a function of energy, volume and composition (which happen to define fully the macroscopic state of the system) called entropy, S which is maximum in thermodynamic equilibrium. A similar property holds in non-isolated systems like for instance systems embedded in an environment maintained at temperature T, but the role of entropy is now played by another function called free energy F, which becomes minimum in the state of equilibrium where the system's temperature is equal to T. We refer to functions like $S, F, \cdots$ as *thermodynamic potentials* and will denote them below collectively by $\phi(\mathbf{x})$.

(ii) The evolution in the immediate vicinity of equilibrium in the absence of systematic nonequilibrium constraints is generated by the thermodynamic potentials ϕ and by a symmetric positive definite matrix $\mathbf{L}$, known as the *matrix of phenomenological coefficients* or Onsager matrix:

$$\frac{\partial x_i}{\partial t} = -\sum_j L_{ij} \frac{\delta \phi}{\delta x_j} \qquad (2.7)$$

where δ reduces to the usual derivative in the absence of explicit dependence of ϕ on the spatial degrees of freedom. Here the elements L_{ij} play the role of control parameters in the sense of Sec. 1.2. They are proportionality constants between the driving forces of the irreversible processes that are switched on once the system is not in equilibrium, expressed by $-\delta \phi / \delta x_j$, and the rates of these processes, expressed by the time derivatives of the $x_i's$.

(iii) Let now the system be subjected to nonequilibrium constraints, X_k, like e.g. a temperature difference across its boundaries, an external electric field, etc. If acting alone, each of these constraints generates a response in the form of a flux, J_k traversing the system, like e.g. a heat flux, an electric current, etc. One may extend the concept of entropy to such situations, provided that the space-time variation of the constraints is much less than the microscopic scale variability. The rate of change of entropy can then be decomposed into a flux term plus a term arising from the irreversible processes inside the system. One obtains in this way a balance equation of the form

$$\frac{dS}{dt} = \frac{d_e S}{dt} + \frac{d_i S}{dt} \qquad (2.8a)$$

where the two terms in the right hand side represent the *entropy flux* and *entropy production*, respectively. Under the above mentioned assumption of slow space-time variability the entropy production can in turn be written as the sum total over the system of local entropy source terms σ, which have the form

$$\sigma = \sum_k J_k X_k$$

$$J_k = \sum_\ell L_{k\ell} X_\ell \qquad (2.8b)$$

where the $L_{k\ell}$'s have the same interpretation as in eq. (2.7). The statement is, then, that in so far $L_{k\ell}$ are state-independent σ takes its minimum value at the unique steady state that will necessarily be established in the system in the limit of long times as long as the system in the absence of constraints is in a well-defined, stable phase

$$\sigma = \text{minimum at the steady state} \qquad (2.8c)$$

Notice that unlike what happens in eq. (2.7) σ does *not* generate the evolution. It just determines the properties of the final regime to which the system is driven by the dissipative dynamics switched on by the constraints.

The second law of thermodynamics stipulates the positivity of $d_i S/dt$ and σ. In an isolated system the entropy flux term in eq. (2.8a) vanishes entailing that S is bound to increase in time until it reaches its maximum when the state of thermodynamic equilibrium is reached - the traditional way to state the second law. But in an open system exchanging energy and matter with the surroundings, the entropy flux is non-zero. Furthermore, there is no law of nature prescribing its sign. As a result S may increase or decrease in time. Now, according to equilibrium statistical mechanics in an isolated system entropy at a certain macroscopic state is an increasing function of the number of microscopic states compatible with the values of the associated macroscopic observables providing, in this sense, a measure of the "disorder" prevailing within the system. In this logic, and to the extent that this microscopic definition of entropy could be carried through to nonequilibrium states, an isolated system is bound to tend irreversibly to a state of maximum disorder whereas creation of order may be a priori expected in open systems subjected to nonequilibrium constraints. This possibility, suggested first in the early 1940s by Erwin Schrödinger and developed subsequently by

Ilya Prigogine, allows one to foresee the beginning of an explanation of biological order in terms of the principles governing physico-chemical systems, since biological systems belong definitely to the class of nonequilibrium open systems. On the other hand, as it will become gradually clear in this book, thermodynamic state functions and entropy in particular do not provide a full characterization of the intricate interplay between order and disorder characterizing complex systems: complex systems are more than just thermodynamic engines and, in this respect, openness and nonequilibrium are to be regarded as two elements in a long list of ingredients and prerequisites paving the road to complexity.

2.3 Bifurcations, normal forms, emergence

The macroscopic level of description is the most familiar level at which complexity is manifested, through the generation of solutions of equations like eq. (2.6) displaying unexpected structural and dynamical properties. As discussed in Sec. 1.3, bifurcation is the "elementary act" of the complexity, as it provides the basic mechanism of appearance of new possibilities starting with regimes corresponding to "simple" behavior. To have access to their structure we need to solve the equations. This task is beyond the reach of the techniques currently available in science as far as exact solutions are concerned, except for certain particular cases. We need therefore to resort to qualitative arguments and to approximations. In this context, bifurcation analysis is a most useful and powerful technique allowing one to construct the solutions in a perturbative way, close to the bifurcation point. The procedure, described here for concreteness for bifurcations where the reference state is a steady-state solution, can be summarized as follows.

Suppose that we have a set of evolution equations of the form of eq. (2.2). By a standard method known as *linear stability analysis*, we can determine the parameter values λ for which a certain reference state $\{x_{js}\}$ switches from asymptotic stability to instability.

As discussed in Section 2.1, stability is essentially determined by the response of the system to perturbations or fluctuations acting on a reference state. It is therefore natural to cast the dynamical laws, eq. (2.2), in a form in which the perturbations appear explicitly. Setting

$$x_i(t) = x_{is} + \delta x_i(t) \tag{2.9}$$

substituting into eq. (2.2), and taking into account that x_{is} is also a solution of these equations, we arrive at

$$\frac{d\delta x_i}{dt} = f_i(\{x_{is} + \delta x_i\}, \lambda) - f_i(\{x_{is}\}, \lambda)$$

These equations are homogeneous in the sense that the right hand side vanishes if all $\delta x_i = 0$. To get a more transparent form of this homogeneous system, we expand $f_i(\{x_{is} + \delta x_i\}, \lambda)$ around $\{x_{is}\}$ and write out explicitly the part of the result that is linear in $\{\delta x_j\}$, plus a nonlinear correction whose structure need not be specified at this stage:

$$\frac{d\delta x_i}{dt} = \sum_j J_{ij}(\lambda)\delta x_j + h_i(\{\delta x_j\}, \lambda) \quad i = 1, \cdots, n \quad (2.10)$$

J_{ij} are here the coefficients of the linear part and h_i the nonlinear contributions. The set of J_{ij} defines an *operator* ($n \times n$ matrix in our case, referred as *Jacobian matrix*), depending on the reference state $\mathbf{x}_s$ and on the parameter λ.

Now a basic result of stability theory establishes that the properties of asymptotic stability or instability of the reference state $\mathbf{x} = \mathbf{x}_s$ (or $\delta \mathbf{x} = 0$) of the system (2.10) are identical to those obtained from its linearized version:

$$\frac{d\delta x_i}{dt} = \sum J_{ij}(\lambda)\delta x_j \quad i = 1, \cdots, n \quad (2.11)$$

Stability reduces in this way to a linear problem that is soluble by methods of elementary calculus. It is only in the borderline case in which $\mathbf{x} = \mathbf{x}_s$ (or $\delta \mathbf{x} = 0$) is stable but not asymptotically stable that the linearization might be inadmissible.

Figure 2.6 summarizes the typical outcome of a stability analysis carried out according to this procedure. What is achieved is the computation of the rate of growth γ of the perturbations as a function of one (or several) control parameter(s). If $\gamma < 0$ (as happens in the figure, branch (1) when $\lambda < \lambda_c$) the reference state is asymptotically stable; if $\gamma > 0$ ($\lambda > \lambda_c$ for branch (1)) it is unstable. At $\lambda = \lambda_c$ there is a state of *marginal stability*, the frontier between asymptotic stability and instability.

In general a multivariable system gives rise to a whole spectrum of γ, just as a crystal has a multitude of vibration modes. We will therefore have several γ versus λ curves in Fig. 2.6. Suppose first that of all these curves only one, curve (1) in the figure, crosses the λ axis, while all others are below the axis. Under well-defined mild conditions, we can then show that at $\lambda = \lambda_c$ *a bifurcation of new branches of solutions takes place*. Two cases can be distinguished:

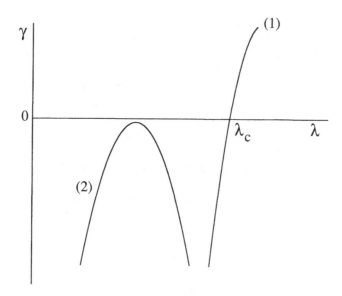

Fig. 2.6. Rate growth, γ, of perturbations as a function of a control parameter λ deduced from linear stability analysis (eq. (2.11)). (1), the reference state is asymptotically stable or unstable according to whether λ is smaller or larger than the critical value of marginal stability λ_c. (2), the reference state remains asymptotically stable for all values of λ.

- If at $\lambda = \lambda_c$ the solutions of (2.11) are nonoscillatory the bifurcating branches at the level of the full equation (2.10) will correspond to steady-state solutions.

- If at $\lambda = \lambda_c$ the solutions of (2.11) are oscillatory, the bifurcating branches will correspond to time-periodic solutions.

In either case, for the description of the bifurcating branches as well as for transient behavior, a suitable set of quantities can be defined which obey a closed set of equations if the parameters lie close to their critical values λ_c. In the first case there is only one such quantity, and it measures the amplitude of the bifurcating branches. In the second case there are two such quantities characterizing the amplitude and the phase of the oscillation. Therefore the original dynamics is effectively decoupled into a single equation or a pair of equations giving information on bifurcation, and $n - 1$ or $n - 2$ equations that are essentially irrelevant as far as bifurcation is concerned. We call the quantities that satisfy the bifurcation equations *order parameters*.

A most important point is that the bifurcation equations turn out to have exactly the same form whatever the structure of the initial laws might be.

In other words, any dynamics satisfying the conditions of either of the two cases above can be cast in a universal, *normal form* close to the bifurcation point. In this respect normal forms can be regarded as a new, "higher" level of description beyond that constituted by eqs (2.2) or (2.10). For dissipative systems the idea of reduction to a few order parameters is frequently associated with the name of Lev Landau, in connection with his theory of phase transitions. In the mathematical literature it is frequently referred to as the Lyapunov-Schmidt procedure, and more recently as the *center manifold theory*.

Inasmuch as the order parameters are the quantities that "perceive" the transition to complexity one is entitled to conclude that they provide a first quantitative measure of emergence - one of the principal signatures of complexity according to Chapter 1. As an example, for a system operating close to a bifurcation of the type depicted in Fig. 1.1, known as *pitchfork bifurcation*, the normal form describing the bifurcation of steady state solutions is

$$\frac{\partial z}{\partial t} = (\lambda - \lambda_c)z - uz^3 + D \nabla^2 z \qquad (2.12)$$

where z is the unique order parameter and u, D are combinations of λ_c and the other control parameters present in the initial equations. Here emergence corresponds to the existence of non-vanishing solutions for z which are inexistent prior to the transition across λ_c. Actually in the presence of symmetry-breaking phenomena like the Rayleigh-Bénard cells in systems of large spatial extent (cf. Sec. 1.4.1), eq. (2.12) has to be amended to account for the presence of a phase variable, specifying the way these cells follow each other in space. The general structure (2.12) is preserved but z becomes now a complex variable, and the term in z^3 is replaced by $|z|^2 z$ where the $|z|^2$ stands for the sum of squares of the real and imaginary parts of z. Another instance in which one deals with a complex order parameter (a convenient and compact way to express the presence of two order parameters through its real and imaginary parts) is the bifurcation of time periodic solutions in systems of large spatial extent. The corresponding normal form equation is

$$\frac{\partial z}{\partial t} = (\lambda - \lambda_c)z - (u_1 + iu_2)|z|^2 z + (D_1 + iD_2) \nabla^2 z \qquad (2.13)$$

differing from (2.12) by the presence of complex coefficients.

More intricate situations can also be envisaged in which several branches cross the λ axis in Fig. 2.6. This leads to interactions between bifurcating solutions that generate *secondary, tertiary*, or even higher order bifurcation phenomena, see Fig. 2.7. The above approach to reduction will still apply, in the sense that the part of the dynamics that gives information on the

Deterministic View

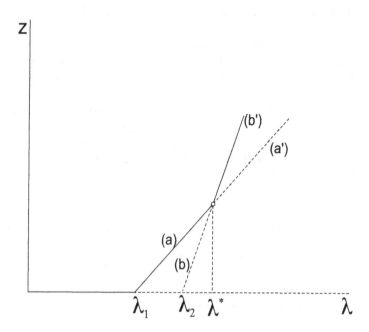

Fig. 2.7. Secondary bifurcation branch (b'), arising from the interaction of the two primary branches (a) and (b), initially stable and unstable respectively, beyond a critical value λ^* of the control parameter.

bifurcating branches takes place in a phase space of reduced dimensionality. However, the explicit construction of the normal form becomes much more involved and its universality can no longer be guaranteed, as discussed further in the next section.

In systems beyond the realm of physical science, using an imaged language, one might say that the macroscopic level of description involves entities like population densities in a coexistence of some sort, whereas the description afforded by the normal form describes the onset of a "revolution" in which these entities are completely reshuffled. The population is in a state of undecidability as several outcomes are possible, yet despite the inherent uncertainties this search takes a generic expression in the sense that details other than the fact that a transition is going to take place no longer matter.

2.4 Universality, structural stability

The possibility to encapsulate the phenomenon of emergence in a closed-form dynamics provided by a normal form signals the tantalizing perspective of recovering the universality lost in the course of the passage to the macroscopic level of description.

The recovery is in fact only partial. It takes its strongest form for bifurcations leading to normal forms involving a single real order parameter, as in eq. (2.12). The point is that in a single equation the right hand side can always be expressed as the derivative of a certain function, U thereby leading to a variational principle:

$$\frac{\partial z}{\partial t} = -\frac{\delta U}{\delta z} \qquad (2.14)$$

As an example, the "potential" U from which eq. (2.12) derives is given by

$$U = -\int d\mathbf{r} \left[(\lambda - \lambda_c)\frac{z^2}{2} - u\frac{z^4}{4} + \frac{D}{2}\left(\frac{\partial z}{\partial \mathbf{r}}\right)^2 \right] \qquad (2.15)$$

where $d\mathbf{r}$, $\partial/\partial \mathbf{r}$ denote integration over space and space derivative, respectively.

We stress that the status of U is completely different from that of the thermodynamic potentials appearing in eq. (2.7). U is generated by the dynamics, as reflected by the presence of parameters like λ or D which are related to the characteristic time and space scales present. For this reason we coin to it the term *kinetic potential*. Nevertheless, one cannot refrain from observing certain formal similarities between U and ϕ. For instance expression (2.15) looks like the expression of the free energy of a fluid in the vicinity of a liquid-gas phase transition, where now λ is related to the temperature, D to surface tension, and the order parameter z to the difference of the densities between the two phases close to the critical point. Such analogies are useful as they can provide inspiration for using ideas and techniques like scaling and renormalization group, which have been applied successfully to phase transitions over the last decades.

There is a whole family of kinetic potentials containing a single real order parameter other than (2.15), see Fig. 2.8. A potential containing quadratic and cubic terms describes a *transcritical bifurcation*, one with a linear and a cubic term a *limit point bifurcation*, one with a nonlinearity beyond the fourth order a degenerate situation in which there is competition between more than two stable states and confluence between two criticalities in which the states can merge by pairs. These situations can be classified in an exhaustive way and analyzed in depth. In the absence of space dependencies this

Deterministic View 47

analysis reduces to *catastrophe theory*, a special branch of nonlinear mathematics and topology. A key concept arising in this context is that of the *universal unfolding*: one can provide the full list of all qualitatively different behaviors that can potentially be generated by nonlinearities up to a certain degree, and determine the number of control parameters that are needed to actually realize them by the system at hand. For instance, in a dynamical system of one variable and nonlinearities up to order three by varying two control parameters one can realize the pitchfork, transcritical and limit point bifurcations, and show that there are no other qualitatively different behaviors that could possibly be encountered.

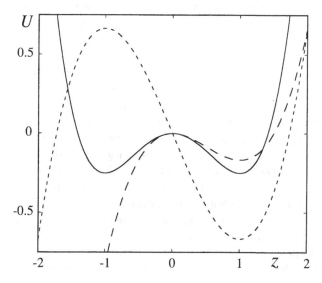

Fig. 2.8. Kinetic potentials corresponding to a pitchfork (full line), transcritical (dotted line) and limit point (dashed line) bifurcations.

Compared to properties (2.7) or (2.8), the range of phenomena belonging to the universality class described by eq. (2.14) is much wider, since there is no restriction concerning the vicinity to thermodynamic equilibrium or the linearity in the flux-constraint relation. On the other hand, so far property (2.14) was guaranteed only for systems reducible to a single real order parameter in the vicinity of bifurcation. It is therefore important to inquire about its possible extension beyond this range. As we saw in Sec. 2.3, transitions involving the interaction between two or more unstable modes are necessarily amenable to a description in terms of more than one order parameter.

Such dynamical systems can derive from a (kinetic) potential and exhibit variational properties only under rather restrictive *integrability conditions*. For instance, in a two-variable system (in its original form or, perhaps, upon a suitable transformation) the derivative of the right hand side of the first equation with respect to the second variable must equal the derivative of the right hand side of the second equation with respect to the first variable. Such conditions are *non-generic*, in the sense that they e.g. impose equalities to be satisfied by the control parameters beyond those defining the bifurcation points. Catastrophe theory addresses with success the inverse problem where, for kinetic potentials involving two variables and nonlinearities up to a certain degree, one can carry out a universal unfolding providing the exhaustive list of all the system's potentialities much like in the one variable case. Unfortunately the class of two variable potentials amenable to this analysis is limited and, in any case, precludes behaviors other than evolutions to steady states. For instance, normal form (2.13) describing the onset of oscillations in a spatially extended system does *not* derive from a potential. Similarly for normal forms involving three variables and describing the interaction between one stationary and one oscillating mode, or normal forms involving four variables and describing the interaction between two oscillatory modes, even in the absence of space dependencies. The situation is further complicated by the fact that independently of whether or not they derive from a variational principle, the *structural stability* of most of the normal forms with three or more order parameters cannot be guaranteed in the sense that one cannot exclude that the addition of a suitably parameterized extra term in their right hand side will not produce qualitatively new behaviors. The breakdown of universality opens therefore the unexpected -and in many respects exciting- perspective of the existence of an unlimited number of evolution scenarios of a complex system. One of the signatures of this versatility is the occurrence of global bifurcation phenomena in the normal form, even though originally the latter may have been established in a small vicinity of the bifurcation from a reference steady state solution. These global bifurcations may lead to, among others, chaotic dynamics as is already the case of eq. (2.13).

In the geometric view afforded by the phase space, universality would amount to achieve a full enumeration and characterization of all possible topologically distinct attractors -or more generally of bounded invariant sets- that can be embedded in phase space. This problem also remains open, and the experience accumulated so far suggests that complexity is accompanied by the generation of a whole "baroque" of forms and evolutions that refuse to conform to the familiar classical archetypes.

In conclusion, complex systems are systems for which the future is open and surprises may occur. In their vast majority they do not satisfy optimality principles of any sort. Their evolution is a kinetically driven "opportunistic" one in the course of which they may get trapped to certain modes of behavior among those potentially available and become then amenable to a universal description before switching to new ones, and we cannot know off hand the full spectrum of these modes. As we shall see in Sec. 2.5 and in the next two chapters universal laws may still emerge beyond this range. These laws are, however, of a new type as they refer to a completely different level of description.

2.5 Deterministic chaos

Contrary to the emergence of steady-state or time-periodic solutions there is no normal form like (2.12) or (2.13) describing the emergence of deterministic chaos in the vicinity of some critical situation and, a fortiori, no universal description as in (2.14) of the evolution towards it. The basic reason is similar to that responsible for the breakdown of universality in the presence of several interacting unstable modes pointed out in the previous section, namely, the intrusion of global bifurcations escaping from known classification schemes.

This fundamental limitation calls for the adoption of a new level of description to apprehend chaos in a quantitative manner. Such a description can be set up by noticing that both experiment and numerical simulation show that chaos appears after a succession of transitions in the course of which periodic or multiperiodic states (rather than steady-state ones) have lost their stability. Since, typically, such states are not known explicitly it would be desirable to map them into fixed point solutions and benefit then from the experience acquired in stability and bifurcation analyses around time-independent states as reviewed in the previous sections. An elegant way to achieve this has been invented by Henri Poincaré. Let γ be the phase space trajectory of a dynamical system and Σ a surface cutting this trajectory transversally. The trace of γ on Σ is a sequence of points $P_0, P_1, P_2, ...$ at which the trajectory intersects the surface with a slope of prescribed sign (Fig. 2.9). If we label these points not by the time at which they are visited (this would lead us to an unsolvable problem since eqs (2.2) cannot be integrated exactly) but, rather, by their order, we realize that the original continuous time dynamics (eqs (2.1)-(2.2)) induces on Σ a discrete dynamics -a recurrence- of the form

$$\mathbf{Y}_{n+1} = \mathbf{g}(\mathbf{Y}_n, \mu) \tag{2.16}$$

where $\mathbf{g}$ and μ are complicated functions of $\mathbf{f}$ and λ. We refer to (2.16) as the *Poincaré map*, Σ being the *Poincaré surface of section*.

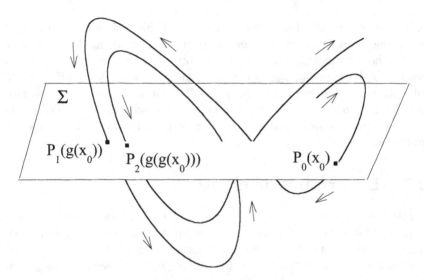

Fig. 2.9. Discrete time dynamics on the Poincaré surface of section Σ (eq. (2.16)) induced by the successive intersections $P_0, P_1, \ldots$ of the phase space trajectory of the underlying continuous time dynamical system with Σ.

The detailed construction of the Poincaré recurrence cannot be carried out quantitatively. Nevertheless, the idea behind it is the origin of a most fruitful approach to chaos. One starts with a particular recurrence, or of a family thereof, arguing that if the functions used are not pathological there is bound to be a family of continuous time dynamical systems embedded in phase space reducible to this recurrence through a judicious choice of the surface of section. Quantitative analysis is performed on this "minimal" model and certain features concerning possible routes to chaos as well as fully developed chaos that would be impossible to unravel from the original description are thus brought out. It is conjectured that the conclusions obtained from this analysis should be generic for systems like (2.2) as well, since the form of the evolution laws on the Poincaré surface were chosen to be non-pathological. The validity of this conjecture is then checked by experiment or by numerical simulation on model equations.

Experience accumulated during the last three decades has fully vindicated the interest of this approach. The reason for this success is in the fact that equations like (2.16) capture global aspects of the dynamics which, as we stressed above, are closely related to chaos.

One might think that the sensitivity to the initial conditions -the principal signature of deterministic chaos, as stressed repeatedly in this book-

Deterministic View 51

would preclude the possibility of its systematic quantitative description. This idea could be further reinforced by the observation that contrary to regular behaviors displaying a limited number of characteristic space and time scales, a system in the regime of chaos displays on the contrary a continuous spectrum of such scales. Yet despite these properties and contrary to certain naive views expressed every now and then, chaos is far from implying the absence of regularities of any sort. It obeys to partial universalities of a new kind of which the Lyapunov exponents and the fractal dimensions are examples. Many of these regularities can only be captured properly by switching to a probabilistic description, as we see more amply in Chapter 3. This level will also allow us to establish a highly interesting connection between chaos and the concept of *information*, by capitalizing the coexistence of both randomness and order within a chaotic regime. Here we summarize certain regularities that emerge in the framework of a purely deterministic analysis.

A most striking example of regularities of this kind which constitutes for many the second most prominent signature of chaos, is the existence of scaling laws governing the successive bifurcations leading to chaotic behavior. In one of these routes the system undergoes successive bifurcations at parameter values $\mu_1, \mu_2, \cdots, \mu_n, \cdots$ each of which marks the appearance of a periodic trajectory of twice the period of the previously prevailing one. These values tend to a finite limit μ_∞ as $n \to \infty$ (in which case the invariant set is a multifractal one describing aperiodic non-chaotic behavior, see Sec. 2.1) such that

$$\frac{\mu_n - \mu_{n-1}}{\mu_{n+1} - \mu_n} \to \delta \qquad (2.17)$$

where $\delta \approx 4.669201...$ is a universal constant presiding in any system that follows this scenario. Similar regularities govern the amplitudes of the successive periodic states as well as other routes to chaos. Their quantitative study uses the method of renormalization which originally was applied with success to the phenomenon of phase transitions in thermodynamic equilibrium, but took subsequently a new form thanks to its cross-fertilization with the ideas of nonlinear dynamics.

It is remarkable that these regularities, including the value of δ and other universal parameters, are captured by very simple models displaying the structure of eq. (2.16). One of the most popular examples is the logistic equation,

$$x_{n+1} = 4\mu x_n(1 - x_n) \qquad (2.18)$$

where x is confined to the unit interval as long as μ is a positive number smaller than or equal to one, and $\mu_\infty = 0.89248....$ This equation contains already the ingredients of the instability of motion and of the reinjection of trajectories tending to escape as a result of it into a finite part of phase space, here the unit interval.

A scaling like in eq. (2.17) implies a power law approach of μ_n to μ_∞, $\mu_n = \mu_\infty - \text{constant}/\delta^n$. The origin of this law is very different from that governing the self-similarity of fractal sets (Sec. 2.1) or the Hurst phenomenon (Sec. 1.4.2), yet it is appropriate to realize that such laws seem to recur in many complexity related phenomena. Some authors hold in fact the strong view that complexity is tantamount to such laws and to the concomitant absence of characteristic scales. This view is naive. In reality, scaling laws governing parameter dependencies of a phenomenon may coexist side by side with dynamical behaviors reflecting the presence of characteristic scales. As an example, for some ranges of values of μ beyond μ_∞ eq. (2.18) admits chaotic behavior with a positive Lyapunov exponent (this exponent vanishes at μ_∞) and exponentially decaying correlations. Notice that scaling behavior may invade the time domain as well in certain limiting cases like intermittent systems or systems containing a large number of variables, as discussed further in Chapters 3, 4 and in Sec. 2.6. In this limit universality properties of yet another form may show up. Furthermore, what we usually qualify as "noise", like e.g. the phenomenon of random walk which displays some remarkable scaling properties, turns out to be in reality a deterministic chaos living in a phase space of very high dimensionality.

The experience accumulated from the study of model and real-world systems suggests that both regular and chaotic behaviors are generic for dissipative systems, of few or many variables. There is nothing like an irresistible tendency to burst into chaos beyond the first few bifurcations. Biochemical regulatory phenomena provide a characteristic example, perhaps not fortuitously so since reproducibility must be an essential feature of biological rhythms. In this respect one may also observe that the control parameters of systems of this kind cannot be tuned arbitrarily since each function present in living cells is the result of a long evolution.

If we lower the temperature of a material under equilibrium conditions, eventually we will reach a phase transition point. But if we control a parameter interfering with the behavior of a generic nonequilibrium system we can observe oscillations, steady states similar to equilibrium, chaos of even something different. One can evolve from equilibrium-like "thermal" disorder to large scale order and then reach this peculiar state where order and disorder coexist that is chaos; or we can witness an opposite evolution or even miss chaos altogether. As stressed from the very beginning of this book it is this inexhaustible baroque-like diversity of evolution scenarios that constitutes one of the principal fingerprints of complex systems.

2.6 Aspects of coupling-induced complexity

So far we focused on the origin of the generic mechanisms presiding in the onset of complexity and on the characterization of the corresponding regimes, starting from the fundamental description afforded by the laws of physics. In this view, the detailed structure of the evolution equations was coming into play only to the extent to which it conferred to the evolution equations certain properties enabling these mechanisms to be activated. On the other hand, as stressed repeatedly, a complex system is typically constituted of a (usually large) number of coupled subunits. It is therefore appropriate to proceed to a more detailed analysis in order to assess the relative roles of the individual characteristics and of the coupling in the observed behavior. In this section we address this question, with emphasis on the macroscopic level description (Sec. 2.2.2).

In all generality, two kinds of couplings can be identified: the coupling between subunits in physical space; and their "functional" coupling arising from their joint participation in processes going on within a given small space region. This separation is nicely illustrated in equations like (2.6), (2.12) and (2.13) where the spatial coupling is accounted for by the second spatial derivative terms and the local processes by the remaining terms. It goes, however, beyond these examples. As seen in Sec. 2.2.2 the evolution of a macroscopic variable x_i can be written as a balance equation of the form
rate of change of $x_i =$
(source term arising from localized processes within a given volume element)
+ (transport of a property associated to x_i across the interface of this element and the surroundings)
In the absence of source terms this relation must reduce to a conservation law, and this forces the second term to be a combination of spatial derivatives (more technically, the divergence) of a vector referred as the *flux* of i. One obtains then the structure

$$\frac{\partial x_i}{\partial t} = \sigma_i(\{x_j\}, \lambda) - div\mathbf{J}_i \qquad (2.19)$$

where $div\mathbf{J} = \frac{\partial J_x}{\partial x} + \frac{\partial J_y}{\partial y} + \frac{\partial J_z}{\partial z}$, J_x, J_y, J_z being the three spatial components of the vector $\mathbf{J}$.

Eqs (2.19) do not constitute a closed description unless $\mathbf{J}_i$ is expressed in terms of the x_j's. Such relations can indeed be found for certain quantities, as seen in Sec. 2.2.2. When the time and space scales governing x_i are much longer than the microscopic scales they have the form

$$\mathbf{J}_i = \mathbf{v}.x_i - D_i\frac{\partial x_i}{\partial \mathbf{r}} \qquad (2.20)$$

where **v** is the velocity of bulk motion, if any, and $\partial/\partial \mathbf{r}$ the vector whose components are the spatial derivatives along x, y, z. In this case spatial coupling is limited entirely to the first neighbors of each spatial region considered separately, as seen by the discretized form of the corresponding part of (2.19)-(2.20) given below (and limited for compactness to a constant coefficient D_i and one space coordinate denoted hereafter by r):

$$\begin{aligned}(\text{spatial coupling})_i &= -\text{div} \mathbf{J}_i \\ &= -\frac{J_i(r + \Delta r) - J_i(r - \Delta r)}{2\Delta r} \\ &= -\frac{\nu(r + \Delta r)x_i(r + \Delta r) - \nu(r - \Delta r)x_i(r - \Delta r)}{2\Delta r} \\ &\quad + D_i \frac{x_i(r + \Delta r) + x_i(r - \Delta r) - 2x_i(r)}{\Delta r^2}\end{aligned} \quad (2.21)$$

Notice that the "advective" term $\mathbf{v}.x_i$ describes non-dissipative, time reversible processes and is nonlinear, whereas the term $-D\partial x_i/\partial \mathbf{r}$ describes dissipative, time irreversible processes and its dominant part is linear.

Contrary to the transport term which has the universal structure just described, the source term in (2.19) is system-dependent except when the system operates close to equilibrium or near a bifurcation in which case it reduces to the forms derived in Secs 2.2.3, 2.3 and 2.4. In fluid mechanics related problems this term arises entirely from the coupling of the system to external forces, like e.g. gravity in the case of the Rayleigh Bénard instability, and is typically linear. In chemical kinetics this term is typically nonlinear (see eq. (2.6)) and accounts for cooperativity - the most important ingredient for the onset of complex behavior.

Let us now summarize the specific effects induced by the source and the spatial coupling terms.

Source terms

These terms are the only ones present when the system is maintained in a state of spatial homogeneity. In chemical synthesis in industrial reactors this is achieved by mechanical stirring; in a suspension of cells in a nutrient medium it may be ensured automatically if mobility is high; and so forth. The resulting structure gives rise to the entire spectrum of bifurcation phenomena surveyed in the preceding sections - multiple steady states, oscillations, chaos. It can also generate interesting transient behavior like *excitability*, whereby the system returns to a steady state after performing a large excursion in phase space along which the relevant variables exhibit overshoots, undershoots, etc. Notice that, typically, source terms confer to the system characteristic time scales, which in the example of eq. (2.6) are given by combinations of rate constants.

Deterministic View

One class of couplings described by source terms that generates universal behavior independent of any bifurcation phenomenon pertains to the coupling of subsystems evolving according to widely separated time scales. This is relevant in many chemistry related problems. For instance, a catalytic reaction under laboratory conditions usually involves catalyst concentrations that are much less than the initial or final product concentrations, and rates that are much higher than the intrinsic rates in the absence of catalyst; as a result, some intermediate steps involving catalytic complexes proceed very quickly. In combustion the activation energies of some of the exothermic reactions are very high, so these reactions proceed, at least in the early stages, much more slowly than do energy and momentum transport. Similar examples can be found in biology, optics, and other fields. On a larger scale the interaction between the atmosphere and the oceans which is responsible for, among others, the breaking of waves, couples also a fast evolving component (the atmosphere) to a much slower one (the oceans).

Intuitively, we expect that during a time interval in which the "slow variables" hardly change, the "fast variables" will evolve a long way until they reach the level expected from the steady state solution of their evolution equations. But there will be one major difference: the values of the slow variables that the fast ones will "see" at this stage will not be their final levels, but rather the instantaneous values predicted by the slow variable equations. Denoting the fast and slow variables by Y and X respectively, we have in this quasi steady state a relation of the form

$$Y = h(X) \qquad (2.22a)$$

playing essentially the role of an *equation of state*. Substituting this expression in the equations for the slow variables, in which both X and Y intervene, we obtain

$$\begin{aligned} \frac{dX}{dt} &= F(X,Y) = F\left[X, h(X)\right] \\ &= f(X) \end{aligned} \qquad (2.22b)$$

which constitutes a closed set of equations for X. We have therefore succeeded in considerably reducing the number of variables involved by eliminating the fast ones. This is analogous to the ideas behind the rate-limiting step and the quasi steady-state approximation familiar in chemistry and, most particularly, in enzyme kinetics. If the number of the remaining slow variables is small this simplifies considerably the analysis and offers valuable insights on the phenomenon at hand, since the spurious effects have been eliminated. This is somewhat similar to the philosophy underlying the reduction to normal forms.

Transport terms

If we think in terms of the space discretized form of eqs (2.19)-(2.20), the simultaneous presence of source and transport terms can be viewed as the first neighbor coupling of local elements each of which generates one of the behaviors summarized above: coupled elements in a single globally stable steady state, coupled bistable elements, coupled periodic oscillators, coupled chaotic oscillators and so forth. We hereafter summarize the principal effects generated by couplings of this sort.

(i) *Space symmetry breaking.* The unique uniform steady state inherited from the unique steady state of the individual elements (supposed to be identical) becomes unstable under the action of the coupling. The new state arising beyond the instability is inhomogeneous in space and exhibits a characteristic length. The most clearcut case where such states are generated belongs to the family of reaction-diffusion type systems (eq. (2.6)). There is, then, a source term providing an intrinsic time scale τ_c, and a spatial coupling term which on the grounds of eqs (2.20)-(2.21) is proportional to D which has dimensions of (length)^2/time. The combination of these terms can then produce a characteristic length, $\ell_c \approx (D\tau_c)^{1/2}$ which is *intrinsic* to the system independent of the size of the embedding space. Compared to this the Rayleigh-Bénard cells described in Sec. 1.4.1 do break the spatial symmetry, but their characteristic length is determined by the depth of the fluid layer. We refer to the above mechanism of space symmetry breaking as the *Turing instability*. Turing patterns have been observed experimentally in chemistry. They are believed to be at the basis of major pattern forming phenomena such as certain structures arising in the morphogenetic processes accompanying embryonic development.

(ii) *Onset of chaos.* The spatial coupling of local elements in a regime of periodic oscillations leads under well-defined conditions to spatio-temporal chaos, the essence of which is already described by the normal form of eq. (2.13). An elegant description of the onset of this phenomenon is obtained by decomposing the order parameter z in this equation into an amplitude part and a phase, $\phi(\mathbf{r}, t)$ part. One can show that under certain conditions ϕ is the leading variable (in the sense of the reduction described in eqs (2.22)). The process is then reduced to a *phase dynamics* of the form (in a one-dimensional system)

$$\frac{\partial \phi}{\partial t} = -(\lambda - \lambda_c) + D\frac{\partial^2 \phi}{\partial r^2} + \mu \left(\frac{\partial \phi}{\partial r}\right)^2 \qquad (2.23)$$

which has been the subject of extensive studies.

(iii) *Synchronization*. In the opposite end of the foregoing one finds synchronization. A first class of systems are locally chaotic elements. When the spatial coupling is sufficiently strong chaotic behavior is wiped out and the system evolves to a periodic attractor. A second class arises in systems whose properties are spatially distributed already in the reference state, like e.g. local oscillators with different frequencies. In a physico-chemical context this may be the case if, for instance, the system is submitted to an external static, but space dependent forcing. The result is, then, that under the effect of coupling the oscillators eventually adopt a common frequency, although there may still be phase differences subsisting across space.

(iv) *Wave fronts*. The spatial coupling of elements possessing two stable steady states or a stable and an unstable one, gives rise to a wave front propagating from the unstable or the least stable state toward the stable or the most stable one. Here relative stability is assessed on the basis of a global criterion involving a kinetic potential of the form (2.15). Such fronts are observed in many experiments involving chemical reactions, electric circuits, or fluids and are at the origin of the propagation of the nerve impulse. In the presence of locally oscillating or excitable elements the wave front may take some unexpected forms, from cylindrically symmetric (target) patterns to simple and multi-armed spiral ones. The presence of such patterns can also be viewed as an indication of the inability of the local oscillators to achieve a regime of full synchronization.

(v) *Scaling regimes and turbulence*. As stressed earlier, temporal or spatiotemporal chaos differs from regular behaviors in time and space by the presence of a continuum of interacting modes, each of which exhibits a different time and space scale. A popular way to visualize this proliferation of scales is to construct the *power spectrum* associated to one of the variables $x(\mathbf{r}, t)$ present. As discussed further in Secs 5.2-5.4, in practice such a variable is available in discrete times and discrete space points as e.g. in experiment or simulation: $t_1, t_1 + \Delta t, \cdots t_1 + n\Delta t, \cdots; r_1, r_1 + \Delta r, \cdots r_1 + j\Delta r, \cdots$ (we again take for concreteness the example of one-dimensional medium). If the total system size is L and the time during which the system is monitored is T, then for fixed windows Δt and Δr n and j vary up to the values $T/\Delta t$ and $L/\Delta r$. The power spectrum $S(k,\omega)$ is then, by definition, the signal $x_n(j)$ multiplied by a sinusoidal function in both n and j and summed over all values of these variables:

$$S(k,\omega) = \frac{\Delta r \Delta t}{LT} \left| \sum_j \sum_n x_n(j) e^{2\pi i \left(\frac{j\Delta r}{L} k - \frac{n\Delta t}{T}\omega \right)} \right|^2 \qquad (2.24)$$

Here we used, as done usually in the literature, the imaginary exponential representation. The bars to the second power indicate the sum of squares of the real and imaginary parts of the enclosed function and the prefactors stand for normalization. Suppose that $x_n(j)$ is constant, as in a spatially uniform steady state. Eq. (2.24) gives then rise to a sharp pulse (technically, a Dirac delta function) centered on $\omega = k = 0$. A periodic signal $x_n(j)$ in time or in space will on the contrary produce a sharp pulse at a non-zero value of ω or k. Finally, a chaotic regime will produce a function $S(k,\omega)$ in the form of a smooth background extending over a range of ω's and k's on which may or may not be superimposed some sharp peaks. In the first case the system will exhibit a small scale variability around a limited number of dominant scales. But in the second case the observer will be unable to identify any trace indicative of a spatio-temporal scale: one is, then, in the presence of a scale-free regime.

There is an important class of spatially coupled, large size systems in which, as the constraints are varied, states where $S(k,\omega)$ displays a power law behavior in a certain range of ω and/or k values are reached. This happens, for instance, in the regime of *fully developed turbulence* realized when a viscous fluid flows over an object of characteristic length ℓ at a very high speed, where it turns out that $S(k) \approx k^{-5/3}$ for the energy spectrum in the idealized limit of homogeneous and isotropic flow. This law reflects the presence of scaling behavior in the sense of Sec. 2.1 and 2.5, the important point being that this behavior arises here as a result of the underlying dynamics. In physical space the counterpart is that the fluid motion splits into eddies of many different sizes thereby giving rise to irregular fractal patterns. In the time domain, the fluid velocity at a given point undergoes a highly irregular evolution.

Most of the natural large scale flows like atmospheric circulation, and of flows taking place in industrial devices especially in connection with combustion, turn out to be turbulent. The repercussions are very important, as many key properties are deeply affected. In particular, turbulence speeds up dramatically the rate of energy transport across different regions.

A full understanding of turbulence on the basis of the evolution laws of fluid dynamics constitutes still an open problem, despite some significant recent progress. Substantial breakthroughs have been achieved in the computational techniques, allowing one to reproduce the salient features of the process. Turbulent-like behavior with its characteristic scaling laws and fractal structures can also be emulated by ad hoc models such as spatially coupled discrete time dynamical systems (see also Sec. 2.7), cascade models, shell models, and so forth. We stress that fully developed turbulence should not be confused with ordinary deterministic chaos of which it constitutes, rather, an extreme limiting form.

Deterministic View   59

Turbulence serves also as an archetype for apprehending a variety of other problems in which irregular behavior in spatially extended systems displaying scaling properties is observed. The assertion is sometimes made that such systems -including turbulent ones- are in a critical state, on the grounds that the most typical instance in which scaling behavior is known to occur is that of a material in the vicinity of a critical point of phase transition (like e.g. the liquid-gas transition). Actually this is no more than an analogy albeit an interesting one, since it can give inspiration for the transfer of ideas and techniques across disciplines. Certainly one should refrain from identifying complexity to this kind of phenomenology alone, as sometimes asserted by some authors.

2.7 Modeling complexity beyond physical science

As stressed throughout Chapters 1 and 2, a necessary prerequisite for the onset of complexity is the instability of motion and the concomitant bifurcation of new states. In physico-chemical systems these properties are built into the fundamental evolution laws underlying the different levels of description. But in many artificial or natural systems, including human societies, the laws are not known to any comparable degree, and yet the behaviors observed are part of the same phenomenology as in physico-chemical systems. To take advantage of the concepts and tools elaborated in a physically-based complexity theory one is then led to proceed by *analogy*.

The first step in this process is to assess the nonlinear character of the underlying dynamics, to recognize a set of suitable control parameters including those mediating possible couplings, and to identify a set of variables capable of undergoing instabilities and bifurcations. This procedure is greatly facilitated by the results surveyed in Secs 2.3 and 2.4 according to which the essence of the qualitative behavior is often contained in simple universal evolution equations, the normal forms, satisfied by a limited set of collective variables, the order parameters. Potentially, therefore, a multivariable system, like most of the systems encountered in nature, technology and society, may reduce to a simplified description near a bifurcation point. For instance, if we know from experiment that the system of interest undergoes a transition from a single to two simultaneously stable steady states as in the recruitment process in social insects (Sec. 1.4.3), we are tempted to use a cubic dynamics (eq. (2.12)) as a zero order description. If, on the other hand, we are in the presence of a transition to sustained oscillatory behavior as observed in a host of biological regulatory phenomena at the subcellular, cellular, organismic and population level one would be entitled to regard eq. (2.13) as a promising starting point. Finally, a chaotic like behavior, be it in the context of atmospheric dynamics or of finance, would suggest that one should first consolidate the evidence by evaluating from the data Lyapunov exponents, fractal dimensions, etc.; and next proceed to modeling using a minimal number of variables compatible with these findings.

These considerations, whose usefulness should not be underestimated, must be used with lucidity, otherwise they may overshadow what must remain an equally important aspect of the problem, namely the specificity of the system considered. Therefore, at each stage of the analysis the modeler must select nonlinearities compatible with the underlying mechanisms before associating them and the corresponding variables to particular types of normal forms and order parameters. He must also make sure that certain constraints imposed by the very nature of the problem are satisfied such as the dissipativity conditions, the positivity of population-like variables, or the time and space scales predicted by the model in comparison with those revealed by the data.

Bearing these points in mind we now proceed to a series of examples of the type of modeling and of the associated nonlinearities encountered in complex systems beyond the strict realm of physical science.

(i) *Competitive growth* An uninhibited growth process leads to exponential explosion. In reality, since the number of resources is finite the process is eventually bound to stop. Between these two extremes growth is regulated and takes a subexponential character. The minimal nonlinearities needed to model this process lead to evolution equations extending the famous model originally proposed by Pierre François Verhulst,

$$\frac{dx_i}{dt} = k_i x_i \left(N - \sum_j a_{ij} x_j \right) \qquad (2.25)$$

where the *carrying capacity* N stands for the resources available and the second term in the parenthesis accounts for the slowing down and the eventual saturation. In some cases spontaneous transitions between the different entities represented by the x_i's become possible like e.g. mutations in connection with biological evolution. Eq. (2.25) must then be augmented by a term of the form $\sum_j m_{ij} x_j$, m_{ij} being the mutation rate, as discussed further in Sec. 6.6.

(ii) *Choice between options* As seen in Secs 1.4.3 and 1.4.4 this process is ubiquitous in, among others, social insects and finance. To account for the sharpness of the choice one usually introduces nonlinearities producing

Deterministic View

S-shaped responses, of the form

$$\text{rate of change of population choosing option i} = \phi \frac{(k+x_i)^\ell}{\sum_{j=1}^{S}(k+x_j)^\ell} \quad (2.26)$$

Here ϕ is the total flow of individuals, k a threshold of x value beyond which cooperativity begins to play, and the exponent ℓ quantifies the degree of cooperativity. Eq. (2.26) may arise from basic chemical kinetics with rate laws in the form of products of concentrations, when certain fast variables are eliminated using the procedure outlined in eq. (2.22) and the catalysts involved have a built-in cooperativity as it happens for the class known as allosteric enzymes. It is also used widely in biochemistry. As a rule, the parameters are fitted from data rather than deduced from first principles.

(iii)*Modeling with delays and discretizing time* In many instances the response to an applied constraint is not instantaneous but is, rather, manifested after a period of latency. This may reflect intrinsic structural-functional changes like refractory periods in neurophysiology, transport along a certain distance, or the simultaneous presence of individuals belonging to several generations within a population. The evolution laws (2.2) acquire then a structure of the form

$$\frac{dx_i(t)}{dt} = f_i\left(\{x_j(t - \tau_j)\}, \lambda\right) \quad (2.27)$$

where τ_j are positive or zero. An elementary example of (2.27) is provided by the delayed Verhulst equation

$$\frac{dx(t)}{dt} = kx(t-\tau)(N - x(t)) \quad (2.28)$$

A different way to account for finite time lags is to use recurrence relations of the form shown in Sec. 2.5. The ubiquity of the logistic law (eq. (2.18)) has prompted several authors to undertake a modeling of stock market evolution based on this equation, where the value of μ is gradually increased to account for changes of the price variable $P(t)$ away from its fundamental value (see discussion in Sec. 1.4.4).

As expected, the main effect of delays and of time discretization is to generate chaotic dynamics in systems amenable to few state variables.

(iv) *Parameterizing the short scale variability* The evolution of a spatially extended system is governed by a set of partial differential equations. In practice, to issue a prediction about the future state of such a system one adopts necessarily a coarse grained view: either the space is discretized and the system is converted into a set of elements coupled in space as in eq.

(2.21); or the variables are expanded in series of known functions satisfying the symmetries and the boundary conditions imposed on the system and the expansion is truncated to a finite order. In doing so one overlooks the processes occurring within the discretization box or beyond the highest mode retained in the truncation which, typically, display space and time scales faster than those retained. A very important class of problems where this procedure is followed concerns atmospheric dynamics and, in particular the numerical models used in operational weather forecasting. Neglecting subgrid variability has here proven to be highly unsatisfactory since it amounts to overlooking important phenomena like convection that are eventually manifested on larger scale as well, on the grounds of the inherent instability of motion. To cope with this difficulty one augments the truncated equations by terms in which the subgrid variability is modeled in some way. This modeling is inspired to a certain extent by closure relations like in eq. (2.20), but appeals to a number of heuristic parameterizations as well. For instance, to enhance numerical stability while accounting for subgrid variability one introduces in (2.20) higher order spatial derivatives in the form

$$\left(\frac{\partial x_i}{\partial t}\right)_{\text{subgrid}} = (-1)^{k-1}\nu_i \nabla^{2k} x_i \qquad (2.29)$$

followed by discretization as above, where the exponent of (-1) guarantees that the dissipative character of the process is preserved.

(v) *Global couplings*

In everyday experience, communication is not always ensured by encounters between adjacent subunits constituting a system. Financial agents competing in the stock market are subjected to long range communication channels. Nowadays, an ordinary person is likewise exchanging information simultaneously with different partners via the Internet. And in biology, a typical neuron in the cortex can be coupled to several thousands of other neurons.

In a physical context long range communication may also arise when the subunits are coupled to each other via an embedding medium. An early example was provided by Christiaan Huygens' remarkable experiment of two pendula coupled via a common wall (more generically, an elastic medium). In a similar vein, distant parts of sea surface or of a solid catalyst of the type involved in chemical synthesis communicate via the atmospheric layer or the gas phase containing the reactants above them. Usually the resulting couplings are treated empirically, by eliminating the embedding medium through a relation similar to (2.22a). To model these situations one replaces the nearest neighbor structure exhibited in eqs (2.19)-(2.21) by terms of the

Deterministic View

form

$$\frac{1}{N}\sum_{j=1}^{S} g(x_i, x_j) \qquad (2.30)$$

where the number S of elements coupled to i, can be as large as the total number, n of variables. The function g is system specific and in many cases (as, for instance, in coupled oscillators where x_i stands for the phase of oscillator i) depends on i and j via the difference $x_i - x_j$.

Compared to a nearest neighbor coupling of the same strength a coupling of the form of (2.30) tends to enhance coherence and to favor synchronization. There is, however, a whole variety of regimes, lying between the incoherent state where each unit behaves in its own way as if it were isolated and the fully synchronized one. Most intriguing among them is the regime of *clustering*: an evolution where the elements split into groups ("clusters") within each of which there is mutual synchronization. Since the different clusters are not synchronized the collective dynamics may still be quite complex giving rise, for instance, to spatio-temporal chaos.

There exist several models for which detailed "state diagrams" have been constructed allowing one to follow these transitions as the parameters are varied. A specially instructive family is that of coupled elements evolving in time in a discrete fashion, such as coupled logistic maps (eq. (2.18)). Their study allows one, in particular, to assess the role of the chaotic or not character of each element in the global behavior.

Chapter 3

The probabilistic dimension of complex systems

3.1 Need for a probabilistic approach

A number of reasons for undertaking a probabilistic description of complex systems have already been alluded to in the preceding chapters. The crux of the argument was that in nature the process of measurement, by which the observer communicates with a system, is limited by a finite precision. As a result, a "state" is in reality to be understood not as a point in phase space (Sec. 2.1) but, rather, as a small region whose size reflects the precision of the measuring apparatus. Additional sources of delocalization also exist in connection, for instance, with incomplete specification of initial data (a situation to which weather forecasters are regularly confronted) or numerical roundoffs.

If the dynamics were simple the difference between the point-like description and the delocalized description described above would not really matter. The situation changes entirely in the presence of instability of motion, where (see e.g. Figs 1.1, 1.2 and 2.5) nearby initial data of a system undergoing bifurcation or being in the regime of deterministic chaos, that are experimentally indistinguishable, may follow quite different histories. To the observer this will signal the inability to predict the future beyond a certain transient period on the basis of the knowledge of the present conditions. This property constitutes a compelling motivation for undertaking a probabilistic description, the only one to reflect this delocalization and to cope in a natural fashion with irregular successions of events.

Additional elements come into play in systems composed of several subunits. Consider first an ordinary physico-chemical system. The observation of its state usually involves an averaging of the instantaneous values of the pertinent variables, either over time or over a volume of supermolecular dimensions. For instance, if we put some 0.33×10^{23} molecules of water in a container of one cubic centimeter ($1\ cm^3$) volume at ambient temperature and pressure, we would conclude that we have a liquid whose number density is 0.33×10^{23} molecules/cm^3 and whose mass density, in grams, is $0.33 \times 10^{23} \times$ (mass of H_2O molecule) $= 1g/cm^3$. Number and mass densities are the sort of variables we have dealt with in most of the preceding chapters, but there we looked primarily on the macroscopic scale. The number density of a liquid in a volume element of the order of, say, one hundred cubic angstroms (or $10^{-22} cm^3$) will continuously deviate from its value in the surrounding macroscopic volume. Molecules will cross the boundaries of microscale volume elements, and because of the random character of their motion the number of particles contained at any moment in each small volume will be essentially unpredictable. Here again, the probabilistic description becomes the natural mode of approach. We refer to the deviations generated by this mechanism as *fluctuations*. Because of fluctuations, physico-chemical systems are capable of exploring the phase space continuously and of performing excursions around the state predicted by the solution of the phenomenological, deterministic equations.

Fluctuations are not limited to physico-chemical systems. They are universal phenomena in the sense that they arise in finite size parts of any system, however large it may be on the whole. In the socio-economic context they reflect, in particular, the role of individual variability in a global decision making process. Finally, a variety of systems of growing concern (microchips, nanotubes, catalytically active crystallographic planes of noble metals or zeolites, vesicles, chromosomes, locomotion devices in living cells, etc.) operate on a nanoscale level, which is intermediate between the microscopic and macroscopic ones. Such *mesoscopic* systems not only generate strong fluctuations on the grounds of their small size, but in certain cases they may literally *thrive* on fluctuations by giving rise to unusual kinetic and transport properties that would otherwise be impossible.

3.2 Probability distributions and their evolution laws

According to the foregoing, in a classical system the central quantity to be considered within the framework of a probabilistic description is the probability $P_{\Delta\Gamma}(t)$ of being in a phase space cell $\Delta\Gamma$ at time t. A quantity of more intrinsic nature, independent of the size and shape of $\Delta\Gamma$, can be defined by taking the limit of small $\Delta\Gamma$ and introducing the corresponding probability density,

$$P_{\Delta\Gamma}(t) = \rho(x_1, \cdots, x_n, t)dx_1 \cdots dx_n \qquad (3.1)$$

A point-like description of a system amounts to stipulating that the initial state $x_{10}, \cdots x_{n0}$ is known with an infinite precision. The corresponding probability density ρ is then an infinitely sharp peak (a Dirac delta function) following the phase trajectory emanating from the initial state. In fact, according to the viewpoint adopted in Sec. 3.1. we shall rather place ourselves in the case where $\rho_0(\mathbf{x}) = \rho(x_{10}, ..., x_{n0}, 0)$ is a smooth function. Obviously, the probability density at later times will then be given by a superposition of the above peak-like evolutions over all the initial states for which $\rho_0(\mathbf{x})$ is non-vanishing. We may therefore write

$$\rho_t(\mathbf{x}) = \int_\Gamma d\mathbf{x}_0 \delta[\mathbf{x} - \mathbf{F}^t(\mathbf{x}_0, \lambda)]\rho_0(\mathbf{x}_0) \qquad (3.2)$$

where δ is the Dirac delta function and $\mathbf{F}^t$ is the quantity defined in eq. (2.1). This relation is known as the *Frobenius-Perron equation*. For continuous time systems, differentiating both sides with respect to t leads to an alternative expression in which ρ satisfies a partial differential equation known as the *Liouville equation*,

$$\begin{aligned}\frac{\partial \rho}{\partial t} &= -\sum_i \frac{\partial}{\partial x_i}(f_i(\mathbf{x}, \lambda)\rho) \\ &= L\rho \end{aligned} \qquad (3.3)$$

where f_i is defined in eq. (2.2) and L is the *Liouville operator*. Notice that in both eqs (3.2) and (3.3) we actually deal with a deterministic dynamical system. It is therefore understood that probabilities are here not given ad hoc but are, rather, to be determined from the underlying dynamics.

By definition, ρ must be non-negative. Furthermore it must satisfy a *normalization condition*, stipulating that the sum of probabilities $P_{\Delta\Gamma}$ to find the system in the different regions $\Delta\Gamma$ must be equal to unity:

$$\begin{aligned} P_{\Delta\Gamma} &\geq 0, \quad \rho \geq 0 \\ \sum_{\Delta\Gamma} P_{\Delta\Gamma} &= 1, \quad \int_\Gamma dx_1...dx_n \rho(x_1,...,x_n,t) = 1 \end{aligned} \qquad (3.4)$$

In quantum systems the phase space description breaks down, as explained in Sec. 2.1. Furthermore, a state can either be part of a continuum (as it always happens in a classical system at the phase space level of description) or be discrete. In both cases eqs (3.2)-(3.3) can be extended, but they now refer to an object which is an operator rather than a function. There are some important aspects of complexity specific to quantum systems, but they will not be developed systematically in this book.

In classical systems a discrete state representation showing some common features with the quantum description becomes natural in the context of a *mesoscopic level* approach, which constitutes an intermediate method of approach, between the microscopic and the macroscopic levels defined in Sec. 2.2. This approach may be summarized as follows.

(i) *Choice of state variables.* The instantaneous state is not described by the variables pertaining to the system's elementary subunits, which in a classical system would be the position coordinates and the velocities (or momenta) of the N particles. Instead, one focuses on a limited number of collective variables $X_1, ..., X_n (n << N)$ related to the variables involved in the macroscopic description (concentrations, temperature, population densities, prices, etc.).

(ii) *Evolution of the state variables.* The macroscopic description is augmented to account for the fact that the X_i's are constantly varying under the action of the microscopic dynamics. As an example, consider a system where a certain constituent is produced or consumed by chemical reactions. The evolution of the number of particles X of this constituent can be decomposed into two types of steps, Fig. (3.1): steps where particles will be in free motion or will simply rebounce as a result of elastic collisions, that will obviously not modify the value of X; and reactive steps where, under the effect of sufficiently violent (necessarily inelastic) collisions, the chemical identity of the molecules involved is modified. As a result the value of X will be modified, typically by elementary steps of one unit. Similar considerations hold for the evolution of competing populations in a medium of limited resources, or for the spread of a mutation in population genetics.

By performing the same type of experiment, under the same environmental conditions and starting with the same initial value X_0, we will find several "trajectories" or *realizations* similar to that of Fig. 3.1., but differing from it by the values of the times t_i at which transitions took place and by the sign (± 1) of the jump undergone by X in the transition. One expects that the value $\bar{X}(t)$ provided by the macroscopic description will be the average of the instantaneous $X(t)$'s over these different "histories". Our objective is to go beyond this approach and obtain information on the deviations from the average, $\delta X = X - \bar{X}(t)$. These deviations, to which we already referred as

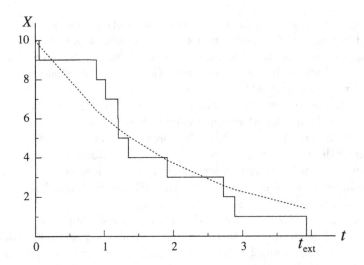

Fig. 3.1. Steplike evolution of the number of particles of a constituent X participating in a chemical reaction that consumes it irreversibly. Under the effect of fluctuations reactive steps occur at randomly distributed times and, between two successive such steps, the value of the variable remains constant (horizontal plateaus in figure). t_{ext} denotes the extinction time and dashed line denotes the macroscopic (mean field) trajectory.

fluctuations, constitute the principal signature of the microscopic dynamics within the framework of a mesoscopic level description (Fig. 3.2a).

As we saw in the previous chapters, at the fundamental level of description the dynamics of a system of many particles is universally chaotic. As a result, $\delta X(t)$ is expected to be a very complex function exhibiting practically unpredictable behavior (Fig. 3.2b). To cope with the presence of such *random variables* we embed the dynamics of the X_i's into a probabilistic description: rather than follow the variation of X_i in all its details we focus on the probability $P(X_1, \ldots, X_n, t)$ to observe particular values of those variables at time t. Similar arguments will hold for systems beyond a physico-chemical context, the principal cause of randomness being now the individual variability in the decision making process and/or the stimuli impinging on the system from the external environment.

The problem is now to deduce the evolution of P on the basis of the dynamics at the level of the elementary subunits. To this end we will establish the balance -in the sense of Sec. 2.6- of processes that lead the system to a given state $X = X_1, \ldots, X_n$ at time $t + \Delta t$. These processes will be related

The Probabilistic Dimension of Complex Systems

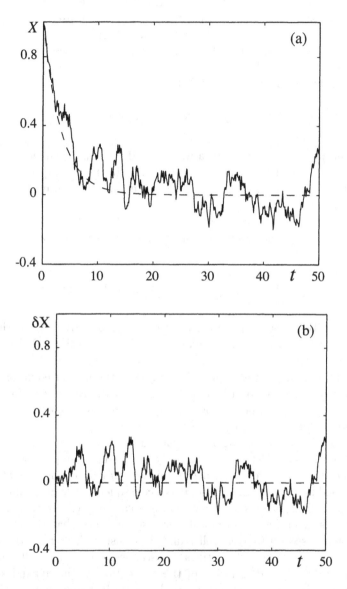

Fig. 3.2. (a) Macroscopic (dashed line) and mesoscopic (full line) trajectories of a typical observable X; (b) fluctuations of X around the instantaneous mean value.

to the *transitions* toward this state starting from states $X' = X'_1, \ldots, X'_n$ in which the system could be found initially (we adopt a discrete time picture since transitions between states are basically discrete events, and suppose that the interval Δt is sufficiently short such that the system will undergo at most one transition). Let $W_{\Delta t}(X|X')$ be the probability to perform such a transition during Δt. One is then led to write

$$P(X, t + \Delta t) = \sum_{X'} W_{\Delta t}(X|X') P(X', t) \qquad (3.5a)$$

In a given problem W will have a well-defined form, independent of P, that will have to be modeled or be determined experimentally. Eq. (3.5a) constitutes therefore the evolution law we were after. It is known in the literature as the *master equation*.

Since $P(X, t)$ must satisfy the conditions (3.4), it follows from (3.5a) that $W_{\Delta t}$ itself must be non-negative and normalized over all final states X for each given initial state X' :

$$W_{\Delta t}(X|X') \geq 0, \quad \sum_X W_{\Delta t}(X|X') = 1 \qquad (3.5b)$$

Notice that $W_{\Delta t}$ is an object displaying two indices, X and X'. It therefore has the structure of a matrix. A matrix satisfying conditions (3.5b) is called *stochastic matrix*.

The plausibility of the arguments leading to the master equation should not mask the presence of a heuristic element in the reasoning. Indeed, we have tacitly assumed that the probability at time $t + \Delta t$ depends solely on the probabilities at a time t just before the (very last) transition, or, in other words, that there is loss of memory of the transitions prior to Δt. We call this class of phenomena *Markov processes*. At first sight these processes seem to contradict the basic premise of deterministic character of the laws of nature as expressed by eq. (2.1) or (2.2) and, as a consequence, the Frobenius-Perron or the Liouville equations (3.2) and (3.3) as well. The contradiction is in fact only apparent, and all elements at our disposal show that the mesoscopic description based on (3.5) is well founded. Considerable progress has been achieved in view of its full justification starting from a Liouville equation involving a full scale description of the fine details of the dynamics of the elementary subunits. Aspects of this problem will be taken up in Secs 3.3 and 6.1. For now we summarize some key elements underlying these attempts, namely, time scale separation. The transitions between two macroscopic states corresponding to two different values of a variable X are sandwiched between a very large number of transitions affecting the system's variables

The Probabilistic Dimension of Complex Systems 71

other than X, which in a physico-chemical context are the velocities or the positions of the individual particles (see horizontal lines in the graph of Fig. 3.1). Everything happens therefore as if, at the end point of this long chain of transitions, the system "restarted" its evolution, in which the updated value X plays the role of an initial value. We recall that time scale separation is also at the basis of the existence of a closed form macroscopic description displaying a limited number of collective variables, as discussed in Sec. 2.2.

In most of the cases of interest involving physical systems one is concerned with the limit of continuous time. To deduce the form of the discrete master equation in this limit we add and subtract $P(X, t)$ on both sides of eq. (3.5a) and let $\Delta t \to 0$. This yields

$$\begin{aligned} \frac{dP(X,t)}{dt} &= \sum_{X' \neq X} [w(X' \to X)P(X',t) - w(X \to X')P(X,t)] \\ &= \sum_{X' \neq X} [J(X' \to X, t) - J(X \to X', t)] \end{aligned} \quad (3.6a)$$

Here $J(X' \to X, t)$ is a *probability flux*, and the transition probability per unit time w is related to the transition probability W of eq. (3.5a) by

$$\begin{aligned} w(X' \to X) &= \frac{W_{\Delta t}(X|X')}{\Delta t} \quad (X' \neq X) \\ &= \frac{W_{\Delta t}(X|X) - 1}{\Delta t} \quad (X' = X) \end{aligned} \quad (3.6b)$$

A very useful (albeit more heuristic than the master equation approach) variant of the mesoscopic level description consists in augmenting the macroscopic level evolution (eqs (2.2)) by *stochastic forcings* R_i accounting for the fluctuations:

$$\frac{\partial x_i}{\partial t} = f_i(x_1, ..., x_n, \lambda) + R_i(x_1, ..., x_n, t) \quad (3.7)$$

In this description, referred to as the *Langevin equation* description, the state variables and the time remain continuous. The principal problem is, of course, how to relate R_i to the dynamics of the elementary subunits. Usually, this is done by decomposing it to a product of a coupling function $g_i(x_1, ..., x_n)$ and a random process $r_i(t)$. The latter, also referred as *random force*, is manifested as a succession of impulses associated, for instance, to microscopic level events like collisions. In view of the ultimately chaotic and short scaled character of these events one expects that $r_i(t)$ will display an erratic dependence on t, taking indifferently positive or negative values at a given time independently of the previously registered values. We call *white noise* a random force satisfying these conditions. A powerful result of the theory of stochastic processes is that the evolution of a system submitted to such a forcing (which would otherwise be deterministic) can be mapped into a probabilistic process corresponding to diffusion in phase space. The associated probability density ρ satisfies then the *Fokker-Planck equation* written hereafter for concreteness in the case of *additive noise*, where the coupling function g_i is constant:

$$\frac{\partial \rho(x_1, ..., x_n, t)}{\partial t} = L\rho + \sum_i D_i \frac{\partial^2 \rho}{\partial x_i^2} \qquad (3.8)$$

Here L is the Liouville operator defined in eq. (3.3) and the D_i's depend on the strengths of the noises r_i, taken to be completely uncorrelated from each other.

It can be shown that eq. (3.8) can be deduced from the master equation (3.6) in the limit where the number N of the constituent units is very large and the system possesses a single attractor. Eq. (3.8) also applies when the process of interest consists in forcing deliberately a system from the environment - for instance, for the purposes of achieving a control of some sort. In this latter case one usually has to account for the presence of a non-trivial coupling function g_i as well.

3.3 The retrieval of universality

A major novelty introduced by the probabilistic approach is that the evolution laws at this level of description (eqs (3.3), (3.5) and (3.8)) are *linear* with respect to the unknown functions ρ or P. This is in sharp contrast with the nonlinearity underlying the deterministic description, which was repeatedly advanced as the major ingredient that makes complex behavior possible. Still, it does by no means imply that the solutions of the equations for the probabilities are straightforward: the complexity of the deterministic laws of evolution is reflected by the presence of nonlinearities in the coefficients of these equations as a result of which their full analysis constitutes an open problem in all but a few exceptional cases.

But there is a second novelty in the probabilistic description which, combined with the previous one, leads to a breakthrough in our understanding of complexity: in most cases of interest the probabilistic description displays strong stability properties - again, in sharp contrast with the deterministic description in which instability is the prime mechanism behind the ability of

The Probabilistic Dimension of Complex Systems 73

complex systems to explore the state space, to diversify and to evolve. As we see presently, the conjunction of these two features of linearity and stability allows one to bring out some unifying trends and thus to retrieve, at least in part, the lost universality of the deterministic description in the sense of Secs 2.3 and 2.4.

We first need to summarize some facts concerning the nature of the solutions of eqs (3.3), (3.5) and (3.8). We start with (3.3) which provides in a sense a "fundamental" description, being generated straightforwardly by the underlying evolution laws. The question we ask is, what kinds of probability densities describe a system that has settled on an invariant phase space region, e.g. an attractor. Such regimes should share with the state of thermodynamic equilibrium the property that, although the system performs continuously transitions between different subregions, the overall properties remain invariant in time. As the latter are generated by the probability density, we conclude that there should be a stationary state solution, $\bar{\rho}$ of (3.3) such that

$$\partial \bar{\rho}/\partial t = 0, \quad \text{or} \quad L\bar{\rho} = 0 \qquad (3.9)$$

This relation is, in fact, the quantitative expression of what a statistician would call the *stationarity* of a process.

The existence of solutions of eq. (3.9) defined on almost all phase space points accessible to the system and possessing, in addition, the property of uniqueness and of being sufficiently smooth in the mathematical sense of the term, is referred as the property of *ergodicity*. In practical terms this means that the phase space of an ergodic system cannot be decomposed into non-trivial invariant subspaces within each of which the trajectories remain trapped during the evolution. Here the terms "almost all" and "non-trivial" stand for the fact that there may be certain excluded sets of points like e.g. integer or rational values of x; such sets do not matter because they are "exceptional" - technically, their "measure" defined as the integral of $\bar{\rho}$ over the set vanishes.

Let us ask next how $\bar{\rho}$ is approached in time when the system starts with an initial smooth distribution ρ_0 of the kind described in Sec. 3.2. In a deterministic description what would happen is that the system's variables tend to their values on the invariant state as time goes to infinity. In the probabilistic description this would show up in the form

$$\int dx_1...dx_n \ A(x_1...x_n)\rho(x_1,...,x_n,t) \to \int dx_1...dx_n A(x_1...x_n)\bar{\rho} \qquad (3.10)$$

for any sufficiently smooth initial ρ_0, where A designates the set of relevant observables. We refer to (3.10) as the property of *mixing*. The strongest form of this property would be that ρ itself tends to $\bar{\rho}$ in the long time limit.

Clearly, ergodicity and mixing together imply uniqueness and asymptotic stability of the solutions of eq. (3.3) in sharp contrast as pointed out already with the non-uniqueness and instability of those of the deterministic description (2.2). It is legitimate to anticipate that in view of such properties the evolution of probabilities should display universality properties. We come to this question soon, but for now we stress that ergodicity and especially mixing hold true only in the presence of sufficiently complex dynamics. For instance for periodic or quasi-periodic behavior ergodicity holds true on the attractor, but the mixing property does not hold. In contrast, chaotic dynamical systems are as a rule both ergodic and mixing. Much of this evidence comes from the analysis of solvable models, from numerical simulations and from experiment. Proving that a given system is ergodic or mixing constitutes a largely open problem.

Consider next the mesoscopic level of description afforded by eqs (3.5) and (3.8). For concreteness we focus on the master equation, but most of the discussion applies to the Fokker-Planck equation as well. Let M be the full (possibly infinite) set of states of the (discrete) variables X. A state is called periodic if the system, once started on it, will return to it as long as the time lag is a multiple of Δt strictly larger than Δt. It is aperiodic, if no such lag exists. A state is called persistent if starting from such a state the system eventually returns to it. Aperiodic, persistent states whose associated mean return times are finite will be called *ergodic states*.

A powerful result of probability theory is that if the set of the available states consists entirely of ergodic states, then there is a unique steady-state solution of the master equation. In addition, starting with an arbitrary initial state this solution is approached irreversibly as time tends to infinity, a result reminiscent of the second law of thermodynamics. The mathematical property behind this powerful result is the *Frobenius theorem*, asserting that a stochastic matrix (in our case, the transition probability matrix $\mathbf{W}$) possesses an eigenvalue $\lambda_1 = 1$ to which corresponds an eigenvector $\mathbf{u}_1$ with positive components (and thus representative of a probability distribution), the absolute value (or modulus in the case of complex numbers) of all other eigenvalues being less than unity, $|\lambda_j| < 1$ for $j \neq 1$. Interestingly, the properties of uniqueness and asymptotic stability are here gained almost for free since they are guaranteed by the Markov character of the evolution.

We will now show, using the master equation, how uniqueness (ergodicity) and stability (mixing) can conspire to lead to a universal property underlying the evolution, in the form of a variational principle. We will be interested in large systems ($N >> 1$ in the notation of Sec. 3.2). One can then show that eq. (3.5) admits solutions of the form

$$P \approx \exp(-NU(X/N,t)) \quad (3.11)$$

where U plays the role of a generalized potential, in the sense that its minimum corresponds to the most probable state of the system. A potential of this sort always exists, and under certain conditions it is analytic in the sense that it can be expanded in powers of $1/N$ and of the order parameters appearing in the normal form description of Secs 2.3 and 3.4 :

$$U = (U_0 + \frac{1}{N}U_1 + ...) \quad (3.12a)$$

$$U_0 = \text{power series of the order parameters} \quad (3.12b)$$

Most important for our purposes is that U_0 governs the evolution toward the attractor, where it attains its minimum value :

$$\frac{d}{dt}U_0(X,t) \leq 0 \quad (3.13)$$

This expression is reminiscent of the extremum properties of the thermodynamic potentials (Sec. 2.2). Two major differences are that U_0 is generated by the dynamics of the macroscopic variables augmented by the effect of fluctuations, and that its range of validity is much wider than the vicinity of equilibrium.

It should be noticed that the validity of a power series expansion in the sense of eq. (3.12b) breaks down whenever the normal form description ceases to be universal in the sense of Sec. 2.4. Universality at the probabilistic level can still hold true in this case as well, provided that one gives up the analyticity property of the generalized potential.

The analysis of this section has brought out strong analogies in the properties of the master and Fokker-Planck equations on the one side, and of the Liouville equation for deterministic dynamical systems displaying complex behavior on the other side. As we will now see the analogy can be pushed further by the establishment, under certain conditions, of a master equation description starting with the Liouville equation.

We have already pointed out in Sec. 3.1 that in the presence of complex dynamics the monitoring of a phase space trajectory representing the succession of states of a system in a pointwise fashion loses much of its interest. One attractive alternative to this limitation is *coarse-graining*: we partition the phase space into a finite (or more generally denumerable) number of non-overlapping cells $C_j, j = 1, ..., K$ (Fig. 3.3) and monitor the successive cell-to-cell transitions of the trajectory. One may look at the "states"

$C_1, ..., C_K$ as the letters of an alphabet. In this view, then, the initial dynamics induces on the partition a *symbolic dynamics* describing how the letters of the alphabet unfold in time. The question is, whether the "text" written in this way displays some regularities (a "grammar" of some sort) and, if so, how these are related to the dynamics. It can be answered for a class of partitions known as *Markov partitions*, which enjoy the property that the partition boundaries remain invariant and each element is mapped by the dynamics on a union of elements. Such partitions exist if the dynamics shows sufficiently strong instability properties, as in the case of fully developed deterministic chaos. If the state of the system is initially coarse-grained, one can then show that eqs (3.2) or (3.3) can be mapped onto a master equation of the form (3.5), in which the transition probability matrix **W** is determined entirely by the indicators of the underlying dynamics. This mapping gives a concrete meaning to the statement made repeatedly that chaos is associated with a "random looking" evolution in time and space. There is no contradiction whatsoever between this result and the deterministic origin of chaos: in the probabilistic view we look at our system through a "window" (phase space cell), whereas in the deterministic view it is understood that we are exactly running on a trajectory. This is, clearly, an unrealistic scenario in view of our earlier comments.

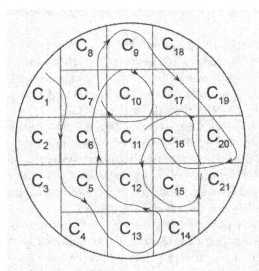

Fig. 3.3. Coarse-grained description in terms of transitions between the cells $C_1, ...$ of a phase space partition, as the trajectory of the underlying deterministic dynamical system unfolds in phase space.

Symbolic dynamics provides one with a powerful tool for classifying trajectories of various types and for unraveling aspects of the system's complexity that would remain blurred in a traditional description limited to the trajectories. An additional motivation for developing the idea of symbolic dynamics is that in many natural phenomena strings consisting of sequences of letters play a central role. For instance, the DNA (Deoxyribonucleic acid) and RNA (Ribonucleic acid) molecules, the principal carriers of information in living systems, are linear strings written on an alphabet consisting of four letters, "A", "C", "G" and "T" (or "U") according to whether the nucleotide subunit contains the bases adenine, cytosine, guanine and thymine (or uracil). Furthermore most of the messages transporting information or having a cognitive value such as books, music, computer programs or electrical activity of the brain are amenable in one way or the other to strings of symbols. These important issues will be taken up in Chapter 4, where explicit examples will also be given.

3.4 The transition to complexity in probability space

Let us come back for a moment to the phenomenology of the Rayleigh-Bénard convection and other similar phenomena discussed in Sec. 1.4.1, this time from the standpoint of the mesoscopic level description outlined in the previous sections. When ΔT is below the critical value ΔT_c, the homogeneity of the fluid in the horizontal direction makes its different parts independent of each other. But beyond the threshold ΔT_c, everything happens as if each volume element was watching the behavior of its neighbors and was taking it into account so as to play its own role adequately and to participate in the overall pattern. This suggests the existence of *correlations*, that is, statistically reproducible relations between distant parts of the system. It is important to note the long-range character of these correlations, which extend over macroscopic distances, as compared to the short range $(10^{-10} m)$ of the intermolecular forces. Put differently, intermolecular forces operate up to a distance equal to about the size of one molecule; but, as stressed in Sec. 1.4.1, a single convection cell comprises something like 10^{23} molecules. That this huge number can behave in a coherent fashion, despite the random thermal motion at the molecular level is one of the characteristic manifestations of the emergence of complex behavior. One of the interests of the probabilistic description is to allow one to formulate and analyze the concepts of coherence and correlation in a systematic manner.

Imagine that a fluctuation inflicts at a certain moment an excess $\delta x_i(\mathbf{r})$ to one of the variables of the macroscopic description - for instance, the value of the vertical component of the velocity at a point $\mathbf{r}$ in the Rayleigh-Bénard problem. What will be the impact of this event in a different point $\mathbf{r}'$? Clearly, in the absence of effect, $\delta x_i(\mathbf{r}')$ will be independent of $\delta x_i(\mathbf{r})$. Since the average value of all possible fluctuations is zero by definition, a natural measure of a possible dependence of $\delta x_i(\mathbf{r}')$ on $\delta x_i(\mathbf{r})$ is provided by the average of their product, where each individual value is weighted by the probability distribution descriptive of the system's regime as provided by the Liouville, master or Fokker-Planck equation, an operation denoted hereafter by angular brackets:

$$g(\mathbf{r}, \mathbf{r}') = <\delta x_i(\mathbf{r})\delta x_i(\mathbf{r}')> \qquad (3.14)$$

We summarize hereafter the properties of the *correlation function* defined by eq. (3.14).

(i) In thermodynamic equilibrium the fluctuations at two different points $\mathbf{r}$ and $\mathbf{r}'$ are uncorrelated, a fact reflected by $g(\mathbf{r}, \mathbf{r}')$ being proportional to a Dirac delta function $\delta(\mathbf{r} - \mathbf{r}')$.

(ii) The presence of nonequilibrium constraints switches on correlations that can be sustained indefinitely. The amplitude of the associated correlation function is proportional to the nonequilibrium constraint, and its range is macroscopic.

(iii) In the vicinity of a bifurcation, the birth of new regimes endowed by emergent properties -such as the Rayleigh-Bénard convection cells- is reflected by spatial correlations whose range tends to infinity in a direction transversal to the constraint, following a law of inverse proportionality with respect to the distance from the threshold. This invasion of the system by the correlations signals the qualitative change of its properties and provides an additional characterization of the phenomena of emergence and complexity.

Expressions like in eq. (3.14) are examples of a set of key quantities referred as *moments*, generated by a probability distribution and providing useful ways to characterize it. A moment of order $k_1 + ... + k_n$ is the sum of products $x_1^{k_1}...x_n^{k_n}$ weighted by the probabilities of their occurrence. If all k's but one are zero and the remaining k is one, one deals with the average $<x_i>$ of the corresponding variable. As seen above one is usually interested in the deviations δx_i of x_i around $<x_i>$ and in the associated moments $<\delta x^{k_1}...\delta x_n^{k_n}>$. If all k's but one are zero and the remaining k is two, one deals with the *variance*, $<\delta x_i^2>$ of the corresponding variable. Knowing the probability distribution is tantamount to knowing the infinite set of its moments of all orders, the converse being also true.

In practice, a drastic reduction becomes possible if the probability happens to be peaked sharply around a single, well defined value of the state variables. Consider for concreteness one particular variable, x. A measure of the dispersion of the probability around the average $<x>$ is provided by the *standard deviation*, defined as $<\delta x^2>^{1/2}$ or, more adequately, by the normalized quantity $\delta = <\delta x^2>^{1/2}/<x>$. A small δ implies a distribution that is sharply peaked about $<x>$, which in turn is close to the most probable value $\bar{x}$, for which the probability takes its unique maximum. From the standpoint of the previous chapters this is expected to be the case prior to the onset of complex behavior, when the system is settled in a single, stable macroscopic state.

In a physico-chemical context the smallness of δ is controlled by the size of the system, or by the number of steps ("trials") it takes to reach a certain value of the variable of interest. This can be expressed in an elegant and general form by two basic results of probability theory known as the *law of large numbers* and the *central limit theorem*. In both cases one is interested in the properties of a variable that is the sum of statistically independent variables having a common distribution, $X = x_1 + \ldots + x_N$. Under the very mild assumption that the averages $<x_k> = m$ exist, the law of large numbers asserts that the probability that the arithmetic mean $(x_1 + \ldots + x_N)/N$ differs from m by less than any prescribed small ϵ, tends to one as N tends to infinity and ϵ to zero. Under the additional assumption that the variances $<\delta x_k^2> = \sigma^2$ exist, the central limit theorem asserts that in the limit of N tending to infinity the probability distribution of $[(x_1 + \ldots + x_N)/N - m]/(\sigma/\sqrt{N})$ becomes as close as desired to the *Gaussian distribution*. It follows that the variance of X/N is of the order of $\sigma^2 N^{-1}$, the variance of X itself being $\sigma^2 N$. In terms of our dimensionless measure of relative dispersion, we have $\delta = (\sigma^2 N)^{1/2}/(mN) = \sigma/(mN^{1/2})$. This is a very small quantity indeed if N is of the order of the Avogadro number. There is, then, a clearcut separation between macroscopic behavior and microscopic scale dynamics. Similar results hold for probability distributions descriptive of independent events associated to discrete variables, such as the binomial and the Poisson distributions.

Results of this kind dominate the literature of probability theory; however, it is easy to imagine counterexamples. Consider, for instance, the situation prevailing beyond the bifurcation point of Fig. 1.1. We are witnessing a transition from a state of a unique most probable value $\bar{x}$ to a state characterized by two such values, say $\bar{x}_+$ and $\bar{x}_-$ for the upper and lower branch, respectively. This "probabilistic analog" of bifurcation is depicted in Fig. 3.4, curve (a). Clearly, we must be violating here the premises of the central limit theorem, and the variable distributed according to the figure can in

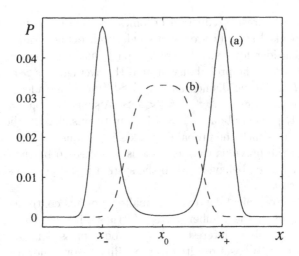

Fig. 3.4. Probabilistic analog of bifurcation. (a) two-humped probability distribution in the regime of two simultaneously stable states. (b) probability distribution at criticality, where the unique state prevailing prior to bifurcation is about to lose its stability.

no way be the sum of statistically independent random variables. This reflects the appearance of coherent behavior within the system. As a corollary, the standard deviation is of the same order as the mean, and the effect of fluctuations reaches a macroscopic level. In the framework of a deterministic description the question of how this coherence emerges and is sustained could not even be formulated. This is one of the basic reasons that the study of fluctuations forms an integral part of the dynamics of complex systems. Another reason is that in a multihump regime beyond bifurcation it is necessary to estimate the relative stability of the most probable states and the times the system spends in their vicinity, before undergoing sooner or later a fluctuation-induced transition to some other simultaneously stable branch.

Curve (b) in Fig. 3.4 sketches the expected form of the probability distribution in the borderline between unimodal and multimodal regimes. We have a flat distribution, which prepares the way for bifurcation by exploring regions of phase space that are increasingly distant from the most probable value. It can be shown that this distribution is non-Gaussian, thereby leading again to the breakdown of the central limit theorem. In this regime of criticality fluctuations now scale as $<\delta X^2> \approx N^{1+\alpha}$, α being a positive number. As criticality is approached from below the variance, which originally scales as N according to the Gaussian law, is enhanced by a factor inversely proportional to the difference between the actual value of the control parameter and its value at the bifurcation point. This behavior is at the origin of property (iii) of the spatial correlation function $g(\mathbf{r}, \mathbf{r}')$ referred earlier in this section. It is formally similar to what is observed close to the critical point when a material in thermodynamic equilibrium undergoes a phase transition. To mark this similarity one sometimes speaks of *nonequilibrium phase transition* in connection with this behavior, although its origin is quite different.

Notice that, as it also happens in ordinary phase transitions, the behavior of the fluctuations depends crucially on the way they couple in space. In a low-dimensional system, like a surface or a macromolecular chain, spatial couplings can modify drastically the way the variance scales both with respect to N and the distance from the bifurcation point. These effects can be evaluated by suitably adapting the technique of renormalization group originally developed in the context of equilibrium phase transitions.

Similar results are obtained for transitions leading to periodic behavior, to chaotic behavior and to spatial patterns associated with symmetry breaking. The main novelty is that rather than exhibiting several humps on isolated values of the variables, the probability distribution takes a crater-like form. The lips of the crater are projected on the corresponding attractor, which can now be a fractal set in phase space, whereas its dip corresponds to an unstable state.

We close this section by summarizing how the transition to complexity shows up at the level of the Liouville equation description (eq. (3.3)). As seen in Sec. 3.3, dynamical complexity is responsible for the properties of ergodicity and mixing. A more detailed view would be to determine how these properties are reflected on the structure of the invariant density $\bar{\rho}$ and on the eigenvalues and eigenfunctions of the Liouville operator L. An important result in this direction is that invariant measures (integrals of densities $\bar{\rho}$) descriptive of large classes of deterministic chaos are smooth along the unstable directions in phase space and fractal along the stable ones. As regards the Liouville operator L, it turns out that even the simplest bifurcation phenomenon (Fig. 1.1) leaves already a clearcut signature on its structure: the eigenvalues which are real, discrete and non-degenerate below the bifurcation point, become degenerate while remaining discrete above bifurcation and continuous right at bifurcation. In contrast, in a typical chaotic system there are both eigenvalues forming a continuous spectrum and eigenvalues forming a discrete set, referred as resonances. The corresponding eigenfunctions have as a rule a singular, or at best a (multi) fractal structure. For this reason their determination requires an embedding into a space of functions quite different from that of traditional functional analysis as practiced, for instance, in quantum mechanics.

3.5 The limits of validity of the macroscopic description

As mentioned in Sec. 3.3, in the probabilistic description the complexity of a system is reflected by the presence of nonlinearities in the coefficients of eqs (3.3), (3.5) and (3.8). A ubiquitous consequence of these nonlinearities is to generate an infinite hierarchy of evolution equations in which the averages are coupled to the variances and/or possibly higher order moments; the variances to the averages, third moments, and/or possibly higher order moments; and so forth.

3.5.1 Closing the moment equations in the mesoscopic description

A first question arising in this context is, whether the infinite hierarchy can be broken to a finite order and, if so, whether the resulting equation for the average value is identical to the macroscopic evolution law. The nature of this closure problem is best seen by using the Fokker-Planck equation (3.8). Let us multiply both sides of the equation by one of the variables, x_i and integrate over all possible values of x_1 to x_n. The left hand side will give rise, by definition, to the time derivative of the mean value $<x_i>$. In the right hand side one performs an integration by parts using the structure of the Liouville operator given in (3.3), and expands the rate functions f_i with respect to the fluctuations δx_j around the mean values of the x_j's. Keeping in mind that the average value of a fluctuation is zero one obtains then the expression

$$\frac{\partial <x_i>}{\partial t} = f_i(<x_1>, ..., <x_n>, \lambda) + \sum_{jk} \left(\frac{\partial^2 f_i}{\partial x_j \partial x_k}\right)_{<x_j>...} <\delta x_j \delta x_k>$$

$$+ \text{ terms possibly involving higher order moments} \quad (3.15)$$

where the subscript in the second derivative of f_i indicates that the corresponding expression is to be evaluated at the mean values of the variables. As can be seen, the first term of eq. (3.15) is identical to the macroscopic law of evolution (eq. (2.2), or eq. (3.7) in the absence of the stochastic forcing term). The presence of the extra terms indicates that the evolution of the macroscopic observables does not depend solely on the full set of these observables -a property referred earlier as the *mean field approximation*- but is also coupled to the fluctuations and hence to the microscopic effects. If one now multiplies eq. (3.8) by products like $x_j x_k$ and integrates over all values of x_1 to x_n one will find that the evolution of $<\delta x_j \delta x_k>$ is coupled to that of $<\delta x_j \delta x_k \delta x_l>$ or of still higher moments, and so forth. This is, precisely, the infinite hierarchy problem mentioned at the beginning of this section.

To assess the range of validity of the macroscopic (mean field) description it is necessary to estimate the order of magnitude of the extra terms in eq. (3.15) which, in typical situations, would be determined by the second order variance $<\delta x_j \delta x_k>$. Depending on the conditions, different situations can be realized.

(i) The contribution in $<\delta x_j \delta x_k>$ is negligible. This is what happens in systems composed of a very large number N of subunits and functioning around a unique stable state (not necessarily time-independent). The term in question -which measures in fact, up to a proportionality factor, the strength D_i of the stochastic forcings- is then of the order of $1/N$ and tends to zero as N gets very large, a limit referred to as the *thermodynamic limit*. This separation between macroscopic values and fluctuations remains applicable in the vicinity of a nonequilibrium phase transition (in the sense of Sec. 3.4) as well, provided that the transition point is approached from the prebifurcation region. Fluctuations are now penalized in (3.15) by a factor of the order of $N^{-\delta}$, where δ is a positive number less than one.

(ii) If the system is in a region of multiple, simultaneously stable states the probability density exhibits several peaks around each of these states, as mentioned already in Sec. 3.4 in connection with the pitchfork bifurcation. Clearly, the mean value $<x>$ is no longer representative: it corresponds to a state of low probability, which may for instance coincide with the unstable state formed by the continuation of the initially unique state past the bifurcation point. Fluctuations become comparable to the averages in eq. (3.15), and a global description limited to the macroscopic observables breaks down. The evolution becomes then essentially probabilistic, entailing that an infinity of observables (the full set of moments) is necessary to describe the state. Notice that a local description may still be possible, provided that one remains in the vicinity of each of the probability peaks. This description, obtained by linearizing the evolution equations around a particular stable state, holds as long as the system does not jump to one of the other stable states as discussed further in the next subsection.

(iii) If the system contains a small number of particles, the standard deviation of the fluctuations may be comparable to the mean. This is what happens in nanodevices whose technological importance is increasingly recognized as well as in biology where, for instance, the number of copies of a particular regulatory unit in the genetic material of living cells is, typically, small. Notice that in such small size systems certain collective phenomena like sustained oscillations tend to be decorrelated by the fluctuations.

(iv) In systems evolving according to widely different time scales, minor fluctuations building up during a slow stage may be amplified by the fast stage to a point that they reach a macroscopic level. This will be reflected by the dispersion of certain dynamical quantities such as the ignition time in a combustion process, the fracture time of a material following the proliferation of defects, the switching between the exponential growth and the saturation regime in a population. Other potential realizations of this mechanism pertain to evolutionary processes. During a slow period of evolution fluctuations endow the system with a variety of states visited with some probability, but the initial state remains dominant. The occurrence of a fast stage, which seems to be compatible with many observations, entrains then some of these states at such a high rate that they soon lose track of the common "ancestor", thereby producing a new clone of their own.

(v) In systems of restricted geometry, limitations in mobility or in cooperativity arising from the decrease of the number of neighbors of a given unit tend to favor the generation of anomalous spatially inhomogeneous fluctuations leading eventually to segregation. This will compromise the effective mixing implicit in the mean field picture. Important classes of such systems are catalytic surfaces, biological membranes, or films.

It should be noticed that throughout this discussion it was understood that the interactions between the subunits are localized. In the presence of long range interactions certain non-trivial amendments are required. As a rule, such interactions tend to enhance the range of validity of the mean field description.

3.5.2 Transitions between states

From the standpoint of the deterministic description, a transition between simultaneously available states at given values of the control parameters can only be realized by the action of an external perturbation bringing the system from the domain of attraction of the initially prevailing state to that of the new state. If these two regimes are separated by a finite distance the corresponding perturbation needs to be quite massive, and hence hardly realizable under normal conditions.

The situation changes when fluctuations come into play. Our starting point is eq. (3.15), which we rewrite in the more general form

$$\frac{d<x>}{dt} = \text{(Macroscopic rate law)} + \text{(contribution of the fluctuations)}$$
(3.16)

Clearly, at some stage of the transition process the value $<x>$ is bound to undergo a finite change. To describe the growth of the "critical mode" responsible for such a change, time-dependent solutions of the Fokker-Planck or of the master equations must be envisaged. As growth goes on, the contribution of the variance becomes increasingly marked. Thus, the second term to the right hand side of eq. (3.16) may be non-negligible in certain volume elements ΔV of small size and, after a sufficient (possibly very long) lapse of time, it can take over and drive the average to a new macroscopic regime.

Figure 3.5 illustrates this *nucleation view* of a transition between (for concreteness) two coexisting steady states, say x_+ and x_-, separated by an unstable one, x_0 (Fig. 3.5a). Initially we are given a probability distribution peaked around one of the (deterministically) stable branches, chosen here to be x_- (Fig. 3.5b, full line). Suppose that ΔV is small enough for the system to be maintained spatially uniform. From our previous discussion we expect that this will be so as long as the characteristic dimension of ΔV is less than the correlation length. Because of the fluctuations, the actual value of x in ΔV will experience spontaneous, random deviations from the initial most probable state, x_-. The vast majority of these fluctuations will have a small intensity and range, still, under their cumulative effect the subsystem within ΔV is bound to be driven sooner or later slightly beyond the "barrier" constituted by the unstable state. At this point it will jump quickly to the other (deterministically) attracting state x_+, thereby constituting a "nucleus" of instability. The nucleus will expand at the expense of its environment if x_+ is stability-wise more dominant (in a sense to be specified shortly) than x_- and, at the same time, other volume elements within the system will experience similar transitions. Eventually, the entire system will become contaminated and the associated probability distribution will have the form indicated by the dashed lines in Fig. 3.5b.

Figure 3.5c shows the plot of the associated *generalized potential*, U. As seen in Sec. 3.3 its minima and maxima correspond, respectively, to the asymptotically stable and unstable solutions of the phenomenological equations of evolution. Its interest in the present context is that it allows one to quantify just what is meant by crossing the "barrier" constituted by x_0 and to actually monitor the rate of transition between x_- and x_+. The simplest situation is when the system involves a single variable or operates in the vicinity of a bifurcation point associated to a normal form with a single real-valued order parameter. Detailed calculations show that the typical time scale for the transition from x_- to x_+ is of the order of

$$\tau_{trans} \simeq \exp(N\Delta U) \qquad (3.17)$$

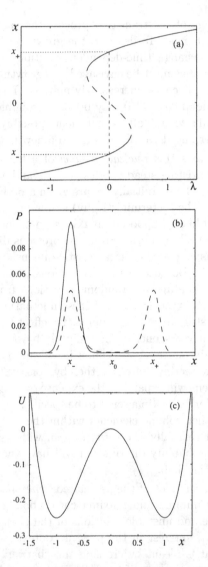

Fig. 3.5. Fluctuation-induced transitions between simultaneously stable steady states. (a) Dependence of solutions of the macroscopic (mean-field) evolution equations on a parameter λ, normalized in such a way that at $\lambda = 0$ state x_+ is equally stable to state x_-. (b) Evolution of an initial probability centered on x_- (solid line) towards a two-humped final distribution centered on x_- and x_+ (dashed line). (c) Kinetic potential U as a function of x at $\lambda = 0$.

where $\Delta U = U(x_0) - U(x_-)$ is the "potential barrier" that has to be overcome by the fluctuations. If a physico-chemical system of large spatial extension for which $N \approx 10^{23}$ is artificially kept uniform τ_{trans} will be extremely large, much larger than any conceivable geological or cosmological time. This shows that in real world situations a transition has to start as a localized event, involving an element of small size. Even so, the characteristic time will remain considerably larger than the characteristic relaxation times displayed by the deterministic laws of evolution, unless the system is very close to the turning points of the S-shaped curve of Fig. 3.5a, at which the value of the barrier becomes very small.

Figure 3.6 describes the kinetics of the fluctuations in the transition between multiple steady states. After a long waiting time, a first sufficiently large germ emerges, whereupon the system jumps almost instantaneously to the new state. The latter becomes in its turn unstable under the formation of a second finite size germ, and so forth. Notice that this second event will be postponed practically indefinitely if the second state is more stable than the first, in the sense of a larger probability peak associated to it.

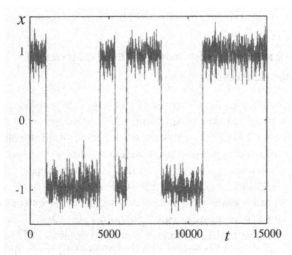

Fig. 3.6. Kinetics of the fluctuations during the transition between two simultaneously stable steady states. Transitions across the barrier associated to the maximum of the kinetic potential (Fig. 3.5c) occur on a time scale given by eq. (3.17).

The kinetics of fluctuation-induced transitions between states can be affected in a most unexpected way by the action of external forcings. We list below two generic mechanisms, which are attracting increasing attention.

(i) The presence of a periodic forcing, even very weak, can favor the transition between states, provided that its period is of the order of the transition time (3.17). This is the phenomenon of *stochastic resonance*, originally discovered to explain the existence of a mean periodicity in the Quaternary glaciations (Fig. 1.5). Its unexpectedness stems from the fact that, contrary to what one would have guessed, noise -here manifested in the form of fluctuations- seems to enhance the system's sensitivity and to be at the origin of orderly behavior. This would be impossible in a purely deterministic setting.

(ii) When the control parameters are varying slowly in time (see also Sec. 2.1) some of the transitions of the type depicted in Fig. 3.5 may become quenched. The system becomes then locked into a "frozen state", even though deterministically the two states involved may have identical stability properties. Time-dependent control parameters arise in, among others, atmospheric and climate dynamics where they stand for anthropogenic effects or for secular variations of the solar constant. On a smaller scale they are related to the process of switching in mechanical, electronic, or optical devices.

3.5.3 Average values versus fluctuations in deterministic chaos

As we saw in Secs 3.3 and 3.4, deterministic chaos is amenable to a probabilistic description provided that certain sufficiently strong ergodic properties are satisfied. However, contrary to the situations encountered in the mesoscopic description, even in the presence of a single attractor the associated probability distributions are, as a rule, delocalized in phase space implying that fluctuations around the mean are comparable to the averages. As a corollary, the average of an observable will not obey to a closed equation, but will be coupled to higher moments in a way similar to eq. (3.16). Eventually, this will give rise to an infinite hierarchy of coupled equations. The question is, whether there exist closure schemes for this hierarchy involving solely the first few moments of the probability distribution, similar to the mean-field limit of the mesoscopic description.

Intuitively, one expects that the answer should be in the affirmative for chaotic evolutions in the form of a small scale variability centered on a clearly defined most probable value. Such situations, reminiscent of the central limit theorem discussed in Sec. 3.4, can indeed arise when a large number of subunits, each one of which is in a strongly chaotic regime, are coupled in space. Since a Gaussian distribution is fully determined by its first two moments, a description limited to the mean values and the variances becomes therefore possible.

On the other hand, in many generic situations complex systems behave more like a collection of "hot spots" loosely coupled in phase (or in real) space, rather than as a more or less uniform background contaminated by a small scale variability. An example is provided already by one of the simplest chaotic systems, the logistic map (eq. (2.18)). Here the probability density takes its largest values at the end points of the unit interval and is minimum at $x = 1/2$, which corresponds to the mean value of the "observable" x. Other significant examples are provided by intermittent systems. At the level of the probability density we now have structures which are either multimodal as in Fig. 3.4 or exhibit a long tail, where they fall like a power law. Can a reduction to a few key observables still be expected under these conditions?

The first idea that comes to mind -simply neglect variances beyond some order- that works so well in macroscopic systems operating around a unique stable state and submitted to weak fluctuations, fails here completely. Rather than neglecting such variances altogether one should thus express them as functions (generally time-dependent) of the low order ones. This stems from the observation that in the regime of long times there exist subclasses of moments varying on a slow time scale given by the eigenvalues of the Liouville operator (eq. (3.3)) that are closest to zero. Any other moment varying on this dominant time scale is thus bound to be expressible in terms of this set, thereby giving rise to a closed set of evolution equations descriptive of the chaotic regime, involving a limited number of variables. This idea has been implemented with success on some of the most representative systems generating deterministic chaos.

Finally, in the presence of highly inhomogeneous chaos yet another possibility would be to reduce the system to a few discrete states corresponding to the "hot spots" linked by transitions as in subsection 3.5.2. Such a mapping, which is reminiscent of the idea of symbolic dynamics discussed in Sec. 3.3 can be carried out rigorously in some simple cases but needs otherwise to be implemented heuristically or justified by numerical experiments. A characteristic feature is that the effective "noise" linking the states can no longer be reduced to a white noise as in eq. (3.6): the overall dynamics is a non-Markovian process, displaying memory effects.

A variant of the above situation would be a transition between simultaneously stable states induced by a deterministic forcing of the chaotic type rather than by the fluctuations, as in subsection 3.5.2. As it turns out, owing to the fact that the probability density of such a forcing is non-vanishing only within a finite region of phase space (defined by the size of the corresponding attractor), the occurrence of transitions requires non-trivial threshold conditions giving rise to a dependence of the rate of transition on the forcing strength which is quite different from the one featured in eq. (3.17). This property in conjunction with the memory effects inherent in the deterministic character of the forcing entails that when possible, transitions occur on a slower time scale. Finally the structure of the probability distribution of the response variable is deeply affected displaying, in particular, a fine structure inherited from the fractal character of the chaotic probability densities along the stable directions as seen in Sec. 3.4.

3.6 Simulating complex systems

In the preceding chapters, we have established the fundamental equations which govern the evolution of complex systems at different levels of description. We have stressed the fact that exact solutions of these evolution laws are known only in a limited number of cases and that, for this reason, one needs to resort to qualitative approaches and to approximation schemes. In view of these limitations it becomes therefore crucial to develop numerical methods enabling one to check theoretical conjectures, to visualize properties that might otherwise look abstract, or simply to address problems for which an analytic approach is beyond the reach of the techniques available. An additional argument in favor of computational methods relates to the lack of universal classification schemes of the various modes of behavior predicted by a typical evolution law, beyond the range of the standard regimes of fixed point or limit cycle behavior.

A first class of computational methods aims to develop algorithms for solving numerically the evolution equations of the variables of interest. There is a huge, highly sophisticated and specialized literature on this important topic. Complex systems research is an intense user of such algorithms, just like any other major scientific field. Here we focus on a second class of computational methods, namely *simulation* methods, addressing directly the process of interest rather than an associate set of evolution equations. This latter type of computational approach has in fact become over the years a major actor in complexity research. One reason for this is that complexity pervades a host of problems beyond the traditional realm of physical sciences, where the structure of the elementary units constituting a system and their interactions may not be known to a degree of detail and certainty comparable to those of a physical law. Simulation methods provide then an appealing alternative, by focusing on a minimal amount of initial information deemed to be essential and by subsequently exploring the different scenarios compatible with this information. In a similar vein, bridging the gap between different levels of description -a question of central importance in connection with the basic property of emergence- is often accompanied by a substantial reduction of the number of variables involved: switching from $N \approx 10^{23}$ microscopic degrees of freedom to a few collective variables, from a large number of coupled processes to a normal form type of dynamics, and so forth. Here again, understanding the main features of this transition is greatly facilitated by a judicious choice of the degree of fine detail to be incorporated at the outset at the most "elementary" level. We hereafter outline some representative approaches along these lines.

3.6.1 Monte Carlo simulation

This approach is an important complement of the mesoscopic description of complex systems based on the master equation (eq. (3.5)) and its continuous time limit (eq. (3.6)). It provides information on the role of fluctuations -or more generally of individual variability- in shaping the overall behavior, on the mechanisms of selection of the different regimes available and on the status of the macroscopic (mean field) description.

Suppose that the state of the system is described by a set of discrete variables taking the values $X_j, j = 1, \ldots$ The idea is that within the range of validity of a Markovian assumption, the evolution of such a system can be viewed as a generalized random walk on a superlattice whose nodes correspond to all possible values of these variables. The system is started initially on a state corresponding to the value X_k. The process is then determined entirely by (a), the probability $W(X_k \to X_\ell)$ that a jump occurs from state k to state ℓ between times t and $t + \Delta t$; and (b), the waiting time, τ in state k. In a Markov process τ is an exponentially distributed random variable with a probability density given by $\pi_k(\tau) = C_k e^{-C_k \tau}$ where C_k is a sum over all transitions leading to a change of the state k, $C_k = \sum_{j \neq k} W(X_k \to X_j)$.

These ingredients form the basis of a Monte Carlo algorithm generating a stochastic trajectory of the system, which can be decomposed into the following steps.

(i) *Initialization*. The current time is set to $t = 0$, the final time t_f is chosen and the sampling interval Δt is fixed. The values of the X's are specified and the transition probabilities W are computed according to a rule depending on the way the processes present are modeled at this level of description. For instance, in a process involving transitions associated to the spontaneous inactivation of a chemical compound one sets $W(X \to X - 1) = kX\Delta t$.

(ii) *Choice among processes present*. If (as it usually happens) processes involving several types of transitions are simultaneously present, only one among them will be activated at a given Monte Carlo step. The choice is based on weighting these processes by appropriate probabilities, determined by the ratios of individual W's to the sum of the W's associated to all transitions present. The particular option, μ that will be realized will then be determined by the outcome of a random number generator.

(iii) *Waiting times.* A particular value of waiting time τ is sampled from the exponential distribution $\pi_k(\tau)$. In practice this involves a second random number generator, following the mapping of the exponential process into a uniform noise defined in the unit interval.

(iv) *Time evolution and updating.* Given μ and τ, the system is evolved by updating time t to $t+\tau$ and $X = X_k$ -the starting value of X- to $X' = X_\mu$, for all X's involved in the process μ selected. The transition probabilities W are recomputed given the new values of X.

(v) A new Monte Carlo step starts unless t exceeds t_f.

Using this procedure several realizations, r of the process corresponding to different choices of initial conditions can be obtained in the form of time series $X(r,t)$, an example of which was already provided in Fig. 3.1. Collective properties such as mean values, variances, correlation functions and higher order moments can thus be evaluated through such relations as

$$< X >_t = \frac{1}{N_{tot}} \sum_{r=1}^{N_{tot}} X(r,t), \quad \text{etc.} \tag{3.18}$$

where N_{tot} is the total number of realizations. Actually, if the process has good ergodic properties one may also obtain the asymptotic (long time) properties by following a single trajectory provided the simulation time is long enough,

$$< X >_s = \frac{1}{N\Delta t} \sum_{j=1}^{N} X(j\Delta t) \tag{3.19}$$

These predictions can be confronted with the experimental and theoretical results available. On the basis of this comparison the rules underlying the transition probabilities W may have to be reassessed, and this can lead in turn to new insights on the problem at hand.

3.6.2 Microscopic simulations

With the growing power of computers, a new way of investigating complex phenomena in physico-chemical systems became available in the mid 1980's.

This method, known as molecular dynamics (MD), consists in solving numerically Newton's or Hamilton's equations of motion (cf. Sec. 2.2.1) for an assembly of interacting particles. An appropriate average over space and time of the basic mechanical variables, i.e. mass, momentum and energy, gives then access to macroscopic quantities such as pressure, temperature, etc. in a way analogous to eqs (3.18)-(3.19). Regarded first as a curiosity by the scientific community, the method eventually became one of the main tools of statistical physics.

For a long time MD was limited to equilibrium phenomena. More recently, it has been successfully extended to the study of nonequilibrium systems as well including a variety of complex phenomena arising in simple fluids such as shock waves, flows past an obstacle or the Rayleigh-Bénard instability (cf. Sec. 1.4.1). Today, it is considered by physicists as a sort of laboratory where they can design a desired "ideal experiment" to check, for example, the validity of a theory in extreme conditions where real experiments are difficult, costly or even impossible to perform, or to bridge the gap between the phenomenological analysis of large scale macroscopic phenomena and their modeling at the microscopic level. We use here the word "ideal experiment" to stress that in no way can MD replace the real world experiment: no matter how cautious one may be, MD remains a numerical approach limited by the finite precision of digital computers, just like any other numerical technique.

The great merit of MD is to keep track of the deterministic origin of the phenomenon considered, without introducing ad hoc probabilistic assumptions. It thus helps to clarify the foundations of complexity, showing how a priori unexpected structures and events can emerge from the elementary laws of microscopic physics. A serious challenge in the full realization of this program pertains to the way one accounts for the constraints acting on the system which, as we saw on many occasions, play a crucial role in the onset of complex behavior. In a physical setting, constraints are manifested in the form of either external fields acting on each of the system's elementary units (e.g. electric field on a system of charged particles), or of boundary conditions describing how the system communicates with the external world through its outer surface (e.g. temperature differences in the Rayleigh-Bénard convection). A full solution of this "thermostating" problem is still missing and, for this reason, extra-mechanical assumptions are sometimes introduced in a MD algorithm to account for constraints in a tractable way. We refer to Sec. 6.1 for some further comments on this point in connection with the foundations of irreversibility.

It is a matter of observation that in many areas of physical science, the macroscopic behavior of a system simulated by MD depends rather weakly on the fine structure of the constituting particles and the details of their interactions. For instance, the hydrodynamic behavior of a fluid whose particles are caricatured as hard elastic spheres is essentially the same as that of a real world fluid provided one is not very close to a liquid-gas transition. Likewise, many of the statistical properties of complex reactive systems are well captured by a "color" or "spin" dynamics, whereby hard spheres change identity ("color" or "spin") upon colliding, mimicking in this way real world reactions. These observations are the starting point of a number of useful variants of a full scale MD, where the underlying "rules" are simplified in one way or the other with respect to the full set of constraints introduced by the laws of physics. Simplifications of this kind may also be accompanied by the replacement of a fully deterministic transition rule by a Monte Carlo step, as in Sec. 3.6.1. Examples of such approaches include lattice gases, where the positions and velocities of the particles are discretized; and the Monte Carlo simulation of the dynamical processes underlying the Boltzmann equation (cf. also Sec. 6.1).

3.6.3 Cellular automata

In the opposite end of the hierarchy of behavioral rules as compared to the full scale MD lie cellular automata simulations, sometimes considered (erroneously) as the exclusive mode of simulation of complex systems.

A cellular automaton (CA) aims to create the conditions for the emergence of self-organization at a higher level, starting from elementary units interacting via a set of simple rules. These units are arranged on the cells of a lattice, usually two-dimensional. They are characterized by a discrete set of internal states, for instance (in the simplest case) by two states that may be labeled by 0 and 1. Time flows in discrete steps and, starting with a given configuration, one computes the state of the whole lattice in the next step by a well-defined, unique rule prescribing how transitions take place on the basis of the present state of each given cell and its neighbors. The rule can be deterministic, in which case the new configuration in the cell will be uniquely determined by its previous state and the state of its neighbors; or probabilistic, in which case a Monte Carlo type step will have to be implemented. In addition of being easily programmable on a computer algorithms of this kind allow for a straightforward and appealing visualization of the evolution, two features that are largely responsible for the popularity of CA.

CA can model with different degrees of realism concrete physico-chemical, biological, social and financial processes such as chemical oscillations, food recruitment in ants (cf. Sec. 1.4.3), or the propagation of rumors and diseases. They also serve as metaphors as in the famous game of life, where a "living cell" is created in the course of encounters of three of its neighbors, or dies from either isolation or overpopulation. In this latter context, it is important to strike the right balance between the real world and its abstract representation, in order to avoid unduly strong claims and induce unnecessary confusions.

Viewed as abstract dynamical systems CA provide interesting insights on the foundations of information and computation theories, which will be addressed more amply in the next chapter. At the origin of these insights is the possibility, first discovered by Stephen Wolfram in the context of one-dimensional models, to classify CA in four classes according to the dynamical behaviors generated. Automata of classes I to III describe processes converging eventually to fixed points and to a succession of regular (periodic) and irregular configurations, respectively. Automata of class IV, to which belongs the game of life, are on the other hand more intricate. They are characterized by long transients and a large diversity of structures that enter in interaction. Such structures can be used to generate binary signals and to construct the elementary logical gates at the basis of the functioning of a computer.

3.6.4 Agents, players and games

Simulations of complex systems, especially in the context of computer science, sociology and finance may also be formulated as a set of entities or "agents" (e.g. the cells in a CA model) that share a common environment and interact directly or via the environment according to a prescribed set of simple rules or "strategies" (e.g., those of the game of life). The key issue is, again, to understand the relationship between individual behaviors and the emergence of collective phenomena. The latter can also be viewed as a "goal", a "problem solving" or a "decision making" that would be impossible to achieve by an individual. The specificity of multi-agent simulations compared to the previously discussed ones is to account for beliefs, desires and intentions of the agents and place emphasis on the adaptive character of the resulting behaviors.

As stressed on many occasions, the evolution of complex systems is a cooperative process with multiple outcomes. A very frequently encountered form of cooperativity is competition between the individual entities and, in this context, the multiplicity of outcomes can be viewed as a set of "options" available to each of them. If, in addition, each of these options carries a "value" of some sort, then the evolution can be simulated as a *game*. A game consists, then, of a set of players, a set of moves (or strategies) available to these players, and a specification of the payoffs resulting from the different strategies, being understood that the ultimate goal is to maximize the players' returns. Since decisions are made in the presence of many interacting players, as a rule the costs and benefits of each option depend upon the choices of other individuals. In games like chess information on these choices is available to each player (games of perfect information), but in many instances (including those where the players act simultaneously rather than sequentially) one deals with imperfect information games.

Originally developed in the context of economics, game theory has since been applied in politics (war strategies) as well as in academic research, from logic and computer science to biology (evolutionary stable strategies), psychology and sociology. There is a highly sophisticated mathematics underlying game theory, which is out of the scope of this book.

3.7 Disorder-generated complexity

So far in this book dynamics -and thus function- was the primary responsible for complex behavior. Now, in many situations of considerable concern one encounters macroscopic or at least supermolecular objects that would definitely qualify as complex, on the sole basis of their intricate structure. What is more, in certain systems of this kind (the DNA being the most provocative example) it seems that it is the structure that dictates the function rather than the other way around, as advanced in our preceding analysis.

The most clearcut examples illustrating these new possibilities are provided by the physics of disordered systems. In a crystal, molecules are fixed on the nodes of a periodic lattice and are thus subjected to identical environments. In an ordinary fluid molecules are continuously moving, yet during their "thermal motion" they scan environments that are still identical though now in a statistical sense. But there is a whole class of materials in which one also observes *glass phases*, where the molecules are frozen in a disordered (though not fully random) structure. For instance, in certain magnetic alloys referred as *spin glasses*, the magnetic moment, or spin of the ions is not pointing in certain prescribed directions in a repetitive fashion but freezes along disordered directions when the temperature decreases below some threshold.

The interest of spin glasses is to provide a paradigm for understanding the principal manifestations of this disorder and, subsequently, for taking advantage of it. This is due to the special character of the basic variables -the spins of the individual subunits $1, ..., N$ constituting the system- which can only take a discrete set of values: a paradigmatic situation indeed, reminiscent of a host of other problems beyond the realm of physics where similar entities come into play, from the sequence of monomers on a biopolymer chain to the opinions of a population of competing agents in the stock market.

The main point to realize is that disorder often reflects the impossibility to find a global configuration minimizing a certain energy-like property under a given set of constraints on the basis of optimizations of the local (e.g. pairwise) structures. This quantity is usually taken to be of the form

$$E = -\sum_{i,j} J_{ij}\sigma_i\sigma_j \qquad (3.20)$$

Here i,j run over the total number, N of subunits, σ_i is the property attributed to subunit i (the spin in the case of the magnetic material) and J_{ij} measures the strength of the i-j interaction. Alternatively one is also led to consider a *fitness function* W, essentially the opposite of E, and require its maximization under the imposed constraints.

It should be emphasized that optimization is a goal that is imposed at the outset to the system at hand. In a material in thermodynamic equilibrium at a given temperature optimization may well follow from the laws of statistical mechanics and thermodynamics, and lead to the minimization of free energy (Sec. 2.2). This is no longer so in more intricate situations.

Be it as it may, let us try to figure out the origin of the incompatibility between global and local optimization. In a molecular system the answer is rather clearcut: incompatibility stems from the nature of the interactions. For instance, in a system of three spins on the nodes of a triangle interacting via negative J's, one can have at most two energetically favorable antiparallel (opposite values of σ) pairs, sandwiching an energetically unfavorable parallel pair. This is the phenomenon of *frustration*. Conversely, let the J's be preassigned and the σ's of the N subunits be initially distributed, e.g. randomly. The subunits are next given, in a sequential fashion, the possibility to shift their σ's in order to decrease the initial value of E. This process goes on as long as needed, until a configuration is reached where no decrease of E is possible by modifying just one σ at a time. Intuitively, one suspects that this configuration should depend on the initial condition and on the values of the J's. This is confirmed by a detailed analysis and constitutes, in fact, one of the major results of the physics of disordered systems: there exists, typically, a large number of locally optimal configurations. Once caught on one of the above configurations the system will remain trapped for ever if it disposes of no mechanism of variability in time. Local optima are thus the analogs of the macroscopic states of our preceding analysis.

In short, disordered systems possess a multiplicity of macroscopic states, thereby satisfying one of the basic premises of the phenomenology of complex systems as discussed in Sec. 1.2. This picture can be extended to account for transitions between local optima much in the spirit of Sec. 3.5.2, by incorporating dynamic variability. The latter can arise from molecular level fluctuations, which will inevitably be switched on as long as temperature is finite; or from "errors" arising from environmental noise. They will be responsible for the escape from an initial local optimum toward a new one. Eventually, after many trials the "absolute" optimum corresponding to the lowest value of E will be reached, but this might require an amount of time so long that would be irrelevant for any practical purpose. In an equilibrium environment the probability of a given configuration in this wandering process is given by a classic formula of statistical mechanics,

$$P(\sigma) = e^{-\beta E(\sigma)}/Z \qquad (3.21)$$

where E is something like (3.20), β^{-1} measures the strength of the fluctuations and Z is the sum of all $e^{-\beta E(\sigma)}$ factors over the possible configurations of the σ variables.

There exists a highly developed theory unveiling a host of interesting properties of the locally optimal states and leading to systematic classification schemes, depending on the ways the coupling constants J_{ij} are assigned in eq. (3.20). For instance, some of the J_{ij} can be zero, others can be chosen completely at random, or sampled from a given probability distribution. We again stress that the fundamental difference with the type of systems considered in the preceding sections is optimization. Once a target of this kind is defined from the outset the behavior is bound to be limited to steady states, all forms of sustained temporal or spatio-temporal activity (periodic oscillations, chaos, etc.) being excluded. In a sense one deals here with a subclass of complex systems corresponding to a highly original variant of catastrophe theory introduced in Sec. 2.4, which has the additional merit to possess a firm microscopic basis thanks to the paradigm of spin glasses.

The concept of frustration and the concomitant existence of a large number of long living metastable states find interesting realizations in biology. A generic class of problems pertains to the motion (in a generalized sense) on a *rugged fitness landscape* provided by a certain quantity playing the role of evaluator of the "quality" of a given configuration, plotted against a set of underlying variables defining the accessible configurations. Consider, for instance, the three-dimensional structures associated with the folding of a protein (the so-called tertiary structure) of a given aminoacid sequence (the so-called primary structure). In an equilibrium environment the probability of occurrence of a configuration is given by eq. (3.21), and the aforementioned evaluator is just minus the free energy. Now, it is well known that minute changes in the aminoacid sequence may strongly affect the folding and, conversely, that quite different sequences may lead to similar three-dimensional structures. The mapping from the "sequence space" to the configuration space is thus expected to be neither monotonic nor even smooth. In other words, the fitness function would possess many local maxima in the sequence space - the set constituted by all possible aminoacid sequences. Moving along this space may transform a local maximum to a global one, and vice versa. These conjectures cannot be fully justified a priori using real world interaction potentials . There exist, however, some very suggestive models. Despite their heuristic character they lead to interesting insights on the structure of these fitness landscapes, which may exhibit in certain cases fractal like properties such as self-similarity.

Considerations of this kind become especially attractive and important in the context of biological evolution, especially when transposed to nucleic acids. In a DNA or RNA molecule *a genotype* is the sequence of bases (Guanine, Adenine, Cytosine, Thymine or Uracil) which carries the genetic information of an organism. It undergoes continuous variations by mutation or recombination, but as such it is not subjected to "evaluation" of any sort. The latter implies first the mapping of the genotype to a *phenotype* through the (generally environment-dependent) folding of the primary sequence, on which natural selection may subsequently operate to differentiate between different variants. The associated fitness functions depend on both thermodynamic and kinetic parameters and are, again, quite involved, displaying a large number of maxima. A great deal of work has been carried out with a fitness function provided by the *mean excess production*, defined as the weighted sum of the net production rates of the species present, where the weighting factor is given by the fractions of the species in question within the total population. Such "kinetic" fitness functions also illustrate the *adaptive character* of the optimization, since they are continuously being redefined by the very process of evolution that they are supposed to command (cf. also Sec. 6.6).

The kinetics of optimization is an intricate phenomenon of great interest. In some cases, under the effect of variability (fluctuations, noise or mutations, depending on the case) one witnesses a smooth, monotonic ascent -apart from small random fluctuations- to the global maximum. But as the constraints become more stringent one may observe instead a jumpwise increase. Furthermore, when variability exceeds a threshold the populations diffuse randomly in sequence space and no optimization is observed. Similar ideas have been used to model aspects of the immune response, economic and financial activities, the propagation of rumors, and so forth.

The evolutionary perspective also allows one to make a synthesis between the function-driven and the structure-driven expressions of complexity. The key point is that an innovation at the origin of a particular structure or function involving a large number of subunits requires an underlying dynamics that has to be sufficiently complex and removed from the state of equilibrium; otherwise, the waiting time to see this event happen would be enormous, owing to the combinatorial explosion of the number of possible arrangements of the subunits. But once such a phenomenon takes place, there is nothing opposing its subsequent stabilization in the form of a static, equilibrium like structure. This is what must have happened when the history of biological evolution was encapsulated in the structure of the DNA: an "aperiodic crystal" following Erwin Schrödinger's famous aphorism, which coexists today with its environment under equilibrium conditions much like an ordinary crystalline material, while being at the same time, thanks to its non-repetitive structure, the carrier of information that allows the switching on of a whole series of complex dynamical phenomena specific to life.

Chapter 4

Information, entropy and selection

4.1 Complexity and information

We have insisted repeatedly on the presence of an element of *choice* underlying a complex system related, in particular, to the multitude of the a priori available states; and correspondingly, on the fact that the outcome of the evolution of such a system will comprise an element of *unexpectedness*, reflected by the difficulty of an observer to localize the actual state in state space on the basis of the data that are available to him. A natural question to be raised is, then, how to construct a quantity serving as a measure of choice and unexpectedness, thereby providing yet another indicator of the complexity of the system at hand.

Now this question is reminiscent of a central problem of information and communication theories, namely, how to recognize a signal blurred by noise. Consider a message source possessing N states, each realized with probability $p_i (i = 1, ..., N)$. For instance, in the early years of telegraphy there were two elements out of which the code could be constructed -current and no current- which one might denote as 1 and 0. Such binary codes are also nowadays ubiquitous in modern theoretical informatics and computer science; in literature, we dispose of the letters of an alphabet -26 in the case of the english language- to which one may also add the pause and other punctuation signs; and so forth. The point is that in any real world information source, living or mechanical, choice is continuously exercised. Otherwise, the messages produced would be predetermined and completely predictable and there would be no need for communication. Corresponding to the choice exercised by the message source in producing the message, there is an uncertainty at the level of the recipient of the message. This uncertainty is resolved when the recipient examines the message. Clearly, the measure we are looking for must increase with the amount of choice of the source, and, hence, with the uncertainty of the recipient as to what message the source may produce or transmit. A major result of information theory is that the *information entropy*

$$S_I = -\sum_{i=1}^{N} p_i \ln p_i \qquad (4.1)$$

introduced by Claude Shannon in 1949, provides this measure as it possesses all the desired properties that one could reasonably assign:

(i) it takes its largest value for $p_i = 1/N$, implying that in the case of equiprobable states the uncertainty about the particular state actually realized -and hence the amount of data necessary to resolve it- is the maximum one;

(ii) adding an impossible event $\alpha, p_\alpha = 0$, does not change S_I;

(iii) The information entropy $S_I(A, B)$ of a composite system AB equals the entropy of subsystem A plus the conditional entropy $S_I(B|A)$ of subsystem B provided that subsystem A is in a given state, a property referred as *additivity*.

In fact, one can show that (4.1) is the only quantity possessing these properties. We emphasize that at this stage one should refrain from identifying information entropy with the concept of entropy as used in thermodynamics and statistical mechanics (see e.g. Secs 2.2 and 6.1). We will come to this point later on in this chapter.

To appreciate the meaning of eq. (4.1) and of properties (i)-(iii) it is useful to introduce the concept of information bit and, correspondingly, to realize that there is a close connection between information and decision. In the classic game of Twenty Questions one is supposed to identify an object by receiving "yes" or "no" answers to his questions. The amount of information conveyed in one answer is a decision between two alternatives, e.g. animal or nonanimal. With two questions, it is possible to decide for one out of four possibilities, e.g. plant -mineral, mammal-nonmammal, etc. Thus, the logarithm (here with base 2, but this is only a matter of normalization) of the possible decisions rather than their number is to be used as a measure of the information involved in discriminating between two elementary events/states. This is, precisely, the bit number and the *raison d'être* of the term in $\ln p_i$ in eq. (4.1). Notice that the logarithmic dependence is also the one responsible for the property ((iii) in our list) of additivity. It is also worth mentioning that property (i) reflects the *convexity* of $-S_I$ or equivalently the *concavity* of S_I viewed as a function of the p_i's, see Fig. 4.1.

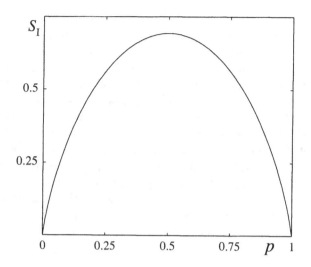

Fig. 4.1. Illustration of the convexity of information entropy S_I as a function of the probability distribution p, for a system possessing two states of probabilities p and $1 - p$.

The probabilistic formulation developed in Chapter 3 allows us to apply straightforwardly the above ideas to complex systems, and to use information entropy as an indicator of complexity. The crux is in the concept of *coarse graining* - the mapping of the dynamics into a discrete set of states. Depending on the case, these can be the states introduced in connection with the mesoscopic level description underlying the master equation (3.5); or the phase space cells of a Markov partition involved in the mapping of deterministic chaos into a symbolic dynamics. The procedure can even be carried out starting from the continuous description afforded by the Fokker-Planck equation (3.8). As an example, consider a system that admits two simultaneously stable steady states. Because of the transitions between states induced by the fluctuations (see Sec. 3.5.2), the system will spend substantial amounts of time near each of these states -the average being given by eq. (3.17)- interrupted by fast transitions to the other state. Let us introduce a more abstract description in which the instantaneous state is labeled "0" or "1", according to whether the system is in the domain of attraction of the first or the second stable steady state. An observer following the evolution at regular time intervals by noticing the attraction basin in which the system is found will therefore detect a sequence like 111001100010110... which provides a description in terms of an alphabet involving only two "letters", 0 and 1.

Conversely, we are entitled on the grounds of the above arguments to view a complex system as an information source. Eq. (4.1) allows then us to connect information entropy with the different evolution scenarios summarized in the previous chapters. Consider, for instance, the bifurcation of Fig 1.1., to which we referred already as one of the "elementary acts" of complexity. Prior to bifurcation ($\lambda < \lambda_c$) the system possesses a unique macroscopic state. In the absence of internal fluctuations or environmental noise one has $N = 1$, $p_1 = 1$ in the notation of eq. (4.1), hence $S_I = 0$. Switching to the other side of the bifurcation point ($\lambda > \lambda_c$) one now has (in the symmetric (pitchfork) bifurcation) two equivalent macroscopic states. In the absence of fluctuations or noise this means $N = 2$, $p_1 = p_2 = 1/2$, hence $S_I = \ln 2$. The presence of variability will modify these results slightly by an amount of the order of the variance of the process, but the essential point is that bifurcation entails a finite jump of S_I plus a small correction that can be discarded to the dominant order,

$$\Delta S_I = S_I(\lambda > \lambda_c) - S_I(\lambda < \lambda_c)$$
$$= \ln 2 \qquad (4.2)$$

This result is universal, as it holds for any dynamical system undergoing this type of bifurcation. It can be extended to other types of bifurcations, including bifurcations leading to periodic or to chaotic behavior. Notice that thermodynamic quantities and other familiar purely macroscopic observables show no universal trends across bifurcation.

On the basis of these results information may be viewed as an emergent property, arising through an evolutionary process leading from the state of "no information" (here $\lambda < \lambda_c$) to the state of "information" (here $\lambda > \lambda_c$). Fluctuations are here the natural carriers of information, as they allow the system to realize the crucial process of transition between the available states.

4.2 The information entropy of a history

So far we have been interested in the case where the information source produces a certain symbol at a give time, whatever the symbols previously produced or to come might be. We will now address the more interesting situation of a source producing a sequence of symbols. Since we are concerned primarily by complex systems as information sources, this means that we are letting the underlying dynamics run for a certain amount of time, map the trajectory through the process of coarse-graining explained in Sec. 4.1 into a discrete sequence of states, and inquire on the restrictions imposed by the dynamics on the succession of symbols such as: constraints of word frequency, constraints of word order or, more generally, "grammatical" rules of some sort.

To handle these situations we divide the message into blocks of symbols of length n. This makes it possible to extend eq. (4.1), corresponding essentially to an entropy per symbol, and define the *entropy per block* of symbols or, equivalently, the entropy associated to a portion of the system's history over a window of length n :

$$S_n = - \sum_{i_1...i_n} P_n(i_1...i_n) \ln P_n(i_1...i_n) \qquad (4.3)$$

Here $P_n(i_1...i_n)$ is the probability to generate the particular sequence $i_1...i_n$ and the sum runs over all sequences compatible with the underlying rules.

As it stands, the value of S_n in eq. (4.3) depends on n. To construct a quantity that is intrinsic to the system we imagine that n is gradually increased. If the blocks are long enough and the system has strong ergodic properties, we expect that only the symbols near the beginning will depend on symbols in the previous block and will anyway form a negligible part of all the symbols in the block. In the limit $n \to \infty$ this will lead to the total information entropy. Dividing through by n one would then obtain the entropy per symbol, a property characteristic of the source and of the coarse-graining algorithm :

$$h = \frac{1}{n} S_n, \quad n \to \infty \qquad (4.4a)$$

If the partition at the basis of the coarse-graining algorithm is Markovian and satisfies some further technical conditions, then h is shown to take the largest value compared to all other partitions. If the underlying dynamical system is deterministic, the limit of (4.4a) for continuous time (time step tends to zero) and for infinitely refined partition exists and is finite. One refers then to h as the *Kolmogorov-Sinai entropy*.

Now, suppose that the source has sent a message in the form of a particular sequence (say the two letters *th* in a message written in english). What is the probability that the next symbol be i_{n+1} (in our example the probability that the next letter be e)? Clearly, we are dealing here with a *conditional* event. Let $W(i_{n+1}|i_1...i_n)$ be the associated probability. By definition, $P_{n+1}(i_1...i_{n+1}) = W(i_{n+1}|i_1...i_n)P_n(i_1...i_n)$. The entropy excess associated with the addition of a symbol to the right of an n block (or "word") is, then

$$h_n = S_{n+1} - S_n$$
$$= -\sum_{i_1...i_{n+1}} P_n(i_1...i_n) W(i_{n+1}|i_1...i_n) \ln W(i_{n+1}|i_1...i_n) \quad (4.4b)$$

One expects intuitively that h_n should have the same significance as h in eq. (4.4.a). This can indeed be shown to be the case for ergodic sources and in the limit of very large n's,

$$h = \lim_{n\to\infty} h_n = \lim_{n\to\infty} \frac{S_n}{n} \quad (4.5)$$

4.3 Scaling rules and selection

The structure of the P_n's, W's and hence of S_n, h_n and h as well, introduced in the preceding section depends crucially on the rules underlying the information processor. If the rules amount to choosing a particular symbol out of an alphabet of, say, K letters independently of the choice made in previous steps (for a two letter alphabet this would be equivalent to tossing a fair coin), then the number of all possible sequences of length n would be $N_n = K^n = \exp[n \ln K]$. Under these conditions, eqs (4.4) - (4.5) lead to

$$P_n = N_n^{-1} = \exp[-n \ln K]$$
$$S_n = n \ln K, \quad h = h_n = \ln K \quad (4.6)$$

As can be seen, the occurrence of each particular word is penalized exponentially with respect to its length n, the rate of penalization being $\ln K$. This is a direct consequence of the adopted rules which lead to a *combinatorial explosion*, associated with the proliferation of words of increasing length. Clearly, an information source of this kind -to which one may refer as a *Bernoulli processor*- is not adequate for generating complex structures such as, say, the Encyclopedia Britannica!

Suppose next that the processor generates the sequences via a Markov process, such as in the mesoscopic level description of the fluctuations afforded by the master equation (eq. (3.5)). The probability of the event $C = i_1...i_n$ is then $P_n(C) = p_{i_1} W(i_2|i_1)...W(i_n|i_{n-1})$ and eqs (4.4)-(4.5) lead to

$$S_n = e + (n-1)h \quad (4.7a)$$

where e is given by eq. (4.1) and the probabilistic analog of Kolmogorov-Sinai entropy h (also referred in this context as "entropy of the source") measures the amount of information obtained when the Markov process moves one step ahead of the initial state i_1

$$h = -\sum_{ij} p_i W(j|i) \ln W(j|i) \quad (4.7b)$$

Notice that this latter value reduces to the value $\ln K$ in eq. (4.6) if (a), two successive steps of the process are completely uncorrelated; and (b), the probabilities of occupying the individual states are equal. What is more, $\ln K$ is an upper bound for (4.7b). This is natural, since constraints like correlations tend to restrict the number of allowable "words". Alternatively, the information necessary to identify a given word out of many that could have been emitted should be less.

One can now prove the following remarkable property. Suppose that we arrange the sequences generated by our information source in order of decreasing probability $P(C)$ and select from this set the first N_n sequences whose sum of the probabilities just exceeds a preassigned finite number between 0 and 1. Then in the limit of large n the number N_n depends only weakly of the above preassigned number, and can be estimated by

$$N_n \approx e^{nh} \quad (4.8)$$

But as we just saw, h is bounded from above by $h_{\max} = \ln K$, a value attained when the successive steps are statistically independent and the states are equiprobable. As a result, for n large,

$$N_n \ll N_{n,\max} \quad (4.9)$$

where $N_{n,\max}$ is given by eq. (4.6). In other words, only a small specific fraction of all the sequences has a sum of probabilities close to unity. How small this fraction is depends on the entropy of the Markov process. We thus arrive at the important conclusion that correlations and deviations from equiprobability act like extremely efficient selectors of privileged sequences out of the huge set of all possible random sequences. Such deviations can be realized if the system is not isolated but instead is put in contact with an external environment. As we have seen repeatedly, such conditions can lead to *nonequilibiurm states* that can be sustained indefinitely because of the exchange of matter or energy between the system and the external world. Nonequilibrium is therefore the natural environment in which selection may take place.

As an example, consider the process of formation of a biopolymer, say a protein $n = 100$ amino acid molecules in length. It is well known that in nature there exist $K = 20$ kinds of amino acids. If biopolymerization were to take place in an isolated system in which all sequences are a priori equally probable, we would have $N_n = e^{100 \ln 20} \sim e^{300}$ equally probable sequences. Hence, any particular arrangement -necessary to fulfill, for instance, a desired biological function- would occur with the exceedingly low probability of e^{-300} ! If, on the other hand, biopolymerization is performed under nonequilibrium conditions corresponding to an entropy of, say, $S_I = 0.1 S_{max}$, then only $N_n \sim e^{30}$ sequences of length 100 would be realized with an appreciable probability. This number is much closer to our scale. It is thus conceivable that evolution acting on such a relatively restricted population in a nonequilibrium environment can produce, given enough time, entities endowed with special properties such as self-replication, efficient energy transduction and so on.

In deriving relations (4.8) and (4.9) we have assumed that the system of interest was giving rise to a stochastic process defined on a discrete (actually finite) set of states and characterized by the Markov property. The first condition is satisfied whether the underlying system is discrete or continuous to begin with, as long as we monitor its evolution through a finite "window" - a mode of description, as we saw earlier, at the very basis of symbolic dynamics. The extent of deviation from equiprobability is then measured by the extent of non-uniformity of the invariant probability in phase space. The second condition is far more restrictive. As seen in Sec. 3.2, it is satisfied in many instances by the internal fluctuations at the mesoscopic level of description. Furthermore, as stressed in Sec. 3.3, in many cases deterministic dynamical systems giving rise to chaotic dynamics with strong instability properties can also be cast in the form of a Markov process when the phase space partition involved in the coarse-graining satisfies certain properties (Markov partitioning, etc.). But what about chaotic dynamical systems that are not unstable in every region of phase space, dynamical systems exhibiting weak chaos in the sense of small Lyapunov exponents, or dynamical systems which, while non-chaotic, give nevertheless rise to intricate evolutions? Such systems are far from being exceptional. For instance, the chaotic attractors generated by continuous time dissipative dynamical systems are not uniformly unstable; weak chaos characterizes the important class of intermittent dynamical systems; simple systems subjected to periodic or quasi-periodic forcings can generate attractors that are fractal, but non-chaotic; and, in addition to these generic situations, one has naturally the critical -and thus non-generic- cases where the system functions at parameter values corresponding to the bifurcation point of a chaotic attractor.

In all the above situations, lumping the continuous dynamics into a discrete (and, a fortiori, a finite) set of states leads, typically, to correlations reflecting the deterministic origin of the process. These are particularly ap-

Information, Entropy and Selection 109

parent when the length, n of the blocks considered is small to moderate, but in some cases they may even persist for as large blocks as desired. One way to accommodate them in the scheme of this section is to stipulate a more general scaling of S_n as a function of n than in eq. (4.7a), of the form

$$S_n = e + gn^{\mu_0}(\ln n)^{\mu_1} + nh \qquad (4.10a)$$

with μ_0 non-negative and strictly less than 1, or $\mu_0 = 1$ and μ_1 negative, leading to (cf. definition (4.4.b))

$$h_n \approx h + \mu_0 g n^{\mu_0 - 1}(\ln n)^{\mu_1} \qquad (4.10b)$$

Such a scaling includes as a special case (i.e. when $h = 0$) a sublinear growth of S_n as a function of n and thus, the subexponential growth of the number of sequences N_n. More importantly, sequences of length n are now penalized subexponentially and thus far less severely than in the Markov or fully chaotic case described by eqs (4.6) or (4.8). The selection of "relevant" sequences achieved by a dynamics of this kind is thus much more efficient, and one may begin to foresee the possibility of realizing complex structures with a non-negligible probability.

As a first example of the foregoing, consider a periodic deterministic system of period p. Clearly, the appearance of a symbol in an admissible sequence on the position $p + 1$ is here completely determined by the preceding p symbols. It follows from eqs (4.3) and (4.4b) that

$$\begin{aligned} S_{p+n} &= S_p & (n \geq 1) \\ h_{p+n} &= 0 & (n \geq 0) \end{aligned} \qquad (4.11)$$

In other words, the addition of symbols beyond a word of length p does not add new information. The selection problem is automatically solved since sequences of arbitrary length are generated with probability one but, as a counterpart, the information conveyed becomes trivial: periodic dynamical systems cannot be used as information sources.

A second example, of special interest in the sense of being intermediate between Markov character (including fully developed chaos) and periodicity, corresponds to setting in eqs (4.10) $h = 0$, $\mu_0 = 1$, $\mu_1 = -1$. This leads to

$$\begin{aligned} S_n &= e + gn(\ln n)^{-1} \\ h_n &= g(\ln n)^{-1} + \text{higher order terms} \end{aligned} \qquad (4.12)$$

Adding a new symbol to a sequence of length n still conveys information, but the new information per added symbol now decreases logarithmically. This implies the existence of long range correlations. Correspondingly, the probability of a word of length n is penalized less severely than in the case where $h > 0$, although the enhancement is rather weak owing to the slowness of the logarithmic dependence.

There exists a class of dynamical systems giving rise to the scaling of eq. (4.12). These can be deterministic dynamical systems of the intermittent type, or their probabilistic Markov counterparts, known as *sporadic systems* to which they can be mapped by a process of coarse-graining involving an infinite (but denumerable) sequence of symbols. A typical example is provided by the mapping

$$x_{n+1} = x_n + Cx_n^2 \quad \text{mod } 1 \quad (4.13)$$

where the mod 1 prescription amounts to reinjecting x in the unit interval when its value exceeds unity, by subtracting 1 from this value. The partition to which the above mentioned infinite sequence of symbols is associated is then provided by the fixed point of this mapping and the sequence of points that are mapped to it after one, two, n, ... iterations, referred to as preimages. Upon performing a further lumping leading to a finite alphabet, one then shows that the resulting sequences generated by this latter alphabet satisfy the scaling of eq. (4.12), reflecting the non-Markovian character of the associated process.

The next step toward a more efficient selection and hence to a less severe penalization of long words would be to have a subexponential law in the form of a stretched exponential. This amounts to setting $h = 0$, $\mu_1 = 0$, $\mu_0 < 1$ in eq. (4.10a), i.e.

$$S_n \approx e + gn^{\mu_0}$$
$$h_n \approx \mu_0 gn^{\mu_0-1}, \quad 0 < \mu_0 < 1 \quad (4.14)$$

The amount of information conveyed per additional symbol falls here even more drastically than in the previous example, suggesting the existence of intricate correlations. Interestingly, this sort of information generation seems to be the one corresponding to written texts or to classical music. From the standpoint of dynamics scaling (4.14) has been demonstrated to occur in intermittent systems of the type of eq. (4.13) where the nonlinearity exponent is higher than 2 as well as in hybrid dynamical systems, that is to say, dynamical systems whose evolution laws switch with a certain probability between two different forms.

The most efficient selection and, correspondingly, the less severe penalization of long words would be to have a power law dependence of the word probability. This would correspond to setting $h = 0$, $\mu_0 = 0$, $\mu_1 = 1$ in eq. (4.10a), i.e.

$$S_n \approx e + g \ln n$$
$$h_n \approx g \frac{\ln n}{n} \tag{4.15}$$

This law is implemented, typically, in dynamical systems functioning at a critical situation as e.g. in the accumulation point of the period doubling cascade described in Sec. 2.5. In this sense it is non-generic, since criticalities occur only when the control parameters take particular (and hence exceptional) values.

We turn now to the case of a typical chaotic attractor generated by a continuous time dynamical system. The mechanism of producing information-rich sequences of symbols that we adopt is as follows. One assumes that when a variable X_i crosses a certain predetermined level L_i with, say, a positive slope a symbol "forms" and is subsequently "typed" (Fig. 4.2). One can envisage in this way a sequence of level crossing variables -symbols- standing as a one-dimensional trace of the underlying multidimensional dynamics. The sequence is by necessity asymmetric (a syndrome that all languages share) as a result of the dissipative character of the process. As an example consider the Rössler attractor, a prototypical system generating deterministic chaos

$$\frac{dx}{dt} = -y - z$$
$$\frac{dy}{dt} = x + ay$$
$$\frac{dz}{dt} = bx - cz + xz \tag{4.16}$$

with $a = 0.38, b = 0.3, c = 4.5$, and thresholds $L_x = L_y = L_z = 3$. A typical sequence generated by this mechanism is

$$zyx\ zxyx\ zxyx\ zyx\ zxyx\ zyx\ zyx\ zx\ zyx\ zyx\ zxyx\ zyx... \tag{4.17}$$

Remarkably, one can verify that it can be formulated more succinctly by introducing the hypersymbols

$$\alpha = zyx, \quad \beta = zxyx, \quad \gamma = zx$$

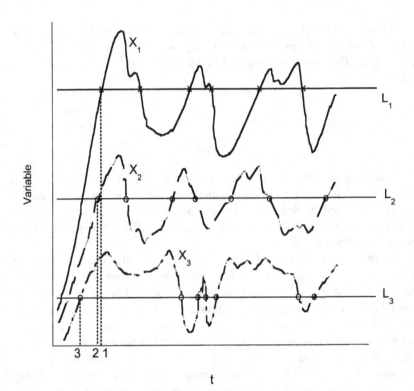

Fig. 4.2. Generation of strings of symbols by a dynamical system such as the Rössler attractor, eqs (4.16). Whenever the state variables $X_1 \cdots$ cross predefined thresholds $L_1 \cdots$ with positive slope a new process is switched on, that allows the transfer of the temporal sequence generated by the system's attractor into a symbolic sequence.

giving rise to

$$\alpha\beta\beta\alpha\beta\alpha\alpha\gamma\alpha\alpha\beta\alpha\cdots \qquad (4.18)$$

A statistical analysis reveals strong correlations in the original sequence which to a very good approximation can be fitted by a fifth-order Markov process. On the other hand, the hypersymbol sequence is a definitely more random first order chain, indicating that the "compression" achieved by the hypersymbols has indeed removed much of the structure of the original sequence.

Figure 4.3 depicts the dependence of the inverse of the number of allowed sequences, N_n^{-1}, of symbols (crosses) and hypersymbols (empty circles). As

Information, Entropy and Selection

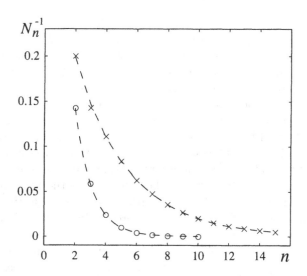

Fig. 4.3. Dependence of the inverse of the number of allowed sequences N_n of symbols (upper curve) and of hypersymbols (lower curve) on the sequence length n generated by the dynamical system of eqs (4.16) according to the algorithm described in Fig. 4.2. Dashed lines correspond to a fitting with an exponential law (eq. (4.8)) and with a law involving a sublinear contribution to the block entropy (eq. (4.10a)) for the upper and for the lower curves, respectively.

can be seen the selection (reflected by the number of excluded sequences) is more efficient in the case of symbols, a result consistent with the presence of correlations in the corresponding sequence. For instance, out of the $3^{10} = 59049$ a priori possible sequences of length 10, only 49 symbol sequences and 8119 hypersymbol sequences are allowed by the dynamics. The extent to which the dependence of either of these curves on n is exponential as in eq. (4.8), or more intricate as in the general scaling (4.10) requires consideration of very long sequences, whose reliability is limited by the very fast proliferation of their number. Still, as seen in the figure (dashed curves), the results on the hypersymbol sequence can be fitted perfectly with a simple exponential, the corresponding value of the entropy of the source h being $h \approx 0.88$. On the other hand, the results on the symbol sequence are fitted quite well by a law featuring the sublinear correction of (4.10a) with $\mu_0 = 1$ and $\mu_1 = -1$, the corresponding h value being now $h \approx 0.28$, a considerably smaller number than before. The deviation from the simple exponential law for N_n^{-1} revealed by this fit reflects, once again, the presence of correlations in the symbol sequence. It is also to be correlated with the small value of h: for small to moderate lengths n of the order of up to h^{-1} sublinear terms may well give substantial contributions, but this type of contribution will inevitably be superseded by the linear one for large n's. In other words, for generic continuous time chaotic flows the scaling in eq. (4.10a) can be understood as a transient behavior prior to the asymptotic limit of $n \to \infty$, whose lifetime is proportional to h^{-1}. Notice that the Lyapunov exponents of model (4.16) are $\sigma_1 \approx 0.12$, $\sigma_2 = 0$ and $\sigma_3 \approx -3.95$. The probabilities of x, y, z are 0.428, 0.267 and 0.306 respectively whereas those of α, β, γ are 0.469, 0.409, and 0.121. The corresponding information entropies are $S_{xyz} = 1.078$ and $S_{\alpha\beta\gamma} = 0.977$. They both are relatively close to the (maximum) value corresponding to equipartition, $S_{\max} = \ln 3 \approx 1.099$. The effectiveness of the selection illustrated by Fig. 4.3 can therefore not be attributed primarily to deviations from equipartition.

In the literature on complex systems statements like "complexity arises at the edge of chaos" have become popular. Our analysis gives a precise meaning to this concept by placing it in the context of selection of information rich sequences of symbols. Fully ordered sequences convey no information. Fully random or Markovian sequences, as well as sequences generated by strongly chaotic dynamical systems with a large value of Kolmogorov-Sinai entropy h, fail to select efficiently a preferred subset of sequences. But in systems whose entropy h is small (which may thus be qualified as weakly chaotic) selection of sequences of small to moderate length proves to be more effective. As seen earlier in this section the effectiveness is further enhanced in systems where $h = 0$. Such systems, although non-chaotic in the sense of having an exponential sensitivity to the initial conditions, may still be quite intricate.

We next proceed to the ranking of the allowed sequences according to the frequency of their occurrence. The result for the symbol sequence is depicted in Fig. 4.4 for a sequence of length $n = 10^6$ and ranks up to 100. One obtains a slowly decreasing overall trend following closely the Mandelbrot extension of Zipf's law (eq. (1.3)), represented by the dashed line in the figure, with an exponent B very close to unity and a cutoff value n_0 of about 10. That laws of the Zipf type, usually associated with socio-economic systems and proposed on the basis of heuristic arguments, may have a dynamical origin is a result of considerable interest. In particular, the dynamics generating them needs not be scale free: this is exactly what happens for the Rössler system (eq. (4.16) and Fig. 4.4) which possesses a chaotic attractor with positive Lyapunov exponent and gives rise to exponentially decaying properties at the level of the original variables x, y, z.

It should be noticed that the data in Fig. 4.4 exhibit a fine structure around the Zipf-Mandelbrot law in the form of plateaus extending over sev-

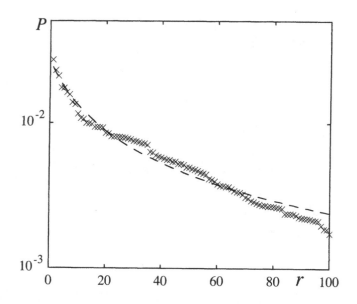

Fig. 4.4. Emergence of a Zipf type law (eq. (1.3)) when the sequences considered in Fig. 4.3 are ranked according to the frequency of their occurrence, for rankings up to $r = 100$.

eral values of the rank r, reflecting the fact that different words of a given length can be produced with very similar probabilities.

4.4 Time-dependent properties of information. Information entropy and thermodynamic entropy

We now extend the description of the preceding sections to account for the ways information is exchanged or produced in the course of time. Our analysis will be carried out in the spirit of Sec. 2.2.3, where a similar question concerning the time variation of thermodynamic entropy was raised. In the present section it will be limited to the information entropy S_I, eq. (4.1), postponing the analysis of the dynamics of time histories to Sec. 4.5.

Since S_I depends on the probability distribution we shall appeal to Sec. 3.2, where evolution equations for this quantity were introduced. In doing so we shall adopt a mesoscopic level description. Consider first a process involving a set of discrete variables X (jump process). Using the continuous time limit of the master equation, eq. (3.6), one may obtain a balance equation for the information entropy,

$$\frac{dS_I}{dt} = -\sum_X \ln P(X,t) \frac{dP(X,t)}{dt}$$

$$= \frac{1}{2} \sum_{X,X'} [w(X' \to X) P(X',t)$$

$$- w(X \to X') P(X,t)] \ln \frac{P(X',t)}{P(X,t)} \quad (4.19)$$

Adding and subtracting the term $\ln(w(X' \to X)/w(X \to X'))$ allows one to split dS_I/dt into two parts,

$$\frac{dS_I}{dt} = J_I + \sigma_I \quad (4.20)$$

where

$$\sigma_I = \frac{1}{2} \sum_{XX'} [J(X' \to X, t) - J(X \to X', t)] \ln \frac{J(X' \to X, t)}{J(X \to X', t)} \quad (4.21)$$

$$J_I = -\frac{1}{2} \sum_{XX'} [J(X' \to X, t) - J(X \to X', t)] \ln \frac{w(X' \to X)}{w(X \to X')} \quad (4.22)$$

We observe that σ_I is non-negative, whereas the sign of J_I is undetermined. It seems therefore appropriate to call σ_I and J_I the information entropy production and the information entropy flux, respectively. It is worth noting that σ_I and J_I vanish when the two fluxes associated to the transitions from X' to X and from X to X' are equal. In physico-chemical systems this condition, known as *detailed balance*, is one of the principal properties of the state of thermodynamic equilibrium. The non-negativity of σ_I implies, then, that information entropy is continuously produced in time until this state is reached. This property is analogous to the second law of thermodynamics.

An even more explicit connection with thermodynamics can be established in the limit where the standard deviation of the fluctuations divided by the average value of X tends to zero (the thermodynamic limit, see Sec. 3.5.1), in which σ_I becomes identical to the thermodynamic entropy production for a wide class of systems. In general, these two quantities will differ by terms related to what one may legitimately call "entropy production of the fluctuations", $\sigma_{I,fl}$. One can show on quite general grounds that in equilibrium, in systems obeying to linear kinetic laws or in systems where the probability distribution P reduces to the Poisson distribution, $\sigma_{I,fl} = 0$. When on the other hand the system functions in the vicinity of equilibrium, it turns out that $\sigma_{I,fl}$ is negative. It is only when the distance from equilibrium exceeds some threshold that $\sigma_{I,fl}$ may become positive.

To address the case of continuous state variables X one needs to switch from (3.6) to the Fokker-Planck equation (3.8). As it turns out, in this limit the information entropy balance is connected to the indicators of phase space dynamics such as the Lyapunov exponents and more generally stability (cf. Sec. 2.1). Specifically, in a conservative dynamical system in the absence of fluctuations S_I is conserved at all times. In contrast, in a large class of dissipative systems the information entropy production is equal in the limit of long times to minus the sum of all Lyapunov exponents (which is negative, otherwise the system will explode to infinity), provided that certain "thermostating" conditions are fulfilled. These conditions guarantee that the extra energy communicated by the constraints (like e.g. the Joule heating in a system of charged particles in the presence of an electric field) is suitably evacuated to the environment, thereby allowing the system to reach a regime in which its statistical properties no longer vary in time. This regime is the (unique) time-independent solution of the Fokker-Planck equation.

4.5 Dynamical and statistical properties of time histories. Large deviations, fluctuation theorems

To reach a more comprehensive view of the connection between information, entropy and dynamics one needs to consider properties related to sequences of states rather than to single states. A first step in this direction was carried out in Sec. 4.2 and is here implemented further.

Our starting point is the stochastic process underlying the continuous time master equation (3.6). We are interested in the *paths* leading from an initial state X_0 at time t_0 to a state X_n at time t_n, via states $X_1, ..., X_{n-1}$, and in those associated with the reverse transition from X_n to X_0 (Fig. 4.5). Keeping in mind the analysis of Sec. 4.4 we introduce the quantity

$$\Phi(t) = \ln \frac{w(X_0 \to X_1)...w(X_{n-1} \to X_n)}{w(X_1 \to X_0)...w(X_n \to X_{n-1})} \quad (4.23)$$

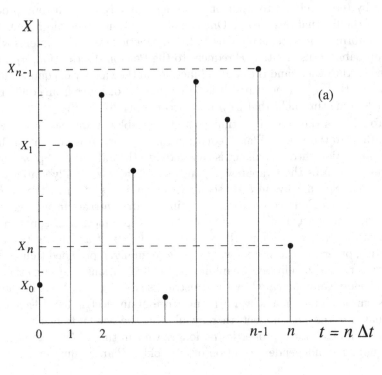

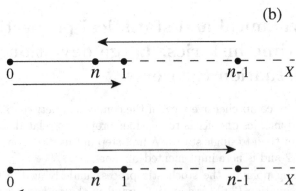

Fig. 4.5. (a) A temporal sequence leading from state X_0 at time t_0 to a state X_n at time t_n. (b) The above path and its inverse, in a representation where states are arranged along the axis according to increasing values of the variable X.

Using eq. (3.6) and the Markovian character of the process one can then establish the following remarkable properties.

(i) The probability that $\Phi(t)$ per time interval t takes a value within some specified range satisfies the equality

$$\frac{\text{Prob}(\frac{\Phi(t)}{t} \text{ belongs to } (\alpha, \alpha + d\alpha))}{\text{Prob}(\frac{\Phi(t)}{t} \text{ belongs to } (-\alpha, -\alpha + d\alpha))} = e^{\alpha t} \qquad (4.24)$$

in the limit $t \to \infty$.

(ii) The average value of $\Phi(t)$ over all possible paths (using the steady-state probability distribution as weighting factor) is given by

$$\frac{1}{t} < \Phi(t) > = \sigma_I \qquad (4.25)$$

in the limit $t \to \infty$. This suggests interpreting $\Phi(t)$ as the information produced along a particular path of the stochastic process. As seen in Sec. 4.4, in the limit where the fluctuations can be neglected this can be further linked to the total entropy produced up to time t. Building on this connection one might be led to associate $\Phi(t)$ to an instantaneous entropy production of some sort. Since nothing guarantees the positivity of $\Phi(t)$ as defined by eq. (4.23), one could then be tempted to conclude at "temporary" violations of the second law of thermodynamics. Such statements have indeed been made in the literature. They are highly misleading, since the principal content of the second law is the irreversible approach to the state of equilibrium in an isolated system or to a stationary nonequilibrium state when constraints are present. This is precisely what happens in the class of systems described by the master equation, itself at the origin of results (4.24)-(4.25).

As pointed out in Sec. 3.3, when the dynamics of strongly unstable systems is projected into a Markov partition, a master type equation governing the transitions between the partition cells is obtained. This holds also true for the full microscopic level dynamics, which is unstable owing to the defocusing character of the collisions between particles as discussed further in Sec. 6.1. Relations like (4.24) and (4.25) are thus expected to be considerably more general than the mesoscopic level description to which eq. (3.6) is at first sight limited, and this is in accord with the results of extensive numerical simulations carried out on interacting many particle systems.

4.6 Further information measures. Dimensions and Lyapunov exponents revisited

So far in this chapter information was associated with the idea of decision underlying the choice between different options, as reflected by the presence of a logarithmic term in the definitions (4.1) and (4.3) of the Shannon and block entropies. In this form it highlights the ubiquitous feature of probability distributions to factorize into a part pertaining to the probability of the state of a subsystem, multiplied by the probability of the state of the remaining parts of the system conditioned by the occurrence of this specific event (property (iii), Sec. 4.1). Now, in certain instances it may be of interest to highlight other features of probability distributions - for instance, penalize or on the contrary favor certain types of events, exceptional or on the contrary frequent ones. The corresponding measures could still be qualified as "information measures", as they reveal different ways to generate and process states or sequences of states by a dynamical system.

An information measure of this kind which, while being close in a sense to the Shannon information, opens some new possibilities is provided by the Renyi information entropy

$$S_\beta = \frac{1}{1-\beta} \ln \sum_{i=1}^{N} (p_i)^\beta \qquad (4.26)$$

where β is a parameter. This quantity still satisfies the additivity property of $S_\beta(A, B)$ being a sum of $S_\beta(A)$ and $S_\beta(B)$ as long as A and B are independent in the sense of $p_{A_i B_j} = p_{A_i} p_{B_j}$, but no longer so if correlations are present. Depending on whether β is larger or smaller than one it may serve as a "magnifying glass" to explore parts of the state space visited with different probabilities. Two noteworthy limiting cases serving as a reference are presented below.

- For $\beta = 0$ one obtains

$$S_0 = \ln N \qquad (4.27)$$

in other words, S_0 grows logarithmically with the number of occupied states. This is reminiscent of the definition of entropy of an isolated system in thermodynamic equilibrium proposed by Boltzmann.

- For $\beta = 1$ definition (4.26) presents an indeterminacy, which may be removed by expansion around $\beta = 1$ (S_β as given by (4.26) is differentiable in β as long as N is finite), yielding

Information, Entropy and Selection   121

$$S_1 = -\sum_{i=1}^{N} p_i \ln p_i \tag{4.28}$$

which is identical to the Shannon entropy.

Coming to general values of β, one sees straightforwardly from (4.26) that S_β is a monotonically decreasing function of β. Furthermore, the function

$$\psi(\beta) = (1-\beta)S_\beta = \ln \sum_{i=1}^{N} (p_i)^\beta \tag{4.29}$$

is monotonically decreasing in β and has the convexity property $\partial^2 \psi(\beta)/\partial \beta^2 \geq 0$.

The next step is to extend the foregoing to sequences of states. As seen in Sec. 4.2, in the context of Shannon's information this extension leads to the concepts of block entropies and of the Kolmogorov-Sinai entropy. Keeping this discussion and definition (4.26) in mind we define the Renyi dynamical entropies of order β as

$$S_{n,\beta} = \frac{1}{1-\beta} \ln \sum_{i_1 \cdots i_n} (P_n(i_1...i_n))^\beta \tag{4.30}$$

For the limiting value $\beta = 1$ one recovers, as expected from the earlier part of this section, the block entropy as defined by eq. (4.3). For $\beta = 0$ one obtains on the other hand $S_{n,0} = \ln N_n$, where N_n is the number of allowed sequences of length N_n (see also Sec. 4.3). This relation extends eq. (4.27).

The Renyi information entropy turns out to be a powerful tool for analyzing the structure of complex patterns like e.g. multi-fractal objects and, in particular, the attractors descriptive of chaotic dynamics (cf. Sec. 2.1). As stressed repeatedly in a complex system these attractors are, typically, highly inhomogeneous in the sense that different parts are visited with quite different probabilities. Because of this, global measures such as the Hausdorff fractal dimension D_0 (eq. (2.5)) are inadequate. To account for the local structure one divides the attractor into equal size cells of linear dimension ϵ. Their number N_ϵ plays a role analogous to the number of states N in (4.26) and, correspondingly, we let p_i designate the probability to visit the ith cell. In the limit $\epsilon \to 0$ (a necessary operation to probe the local structure) N_ϵ tends to ∞ as an inverse power of ϵ and S_β diverges. On the other hand the following set of quantities, to which one refers as Renyi dimensions,

$$D_\epsilon(\beta) = \frac{-S_\beta}{\ln \epsilon} = \frac{1}{\ln \epsilon} \frac{1}{\beta - 1} \ln \sum_{i=1}^{N_\epsilon} p_i^\beta \tag{4.31}$$

usually remain finite as $\epsilon \to 0$. For $\beta = 0$ one recovers the Hausdorff dimension D_0. For $\beta = 1$ one obtains a generalized dimension incorporating the probability of visiting different attractor regions,

$$D_1 = \frac{1}{\ln \epsilon} \sum_i p_i \ln p_i \qquad (\epsilon \to 0) \qquad (4.32)$$

which can be viewed as a measure of how the Shannon information increases when the resolution in the system's monitoring becomes sharper. For $\beta > 1$ eq. (4.31) allows one to capture the effect of correlations between different phase space regions. The limits $\beta \to \infty$ and $\beta \to -\infty$ are likewise of interest, as they describe the behavior of the regions in which the probability is most concentrated and most rarefied, respectively.

Coming back to the Renyi dynamical entropies, the discussion in Sec. 4.3 suggests that they should provide information on the randomness and the associated predictability properties of the underlying dynamics. A first result of this kind concerns already $S_{n,1}$, the block entropies which, as seen in Sec. 4.2 and 4.3, allow one to discriminate between different families of dynamical systems and in particular between deterministic dynamics and stochastic processes of the Markov type. Furthermore, if the system generating the symbol sequence is a strongly chaotic bounded deterministic system possessing at each phase space point expanding and contracting directions, and if the partition used is a Markovian partition, then the Kolmogorov-Sinai entropy h associated to $S_{n,1}$ (eq. (4.7b)) can be related to the Lyapunov exponents σ_i by the Pesin equality

$$h = \text{sum of all positive Lyapunov exponents} \qquad (4.33)$$

The above results still concern global behavior. Concerning now the local instability properties, rather than being of direct use, the Renyi dynamical entropies provide inspiration for building a closely related formalism for extracting precisely this sort of information. The idea can be illustrated on a chaotic dynamical system involving a single variable like in eq. (2.18) as follows. Let $\lambda(x)$ be the rate at which the system is expanding at phase space point x. The quantity $\sigma(x) = \ln \lambda(x)$ can then be qualified as the local Lyapunov exponent. The corresponding (global) Lyapunov exponent is $\sigma = <\ln \lambda(x)>$ where, depending on the case, the average is taken with the time-independent solution $\rho(x)$ of the Frobenius-Perron or Liouville equation. In an ergodic system σ is a constant. To account for possible local inhomogeneities one introduces then a set of quantities (to be compared to eq. (4.30))

$$\sigma_\beta = \frac{1}{\beta N} \ln \int dx_0 \rho(x_0) \lambda(x_0, ..., x_{N-1})^\beta \qquad (4.34)$$

in the limit $N \to \infty$, where $\lambda(x_0, \ldots, x_{N-1}) = \lambda(x_0) \cdots \lambda(x_{N-1}) = \exp \sum_{i=0}^{N-1} \ln \lambda(x_i)$ and the x_i's denote the attractor points visited consecutively by the system in the course of time. For $\beta = 0$ one recovers the global Lyapunov exponent. On the other hand, by differentiating (4.34) successively in β and setting $\beta = 0$ afterwards one generates the whole set of moments of the expansion rates and variances thereof, thereby obtaining the desired quantifiers of local variability. As an example for system (2.18) at $\mu = 1$ one has $<\sigma_1> = \sigma = \ln 2$, and a standard deviation $<\delta\sigma_1^2>^{1/2} = \pi\sqrt{3}/6$, comparable to σ.

Eqs (4.31) and (4.34) are the starting point of an elegant "thermodynamic" formalism of multifractals, which builds on a series of analogies with classical equilibrium thermodynamics in which $-\ln \epsilon$ plays the role of the volume and β the role of the reciprocal temperature. Furthermore, it can be extended to account for a division of phase space into cells of variable size. This provides a natural access to the role of the scales underlying the overall structure that may be present in the problem.

One further step in the search of information measures beyond the Shannon like entropies is to give up the property of additivity altogether, even when the subsystems constituting the overall system of interest are independent. A measure of this kind is provided by the entropy

$$S_q = \frac{1}{q-1}(1 - \sum_{i=1}^N p_i^q) \qquad (4.35)$$

This quantity is easily seen to be non-negative. It reduces to both the Shannon and Renyi information entropies for $q = 1$ but differs from either of them for $q \neq 1$. Furthermore, when instead of a one-state probability one introduces in (4.35) the probability of a time history one is led to yet another quantity generalizing the block, Renyi dynamical and Kolmogorov-Sinai entropies. Such generalizations may be relevant in analyzing data generated by complex systems displaying fractional nonlinearities. The latter may be viewed as "effective" nonlinearities manifested at some level of description (see, for instance, Sec. 2.7), whose connections with the underlying "fundamental" dynamical laws (where additivity is expected to hold) cannot be established in a direct manner.

4.7 Physical complexity, algorithmic complexity, and computation

We live in an era of information technology and communication. Rather than being described in qualitative terms, natural objects are represented as sequences of 0's and 1's stored in the memory of a computer (think for instance of a digitalized photograph). Likewise, statements on how things are to follow logically each other are expressed as computer programs and hence also as sequences of 0's and 1's. One is thus gradually led to the view that natural laws are nothing but algorithms implementable in the form of a computing program, that computes sets of digits as an output starting with a reference set (the initial data) provided by the experimental observations. In short, the world would function according to a "computational dynamics" whereby the observed phenomena are the outcome of a succession of steps in which the structure and the interactions between the constituents of matter manifest themselves primarily through the fact that they are "computing".

In the above summarized philosophy, an object is quite naturally deemed complex when there is no short description of it. More precisely, the algorithmic complexity -also referred as Kolmogorov-Chaitin complexity, $C(K)$- of an object in its digitalized expression of binary sequence of length K, will be defined as the size of the smallest computer program (measured again in number of bits) generating it, *no matter how slow* this program might be.

Let us give a few examples of the foregoing. Consider a binary sequence K data long displaying an overall regularity, e.g. 100100100... Clearly, the message in it can be considerably compressed. For instance, it could be transmitted to a computer by a very simple algorithm "Print 100 ten (or 100, or a million, ...) times". The number of binary digits K^* in such an algorithm is a small fraction of K, and as the series grows larger the algorithm size increases at a much slower rate such that K^*/K tends to zero as K tends to infinity. According to the definition this therefore implies limited algorithmic complexity. Consider next the opposite limit of a sequence of length K whose digits are set to 0 or to 1 depending on the outcome "heads" or "tails" of a (fair) coin tossing game. Clearly, this sequence can be recorded or transmitted only by reproducing it from beginning to end: its algorithmic complexity K^* is equal to its size K. Let us now bring together all binary sequences of length K, from the most ordered to the fully random ones. Their total number is 2^K. Among them there are sequences that can be compressed in the more compact form of binary programs of lengths 1, 2,... up to $m-1$, strictly less than a certain $m < K$. The total number of these programs is less than or equal to $1 + 2 + 2^2 + \cdots 2^{m-1} = 2^m - 1$, i.e., less than 2^m. It follows that among the original 2^K sequences at most 2^m will have a program of length strictly less than m: stated differently, the proportion of sequences of length K having a Kolmogorov-Chaitin complexity strictly less than m, is less than or equal to $2^m/2^K = (1/2)^{K-m}$. For instance, of all binary sequences of length $K = 1\,000$ at most one sequence out of $2^{1\,000-990} = 2^{10} = 1\,024$ has a complexity less than $m = 990$, and at most one out of $2^{20} = 1\,048\,576$ has a complexity less than $m = 980$.

In short, most of the sequences are incompressible and have a maximum algorithmic complexity. How many? To answer this question one needs to compute $C(K)$, and the quest for a result of this kind leads to a major surprise: there is no general procedure for computing it in a finite time. Far from being a temporary drawback related to the lack of sufficiently sophisticated mathematical techniques this conclusion reflects, in fact, the deep concept of undecidability discovered by Kurt Gödel in 1931: any formal system built on a set of axioms comprises statements that cannot be confirmed or rejected on the sole basis of these axioms. Translated in terms of computing -a giant step by itself accomplished by Alan Turing in 1936- this implies that there is no mechanical procedure allowing one to know whether a computer program chosen at random will or will not be executed in finite time. Gregory Chaitin reformulated this "halting problem" in terms of a halting probability, Ω and showed that Ω is an uncomputable number that cannot be compressed in any way whatsoever. One provocative way to interpret this result is that there is an infinite number of unprovable mathematical statements. This "openness" in a field usually considered to be the prototype of definiteness, is not unlike the situation described in Sec. 2.4 that there exists an unlimited, non-classifiable number of evolution scenarios of a complex system.

To what extent is algorithmic complexity tantamount to the complexity one is concerned with in nature? Suppose that we want to encode the dynamical laws generating the Rayleigh-Bénard instability and the patterns along a horizontal direction arising for different parameter values (cf. Sec. 1.4.1). The complexity of the computer program for the first task would seem to be exactly the same whether the system functions below or beyond the instability threshold, since in both cases a generic initial condition would lead to a final structureless or patterned state with probability one - at least as long as one deals with a small size cell and the parameters are near their threshold values. But to describe the gross features of the pattern itself different specifications will be necessary :

- Below the instability one simply has to state the value of $\Delta T = T_2 - T_1$, and the fact that the vertical velocity w is zero everywhere.

- Beyond the instability one still needs to specify the value of ΔT. But the single "instruction" $w = 0$ has now to be replaced by specifying $w = w_0$ at some reference point, and next by adding that this value is repeated in the horizontal direction according to a periodicity whose wavelength needs to be added as an extra "instruction".

In other words, as we move across the threshold more instructions become necessary in order to specify the state generated by the system. This is in accord both with the idea expressed repeatedly in this book that bifurcation is the "elementary act" toward complexity, and with the behavior of the Shannon information entropy across a bifurcation point deduced in Sec. 4.1. Suppose now that the system is moved further and further from the threshold to reach the region of fully developed turbulence as described in Sec. 1.4.1. Obviously, the description of such a state requires a lot of specifications or, in its digitalized form, a long sequence of 0's and 1's. On the other hand, the computer program is not experiencing a similar complexification: the evolution laws remain the same (the Navier-Stokes and the energy balance equations) and the initial conditions leading to the turbulent state need no special specification in the sense that they are generic. In other words, the theory is simpler (in the algorithmic sense) than the data it explains - just the type of theory sought after in physics. The argument goes through for a host of other problems. For instance, eq. (2.18) can be condensed in a few lines of a computer program. For $\mu = 1$, a randomly chosen initiation condition (thereby excluding those that would follow one of the unstable periodic orbits) would lead to a time behavior in the form of fully developed chaos with a Lyapunov exponent equal to $\ln 2$. If one chooses a Markov partition provided by the left and right halves of the unit interval, the symbolic sequence generated by this chaotic trajectory can be shown to be equivalent to a (fair) coin tossing and turns to possess a maximum algorithmic complexity.

Despite the plausibility of these arguments one should be aware of a subtlety related to the information needed to specify the initial conditions. The initial condition for reaching a complex regime of interest may well be generic, but quite frequently the regime in question coexists with a host of other, generally unstable solutions. Suppose that the latter are realized for initial conditions provided by, say, the set of rational numbers. Such conditions are exceptional, since numbers are irrational in their overwhelming majority. To complete the computer program with the specification of a generic initial condition one thus needs to provide an algorithm generating the digits of an irrational number. The trouble is that most of the irrational numbers are incompressible, i.e. random in the sense of algorithmic information. If the regime of interest can be reached only through such initial data, the complexity of the program would thus be comparable to that of the outcome. It may of course happen that the regime can be reached through

Information, Entropy and Selection   127

irrational but compressible data as well, like e.g. those necessary for the specification of $\sqrt{2}$. This would secure, then, the existence of a theory that is less complex than the data it wants to explain.

The discussion above suggests that algorithmic complexity captures certain features of natural complex systems as we have identified them so far in this book. On the other hand, in its basic philosophy it is fundamentally different from the complexity one is concerned with in nature where one seeks to identify emergent patterns, concerted behavior and evolution. True, full orderliness in the form of a complete lack of variability is an extreme case of coherence in which the object is like a fossil, and its behavior can hardly be characterized as complex. On the other hand, the strong variability represented by a random sequence -a random noise- and the concomitant lack of any form of correlations is another, equally non-representative form of organization. In reality, physical complexity must somehow be sandwiched between these two extremes, and thus should not be fully identified with algorithmic complexity.

A still more fundamental reason for differentiating between the two concepts pertains to the role of dynamics and, in particular, to the issue of selection raised in Sec. 4.3. Algorithmic complexity is insensitive to the time needed to accomplish a program (assuming that the latter will eventually halt). But in nature it is important to produce certain forms of complexity as the system of interest evolves in real time. The probability to produce a prescribed pattern/sequence out of the enormous number of the a priori possible ones is usually exceedingly small. But in a dynamical system generating complex behavior such states may be produced with probability one, being the result of a physical mechanism built into the evolution law: the problem of selection simply does not arise. In a way, under appropriate conditions dynamical systems are capable of exploring their state space continuously thereby creating information and complexity; at the same time they act like efficient selectors that reject the vast majority of possible patterns/sequences and keep only those compatible with the underlying dynamics. Furthermore, dissipation allows for the existence of attractors that have asymptotic stability and thus reproducibility. In concluding, it therefore seems legitimate to state that algorithmic complexity is a static, equilibrium like concept whereas physical complexity takes its full significance in a dynamic, nonequilibrium context. To tackle physical complexity, one needs a nonequilibrium generalization of classical information theory. The ideas developed in Secs 4.3 to 4.6 constitute steps in this direction.

Attempts at a compact definition -or at least measure- of physical complexity beyond its algorithmic aspects as formalized by the Kolmogorov-Chaitin complexity have been reported in the literature. An interesting measure, on the grounds of its relation to prediction, is the amount of information necessary to estimate optimally conditional probabilities of the type $W(i_{n+1}|i_1\cdots i_n)$ given the n-fold probabilities $P_n(i_1\cdots i_n)$ (cf. Sec. 4.2). In a quite different vein one associates complexity to "value" of some sort, for instance, the time required to actually retrieve a message from its minimal algorithmic prescription. In this view a message is complex, or *deep*, if it is implausible and can only be brought to light as a result of a long calculation. This introduces the time element that is so conspicuously absent in the Kolmogorov-Chaitin complexity. Alternatively, the complexity of a system in a certain state is identified to the logarithm of the probability that this state has been reached through a particular path from time step $-n$, in the past, to the present time zero. While capturing certain aspects of physical complexity, none of these definitions/measures manages to fully encompass its multiple facets. The question, how to define complexity is thus likely to remain open for some time to come. It may even turn out to be an ill-posed one: after all complexity does not reflect any built-in, immediately recognizable, structure as is e.g. the case of nonlinearity; it is, rather, a set of attributes that spring into life unexpectedly from the laws of nature when the appropriate conditions are met.

4.8 Summing up: towards a thermodynamics of complex systems

Science is at its best when dealing with uniqueness and stability. It provides us under these conditions with universal evolution criteria in the form of variational principles. In particular, thermodynamics drew a well-deserved reputation of efficiency and elegance for its ability to make predictions of this kind on a surprisingly large variety of natural systems, independently of their detailed structure and of the details of the ingoing processes.

As we saw in Sec. 2.2, the universality of this remarkable *tour de force* is limited to thermodynamic equilibrium and to situations close to it. Since one of the main attributes of complex systems is to generate multiplicity and instability and to function as a rule far from thermodynamic equilibrium the question was raised repeatedly in this book, as to whether there exists an extension of thermodynamics capable of coping with these situations while keeping at the same time its main prerogatives of universality and compactness.

The probabilistic approach to complex systems and the developments outlined in the present chapter provide an affirmative, albeit unexpected answer to this question. They show that rather than being extended in the traditional sense, keeping the same basic concepts and tools, thermodynamics is undergoing a profound transformation. The introduction of new ideas from nonlinear dynamical systems and stochastic processes leads to a considerable enrichment of the traditional framework and provides efficient methods for characterizing complex nonlinear systems: variational principles featuring generalized potentials, incorporating aspects of the microscopic dynamics through the fluctuations around the macroscopic variables (eqs (3.10)-(3.12)); or kinetic potentials replacing the traditional thermodynamic ones in the vicinity of a bifurcation, at least when the normal form features a single order parameter (eq. (2.14)). More significantly perhaps, thermodynamically inspired formalisms pervade nowadays many fields outside traditional thermodynamics such as nonlinear dynamics, chaos theory, stochastic processes and information theory. Although the state variables and functionals featured in these approaches bear no direct relation to the traditional thermodynamic variables, the spirit of thermodynamics is very much present in that the goal set is to arrive at universal relationships between the relevant quantities, independent of the detailed structure of the underlying system. The Shannon information and block entropies are of special interest in this respect. They are connected with the indicators of the underlying dynamics such as the distance from bifurcation (eq. (4.2)), the Lyapunov exponents and the Kolmogorov-Sinai entropy. And at the same time, when appropriate limits are taken, they reduce to quantities playing a central role in traditional thermodynamics such as the entropy production (eq. (4.25)). Furthermore, as seen in Sec. 4.6, thermodynamics inspired ideas are at the basis of an elegant description of fractal structures or more generally of complex data sets, based on Renyi entropies, generalized dimensions and local Lyapunov exponents.

Surprisingly the more obvious question, what happens to the traditional thermodynamic functionals like entropy and entropy production when a system undergoes transition to complex dynamics remains largely open. Early attempts in this direction were concerned with the evaluation of the mean entropy or entropy production on attractors of e.g. the limit cycle or the chaotic type, and the comparison with its value in the (unstable) reference state from which these attractors had bifurcated. They did not produce sharp results, suggesting that global thermodynamic quantities do not provide adequate measures of complexity. A more promising approach that has recently led to encouraging results is to monitor local and instantaneous values of entropy production and the like quantities. Some interesting connections with the indicators of dynamical instability and randomness such as the local analogs of Lyapunov exponents and Kolmogorov entropy are in this way brought out, but so far there exists no formal relation assessing the generality of these results. The issue will be taken up later on, in Sec. 6.2.

Chapter 5

Communicating with a complex system: monitoring, analysis and prediction

5.1 Nature of the problem

To an observer, complexity is manifested through the intricate way events unfold in time and organize in space. As stressed in the preceding chapters non-repetitiveness, a pronounced variability extending over several space and time scales, sensitivity to the initial conditions and to the parameters are some of the characteristic signatures of such spatio-temporal complexity. What is more, a given system can generate a whole variety of dependencies of this kind associated to the different states that are, as a rule, simultaneously available. Clearly, the situation is far from the standard picture of a system living around a more or less well-defined configuration in which variability results primarily from contamination by a weak background noise. How to observe, to analyze and to predict a system in a way that does not miss essential features of its complexity and, conversely, how on the basis of the data so collected one may gain access to some key indicators of this complexity is the principal theme of this chapter.

5.2 Classical approaches and their limitations

The basic problems one has to cope with when confronted with data are, how to explore the available information and sort out regularities linking the data points thereby differentiating them from random noise; given a number N of data points, how to infer in the best possible way the outcome of the next few trials; and, given that the observations generating the data are subjected to errors and the models devised to interpret them are incomplete, how to sample best the relevant variables and estimate the "true" state.

5.2.1 Exploratory data analysis

This type of analysis is carried out in a statistical perspective, in which the dynamical origin of the phenomenon and its deterministic aspects are not addressed. In its most familiar version it refers to stationary univariate data, where a single quantity is sampled and long term trends are absent. Two obvious questions asked in the first place pertain to their location and spread. In a classical linear perspective these are provided by the mean and the standard deviation (cf. Sec. 3.2)

$$\bar{X} = \frac{1}{N} \sum_{j=1}^{N} X_j \quad (5.1a)$$

$$\sigma = \left(\frac{1}{N} \sum_{j=1}^{N} (X_j - \bar{X})^2 \right)^{1/2} \quad (5.1b)$$

A more detailed information is provided by the histogram - essentially, the probability distribution underlying the data: the range of variation of the data is divided into regions (bins) and the number of points falling into each bin is counted and put in the ordinate axis. The bin width, δ must be selected in such a way that in the one extreme there are no empty regions and, in the other, essential features of the structure are not missed. A rule of thumb is that δ is inversely proportional to $N^{1/2}$ and proportional to the so-called interquartile range, defined as the difference between the points lying in the 3/4 and the 1/4 of the interval spanned by the values of the data taken in ascending order. The importance of the histogram is to hint at possible limitations of the classical description. If its structure consists of a single well-defined peak with a small dispersion around the mean one may expect that a linear view will be a reasonable first order description. But if it is multihumped, the underlying system might be closer to a multistable system jumping between coexisting attractors under the effect of fluctuations. This is, for instance, the case for the precipitation data depicted in Fig. 1.6. A more sophisticated analysis integrating tools from complex systems will then be necessary.

Having an empirical probability distribution at one's disposal the next step is to compare it with particular mathematical forms, usually taken among the classical distributions of probability theory and referred in this context as "parametric" probability distributions. There is a highly sophisticated statistical theory of how to test the hypothesis related to the adoption of a particular form and how to fit from the data the associated parameters, which is out of the scope of this book.

In addition to the above "static" information, it is useful to extract indications on how successive data points are related to each other in the order they are obtained. A classical measure of such a relationship is provided by the autocorrelation function, defined as

$$C(m) = \frac{\sum_{j=1}^{N-m}(X_j - \bar{X})(X_{j+m} - \bar{X})}{\sum_{j=1}^{N}(X_j - \bar{X})^2} \qquad (5.2)$$

If the j's refer to the successive values of the variable X registered at regularly spaced instants, then (5.2) is the analog, in the time domain, of the power spectrum introduced in Sec. 2.6, which provides a similar information as (5.2) in the frequency domain. In plain terms one seeks the answer to the following question: supposing that initially ($m = 0$) a perturbation is inflicted on the system, how is this expected to affect the relevant variable at later times and, in particular, how long is the system going to keep the memory of this event? The autocorrelation function provides a valuable answer as long as its envelope is found to decay exponentially, in which case the inverse of the factor multiplying the time argument in the exponential can be identified to the persistence time. If this does not happen to be the case and, in particular, if long tails indicative of intricate dynamics are present, the autocorrelation function analysis will need to be completed by more advanced information as seen later on in this chapter. It should be stressed that the correlation function provides no physical explanation about the relationship between the successive values of X, and is insensitive to the deterministic or the random character of the underlying process. Notice that definition (5.2) is representative for m roughly less than the integer part of $N/2$. Otherwise the number of data lost is much too large.

We next consider the case of data pertaining to several variables $X_1, ..., X_n$ say surface pressure records at different geographic locations or temperature and vertical component of the velocity at a given point in the Rayleigh-Bénard experiment (cf. Sec. 1.4.1). A quantity giving a first idea of possible relationships between these variables is the covariance matrix, a symmetric $n \times n$ matrix which constitutes the multivariate generalization of the variance (eq. (5.1b)):

$$V(X_\ell, X_m) = \frac{1}{N}\sum_{j=1}^{N}(X_{j,\ell} - \bar{X}_\ell)(X_{j,m} - \bar{X}_m) \qquad (\ell, m = 1, ...n) \qquad (5.3)$$

If only two variables, say X and Y, are involved, the non-diagonal element of (5.3) normalized by the variances σ_X and σ_Y, referred as the Pearson correlation

$$r_{XY} = \frac{1}{\sigma_X \sigma_Y}\frac{1}{N}\sum_{j=1}^{N}(X_j - \bar{X})(Y_j - \bar{Y}) \qquad (-1 \le r_{XY} \le 1) \qquad (5.4)$$

describes the part of the variability of one of the two variables that is also shared by the other. To extract a similar information in the general case of an arbitrary number of variables one resorts to the concept of "principal components". The idea is that if the variables $\{X_j\}$ are highly correlated, then their full set contains redundant information and one should be able to reduce it to a data set featuring fewer variables that contain most of the variability present. These new variables are sought in the form of linear combinations, $Z_1, ...$ of the original ones, and it is required that they are mutually uncorrelated. Now, as well known, the eigenvectors of a symmetric $n \times n$ matrix like the covariance matrix are mutually orthogonal. This suggests that the desired linear combinations should be the projections of the data vector $\mathbf{X} = (X_1, ..., X_n)$ onto the successive eigenvectors $\mathbf{u}_m$, of $\mathbf{V}$:

$$Z_m = \sum_{k=1}^{n} u_{mk} X_k, \qquad m = 1, ...K \qquad (5.5)$$

The first principal component, Z_1 is the linear combination in (5.5) having the largest variance. This implies that the corresponding eigenvector $\mathbf{u}_1$ is the one associated with the largest eigenvalue of $\mathbf{V}$. The subsequent principal components $Z_2, Z_3, ...$ are the linear combinations having the next largest variance, entailing that the eigenvectors $\mathbf{u}_2, \mathbf{u}_3, ...$ intervening in eq. (5.5) correspond to the second, third, ...largest eigenvalues of $\mathbf{V}$. If there is a group of K rather closely lying eigenvalues separated by the remaining $n - K$ ones by a large gap, then one may limit the analysis to the first K principal components. Alternatively one may prescribe a certain fraction of the total variability to be accounted for, and truncate the set of Z_m's accordingly. Once this is settled then the original variables can be expressed in terms of the principal components by inverting eq. (5.5). The principal component analysis has proven to be invaluable in the description of atmospheric fields such as pressure, temperature, or precipitation. For instance, four principal components turn out to account for nearly half of the total variability of the winter monthly-mean heights at which the prevailing pressure is at the $500\,hPa$ level (a popular quantity in meteorology) in the northern hemisphere. In this context, eq. (5.5) and the analogous relation expressing $\{X_j\}$ in terms of $\{Z_m\}$ in truncated form are referred, respectively, as the "analysis" and "synthesis" equations.

5.2.2 Time series analysis and statistical forecasting

We turn next to the more ambitious question, how to characterize on the basis of the data gathered at successive times $t_1, ..., t_j, ...$ the way the relevant variable (or a set thereof) varies in time. This is usually achieved by representing the data set in terms of a mathematical model involving a limited set of parameters, to be fitted in such a way that the model statistical properties are identical to those determined by the data (cf. preceding subsection). For a stationary process this also allows future values to be inferred from the (inevitably limited) set of data available.

In classical data analysis, the prototype of mathematical models of this kind are the *autoregressive models*. They can be regarded as "filters" converting an uncorrelated process, R of zero mean (a background noise of some sort) to a correlated one, X - the process responsible for the time variation of the variable of interest. For a single variable the simplest such connection is the one linking two successive values of X through the first order, or "AR(1)" model

$$X_{i+1} - \bar{X} = \phi(X_i - \bar{X}) + R_{i+1} \qquad (5.6)$$

where for convenience the average value $\bar{X}$ has been subtracted throughout and ϕ is a proportionality coefficient. This coefficient as well as the variance of R are to be fitted according to the above described procedure. Multiplying (5.6) by $X_{i+1} - \bar{X}$ and by $X_i - \bar{X}$ and taking the average over all values yields

$$\phi = C(1) = \rho, \quad \sigma_R = \sigma_X(1 - \rho^2)^{1/2} \qquad (5.7)$$

where ρ is the autocorrelation function (5.2) at time lag $m = 1$. If, in addition, R is a Gaussian process the X process is fully specified. The procedure can be extended straightforwardly to several variables and to higher order autoregressions, referred as "AR(K)", where X_{i+1} is linked to the K values $X_i, ..., X_{i-K+1}$ preceding it. Further extension leads to the autoregressive-moving average models, ARMA(K, M) where, in addition, several random noise terms lagged by up to M time units are added to R_{i+1} with appropriate weighting coefficients. Non-stationarity connected with a systematic time dependence of the mean can also be accounted for by integrating the output of an ARMA model from a distant past up to the running time i.

Eq. (5.6) and its various generalizations specify completely the statistical properties of the X process. In particular, they allow one to compute conditional probabilities for having a certain value of X at time $i+1$ conditioned on observed past behavior at the $i, ..., i-K$ time step. For an AR(1) model, the conditional mean

$$f(X_i) = \int dy\, y\, W(y, i+1 | X_i, i) \tag{5.8}$$

can be shown to provide the best estimate, $\hat{X}_{i+1}$ of such a value that can thus be used for forecasting, in the sense that the mean square deviation between the actual values and $f(X_i)$ will be minimum. Since eqs like (5.6) are linear all probability distributions pertaining to X, including W, will be Gaussian as long as R is a Gaussian process. This entails that $f(X_i)$ is a straight line, also referred as *linear regression* line, X_i and $\hat{X}_{i+1}$ being the predictor and the predictant, respectively. It must be realized that in view of the presence of random inputs in eq. (5.6) and its various generalizations, which can take quite different values depending on the particular realization considered, a prediction based on eq. (5.8) will make only statistical sense: the detailed time course associated to each of these realizations will lead to widely scattered values.

In a similar vein, one may also devise a linear prediction algorithm on the sole basis of the data available. The target is to arrive at the prediction $\hat{X}_{N+1}$ which minimizes (on average) the squared prediction error, $(\hat{X}_{N+1} - X_{N+1})^2$. We estimate this value by its average over the available values of the data set,

$$\sum_{n=K}^{N-1} (\hat{X}_{n+1} - X_{n+1})^2 \tag{5.9}$$

whereby the predictant is expressed as a weighted sum of the last K measurements (being understood that the mean value is subtracted throughout):

$$\hat{X}_{n+1} = \sum_{j=1}^{K} a_j X_{n-K+j} \tag{5.10}$$

Inserting (5.10) into (5.9) and requiring that the derivatives with respect to a_j vanish leads then to a set of K linear equations for a_j whose coefficients and constant terms are expressible in terms of the covariance matrix of the data (eq. (5.3)), from which these weighting factors can be evaluated . The procedure can be generalized to "lead r" predictions, in which $\hat{X}$'s up to $n+r$ are considered. It should be stressed that at this stage there are no a priori criteria for choosing the value of K and the order of the autoregressive model. Related to this is also the important question of robustness of a prediction against these values. There is a highly specialized part of the literature on mathematical statistics in which these problems, which are out of the scope of this book, are addressed in detail.

We have argued as if the variables X were directly available from experiment, but in actual fact this is not the case. First, experimental measurements are subjected to errors; and, second, the quantities Z actually measured may not be X but combinations of them whose number may even be less than the total number of the X's. For instance, a river discharge is a measurable quantity but the water storage of a basin is not. Assuming that Z are linked to X by linear relationships and that the measurement errors, V are of an unbiased random nature fully uncorrelated from the noise R in eq. (5.6) leads to

$$Z_i = \sum_j H_{ij} X_j + V_i \quad (5.11)$$

where it is understood that X satisfies (5.6) or extensions thereof. The problem is now that of deriving an estimate, $\hat{X}$ of the linear stochastic process X on the basis of Z in a way minimizing the mean quadratic error and restituting the correct average values. This problem can again be formulated as a filtering problem, known in this context as Kalman filter and leads to an answer of the form

$$\hat{X}_{N+1} = \underbrace{\phi \hat{X}_N}_{\text{estimate in absence of measurement}} + \underbrace{G_N(Z_{N+1} - H\phi \hat{X}_N)}_{\text{correction due to the measurement}} \quad (5.12)$$

where the "gain matrix" G is determined by ϕ and the covariance matrices of V and R. This equation can be solved recursively, the initial conditions being determined by Z and the scheme itself in a self-consistent way. Notice that if the random noises are Gaussian the conditional probability of X is also Gaussian, and $\hat{X}$ in (5.12) represents the conditional average computed with this probability.

As seen in the previous chapters in many instances it is natural to resort to coarse-grained states, corresponding to the lumping of higher resolution states. The state variable becomes now discrete, and the transitions between its different values is a finite state stochastic process. There is a large literature in data analysis, especially in the context of atmospheric physics and hydrology, in which this process is at the outset assumed to be Markovian. In general this is not a legitimate procedure: as a rule, lumping introduces memory effects unless the existence of a Markov partition (Sec. 3.3) can be established. Care should thus be taken to test the Markov hypothesis by comparing higher order transition probabilities to those linking two successive states. If the test fails probabilistic models with a finite number of states can still be devised on the basis of the two-time conditional probability *and* the statistics of residence and escape times from each of the states, and will as a rule give rise to memory effects.

5.2.3 Sampling in time and in space

Experimental data are discrete samples of signals that usually are in their original form continuous in time and in space. Continuous signals can be advantageously visualized in the frequency domain, by representing them as infinite sums or as integrals (Fourier series and Fourier integrals, respectively) of sinusoidal functions with appropriate weighting factors. Monitoring such a signal in intervals Δt over a total time period T yields $N = T/\Delta t$ samples. The above infinite sums will then be truncated to a finite number differing by up to 1 from the integer part of $N/2$, to account for the fact that the number of Fourier coefficients must equal the number of data points (cf. also Sec. 2.6). In this context an important mathematical result, known as *sampling theorem*, is of special relevance. It states that a continuous time signal can be represented completely by and reconstructed from a set of samples of its amplitude at equally spaced times, provided that the interval between samples is equal to or less that one-half of the period associated to the highest frequency present in the signal. As a rule for this condition to be satisfied the samples must be taken much too frequently, which poses practical difficulties. For instance, in a signal containing frequencies up to $1\,000 sec^{-1}$ we must use $2\,000$ samples per second. It follows that a signal will rarely be described or reproduced perfectly. Additional problems come from the fact that strictly speaking many signals of interest, including signals generated by intermittent processes, are rarely "band limited" in the sense of having zero energy above a limiting frequency ω_{max}. Random sampling can provide a valuable alternative for coping with these difficulties. For instance, the regularly spaced sampling times $t_j = j\Delta t$, $j = 0, 1, ...$ may be replaced by random times distributed according to a Poisson process with average density τ. One can show that in this way one can detect spectra of any frequency, provided that the observation interval T is large enough. Furthermore, for a deterministic process the mean of the estimate of the power spectrum of the process equals the power spectrum itself.

Most of the above ideas can be transposed to the space domain where they also provide input to the important question, especially in the context of atmospheric physics and hydrology, how to design an observational network capable of representing adequately the spatial characteristics of the quantities of interest. Suppose, for instance, that we want to capture by our measurements spatial scales up to a characteristic length ℓ_{min}. Since by the sampling theorem we need a minimum of two samples per wave the maximum spacing between the network stations must be $\Delta x = \ell_{min}/2$, entailing that the number of stations in a one-dimensional section of the monitored area must be $N \approx 2L/\ell_{min}$ which might be much too high. If so, random sampling is again to be envisaged as an alternative and this idea has indeed been explored in recent years in the context of meteorology and hydrology (see also Sec. 5.4).

5.3 Nonlinear data analysis

In the analysis outlined in the preceding section the regularities contained in a set of data were identified through the structure of correlation function-like quantities as inferred from the data, and prediction was carried out by means of linear models fitted to reproduce these structures. In this statistical and linear perspective the nature of the underlying dynamics was overlooked, and the only allowed possibilities were a monotonic or oscillatory decay towards a mean value around which the system was exhibiting an aperiodic variability generated entirely by the presence of an externally introduced noise. In particular, there could be no question of discriminating between deterministic dynamics (possibly perturbed by some amount of noise present inevitably in the real world) and a random process, nor between different kinds of attractors (or more generally of invariant sets) descriptive of the complexity of the system at hand. These questions, which are of paramount importance in the context of complex systems research, will be addressed in the present section.

5.3.1 Dynamical reconstruction

As in Sec. 5.2 we start with a time series providing the successive values of a certain variable sampled at regular time intervals Δt. At first sight, it might be argued that such a series is restricted to a one-dimensional view of a system which in reality contains a large number of interdependent variables providing, for this reason, a rather limited amount of information. Here we show that a time series is far richer than that in information: it bears the marks of all other variables participating in the dynamics, and allows us in principle to identify some key features of the underlying system independent of any modeling.

Let $X_1(t)$ be the time series derived from the experimental data and $\{X_k(t)\}, k = 2, ..., n$ the other variables expected to take part in the dynamics. As we do not rely on any particular model, we want to reconstruct this dynamics solely from the knowledge of $X_1(t)$. Imagine the phase space spanned by the variables $\{X_k\}, k = 1, ..., n$. As we saw in Chapter 2, an instantaneous state of the system becomes a point in this space, whereas a sequence of such states followed in time defines a curve, the phase space trajectory. As time progresses the system reaches a state of permanent regime, provided its dynamics can be reduced to a set of dissipative deterministic laws. This is reflected by the convergence of families of phase trajectories toward a subset of the phase space, the *attractor*. Our principal goal is to find an answer to the following questions:

1. Is it possible to identify an attractor from a given time series? In other words, can the salient features of the system probed through this time series be viewed as the manifestation of a deterministic dynamics (possibly a very complex one), or do they contain an irreducible stochastic element?

2. Provided that an attractor exists, to what extent is it representative of the attractor in the system's original, full phase space, and what is its dimensionality D? We know from the analysis in Chapter 2 that the dimensionality provides us with valuable information on the system's dynamics. For instance, if $D = 1$ we are dealing with self-sustained periodic oscillations; if $D = 2$ we are in the presence of quasi-periodic oscillations of two incommensurate frequencies; and if D is noninteger and larger than 2 (the case of a fractal attractor, Sec. 2.1), the system is expected to exhibit a chaotic oscillation featuring sensitivity to initial conditions, as well as an intrinsic unpredictability. In this latter case one would also like to have access to indicators of more dynamical nature, such as Lyapunov exponents or Kolmogorov-Sinai entropies.

3. What is the minimal dimensionality, d, of the phase space within which the above attractor is embedded? This defines the minimum number of variables that must be considered in the description of the underlying dynamics. Note that D is necessarily smaller than d.

Our first step is to identify a suitable set of variables spanning the phase space. A very convenient way is to unfold the original time series $X_1(t)$ by successively higher shifts defined as integer multiples of a fixed lag τ ($\tau = m\Delta t$). Taking, in addition, N equidistant points from the data set, we are led to the following set of discretized variables:

$$X_1 : \qquad X_1(t_1), \cdots, X_1(t_N)$$
$$X_2 : \qquad X_1(t_1+\tau), \cdots, X_1(t_N+\tau)$$
$$\cdots$$
$$X_n : \; X_1[t_1+(n-1)\tau], \cdots, X_1[t_N+(n-1)\tau] \qquad (5.13)$$

For a proper choice of τ these variables are expected to be linearly independent, and this is all we need to define a phase space. Still, all these variables can be deduced from the single time series pertaining to $X_1(t)$, as provided by the data. We therefore see that, in principle, we have sufficient information at our disposal to go beyond the one-dimensional space of the original time series and to unfold the system's dynamics into a multidimensional phase space. This information allows us to draw the *phase portrait* of the system -more precisely, its projection to a low-dimensional subspace of the full phase space- or the analog of a Poincaré type mapping, see Sec. 2.5. This view is often helpful and suggestive of what might be going on but, as a rule, the data are too coarse to draw any useful conclusion from this representation alone. Fortunately, using the techniques of nonlinear science, one is in the position to arrive at a sharper characterization of the complexity of the dynamics. Let us introduce a vector notation: $\mathbf{X}_i$ stands for a point of phase space whose coordinates are $\{X_1(t_i), \cdots, X_1[t_i+(n-1)\tau]\}$. A reference point $\mathbf{X}_i$ is chosen from these data and all its distances $|\mathbf{X}_i - \mathbf{X}_j|$ from the $N-1$ remaining points are computed. This allows us to count the data points that are within a prescribed distance, r, from point $\mathbf{X}_i$ in phase space. Repeating the process for all values of i, we arrive at the quantity

$$C(r) = \frac{1}{N^2} \sum_{i,j=1}^{N} \theta(r - |\mathbf{X}_i - \mathbf{X}_j|) \qquad (5.14)$$

where θ is the Heaviside function, $\theta(x) = 0$ if $x < 0$ and $\theta(x) = 1$ if $x > 0$. The nonvanishing of $C(r)$ measures the extent to which the presence of a data point $\mathbf{X}_i$ affects the position of the other points. $C(r)$ may thus be referred to as the integral *correlation function* of the attractor.

Suppose that we fix a given small ϵ and use it as a yardstick for probing the structure of the attractor. If the latter is a line, clearly the number of probing points within a distance r from a prescribed point should be proportional to r/ϵ. If it is a surface this number should be proportional to $(r/\epsilon)^2$ and, more generally (cf. Sec. 2.1), if it is a D-dimensional manifold it should be proportional to $(r/\epsilon)^D$. We therefore expect that for relatively small r, $C(r)$ should vary as $C(r) \sim r^D$. In other words, the dimensionality D of the attractor is given by the slope of the $\ln C(r)$ versus $\ln r$ in a certain range of r:

$$\ln C(r) = D \ln r \qquad (5.15)$$

The above results suggest the following algorithm:

- Starting from a given time series construct the correlation function, eq. (5.14) by considering successively higher values of the dimensionality n of phase space.

- Deduce the slope D near the origin according to eq. (5.15) and see how the result changes as n is increased.

- If the D versus n dependence is saturated beyond some relatively small n, the system represented by the time series should be a deterministic dynamical system and possess an attractor. The saturation value D is regarded as the dimensionality of the attractor represented by the time series. The value d of n beyond which saturation is observed provides the minimum number of variables necessary to model the behavior represented by the attractor.

Notice that in signals generated by Gaussian white noise processes D would increase linearly with n. In correlated noise processes there would be no tendency to saturate either. Still, the presence of highly intricate correlations in conjunction with the limited amount of data may be at the origin of a saturation-like behavior. Similar pitfalls arise when the attractor dimensionality is much too high to be handled by the data available. They reflect the limits of the method of dynamical reconstruction based on delay embedding.

An important question pertains to the expected relationship between D and d. An answer is provided by the *delay embedding theorems*, which originate from Floris Taken's seminal work. Specifically, if $\bar{D}$ is the dimensionality of a smooth compact manifold containing the reconstructed attractor ($\bar{D} \geq D$), then $d = 2\bar{D} + 1$ provides an embedding preserving the salient features of the dynamical system provided that the dynamics is sufficiently smooth and generic (in the sense of coupling all of the variables present). An interesting variant asserts that $d > 2D$, provided that there are no low order cycles of periods equal to the first few multiples of the sampling step Δt. In many instances, however, the n value where saturation is actually realized turns out to lie between D and $2D$.

There is a highly specialized literature on how to optimize the reconstruction algorithm, including the choice of time lag τ, and how to handle best data contaminated by noise. Comparison with the technique of principal component analysis (cf. Sec. 5.2.1), which is somewhat similar in spirit, also allows one the cross-check the reliability of certain conclusions.

Communicating with a Complex System 143

As the reader must have recognized by now, the correlation sum in eq. (5.14) gives access to the generalized dimension D_2 in the classification of Sec. 4.6. The algorithm can be adapted to compute the other dimensions as well. It can also be extended to evaluate the spectrum of the Lyapunov exponents and of Kolmogorov-Sinai entropies from time series data, if the attractor dimension analysis turned out to reveal the presence of chaotic dynamics. The latter procedure is more demanding as far as number of data is concerned and is for this reason subjected to more serious limitations.

The above developments open the tantalizing perspective to capture the essential features of a complex system by a limited number of variables. The nature of these variables is not specified at this stage: time series analysis is not a substitute to modeling, although it provides a most valuable input on the kind of description one should be after. This description should be understood as a coarse-grained view of the full scale system under consideration, rather than as a detailed one. One promising possibility is to resort to modeling based on generic expressions such as normal forms (cf. Sec. 2.3) chosen according to the conclusions drawn from the data analysis and with parameter values fitted from comparison with experiment.

5.3.2 Symbolic dynamics from time series

The ideas developed in Chapter 4 and in Sec. 4.3 in particular, suggest an alternative approach to nonlinear data analysis independent of any phase space reconstruction algorithm, especially suited to information-rich signals generated by chaotic systems: map the original dynamics into a sequence of discrete symbols and identify the nature of the resulting symbolic dynamics using such tools as the scaling of the block entropies S_n with the sequence length n, or even by appealing to standard statistical tests. We have already seen the kind of information provided by this approach on a generic mathematical example of low-dimensional chaos, the Rössler model (eq. (4.16)). In this subsection we outline an application on a set of data pertaining to atmospheric dynamics.

The atmosphere tends to organize in relatively persistent and recurrent circulation patterns, identified and classified according to the space and time scales of interest and to the degree of refinement desired. An analysis of this type, where the method of principal component analysis described in Sec. 5.2.1 plays an important role, has led to the 21 "Grosswetterlagen" of the European North Atlantic region. Less refined classifications are also carried out. Fig. 5.1 provides an example based on the compilation of daily weather regimes by the Swiss Meteorological Institute between 1945 and 1989. The classification scheme consists of 3 main clusters (weather regimes) known as

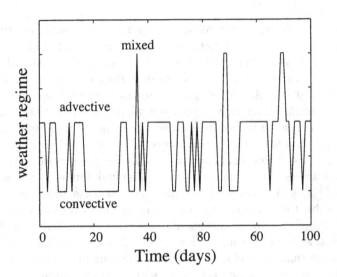

Fig. 5.1. Typical time sequence of the convective, advective and mixed weather regimes over an area covering a radius of 222 km centered somewhere in the Alps.

"convective", "advective" and "mixed". We notice that the time evolution presents a markedly intermittent character. The convective regime is the most frequent one, occurring with a probability $p_c \approx 0.50$. It is followed closely by the advective one ($p_a \approx 0.44$), the mixed one being almost an order of magnitude sparser ($p_m \approx 0.06$). A more detailed information provided by the time series pertains to the histogram of the *exit time distributions* from each state, measuring the persistence of this state if the evolution is started in its immediate vicinity. Fig. 5.2 depicts the results for one of the weather clusters of the data set. One is struck by a rather slow decay of these distributions, a feature that is also reflected by the fact that a fitting by an exponential turns out to be a rather poor one giving rise to deviations that are larger than 10%. Now, exponential decay of exit time distributions is one of the characteristic signatures of a first order Markov process. The results of Fig. 5.2 cast therefore doubt as to the Markovian character of the succession of weather regimes.

The presence of non-Markovian dynamics and, concomitantly, the existence of long range correlations suggested by the above result is further confirmed by the study of the actual frequencies of finding strings consisting of n symbols (here individual clusters) in a prescribed order. For instance, in an 8-day record more than 800 sequences of the type 11111111 instead

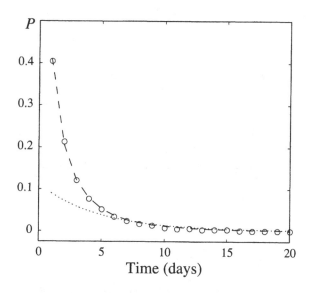

Fig. 5.2. Exit time probability distribution of the convective regime. Dotted and dashed lines represent best fits with an exponential and a stretched exponential, respectively.

of 440 predicted by the Markov model are found, the next most frequent sequence being 22222222 present in about 400 copies, about three times as much compared to the Markov model. Notice that the vast majority of "words" appear with low probability, whose values are comparable for all these sparse sequences.

We now perform an analysis of the symbolic sequence generated by the data based on the scaling properties of the block entropies, S_n. In doing so care must be taken in relation to the limited amount of data available: since the probabilities involved in the definition of S_n (eq. (4.3)) are deduced from observed relative frequencies, the latter are less and less adequately estimated as the length n of the string increases. At the level of S_n this shows up as a saturation of the S_n versus n curve beyond a moderate value of n, even if the process is known to be random to begin with. At the level of the excess entropy h_n it will result in a decrease with n rather than a tendency toward a limiting h as stipulated by eq. (4.5). Methods for overcoming these limitations have been proposed, leading to length corrections necessary to remove such spurious effects. The results of the evaluation of S_n and h_n accounting for these corrections are shown in Figs 5.3a,b. As can be seen the excess entropy h_n is decreasing with n even after the length corrections

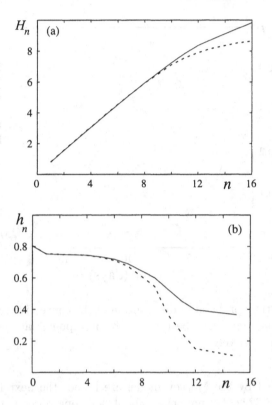

Fig. 5.3. (a) Block entropies for sequences of length n of the weather regimes (Fig. 5.1). The dashed line corresponds to the observed entropy based on the actual frequencies whereas the full line takes into account length corrections. (b) Excess entropies, uncertainties of the next outcome of a sequence consisting of n regimes (dashed line-observed, full line-corrected).

are performed implying a sublinear scaling of S_n with n, indicative of a non-Markovian process.

One of the merits of the symbolic analysis of time series is to provide hints at a probabilistic modeling of the process, enabling one to carry out statistical predictions beyond the predictability time of the fine scaled variables. We briefly outline how this can be achieved for the data set analyzed above. The idea is to represent the transitions between the three regimes as a 3-state continuous time stochastic process. The evolution of the different realizations of such a process can be viewed as a succession of halting periods during which the trajectory remains in a particular state, interrupted by rapid jumps between states. A full specification of the process requires, then, the transition probabilities between states, and the sojourn probabilities for each state, all of which are available from the data. To account for the non-Markovian character revealed by Fig. 5.2 we consider an exit time probability in the form of a stretched exponential, $p(\tau) = K\exp(-b\tau^{1/2})$, leading to a fit of the data much better than the exponential one (dashed line of Fig. 5.2). With this information one may now generate realizations of the underlying process. This simulation can be achieved by a Monte Carlo type approach as follows: (i) start from a given initial state i (cluster); (ii) next, draw a random number τ_i subject to the distribution appropriate to that state. The nearest integer $[\tau_i]$ to this number provides the time that the system will stay in this particular cluster; (iii) at $t = [\tau_i]$, state i is left. The particular state that will occur after the transition is chosen according to the values of the transition probabilities W_{ji} ($j \neq i$). Specifically, a uniformly distributed random number in the interval (0, 1) is drawn. If its value is between 0 and $W_{ji}/\sum_j W_{ji}$ a transition to state j is performed; once in state j, the above algorithm is repeated until a good statistics is realized.

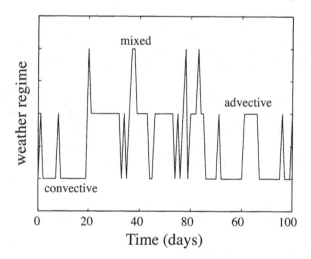

Fig. 5.4. Typical time sequence of the convective, advective and mixed weather regimes generated from Monte Carlo simulations by adjusting from the data the probability of exit times (Fig. 5.2) as well as the transition probability matrix **W**.

Having the realization of the process at our disposal we are in the position to make a forecasting of the evolution starting with a given initial state. Fig. 5.4 depicts a time series generated by the above method. The simulation reproduces the salient features of the data (Fig. 5.1), especially the intermittent character of the transitions between states. The procedure may also be used to generate the block entropies S_n and h_n, which turn out to be quite close to the forms constructed directly from the data.

The issue of correlated versus random symbolic sequences generated by a data set is also of central importance in biology, in connection with the analysis of the data on genome structure that have become available in recent years. We defer a discussion of this important topic to Sec. 6.6.4.

5.3.3 Nonlinear prediction

The procedure outlined in Sec. 5.3.1 allows one to develop prediction methods for nonlinear systems on the sole basis of experimental data, thereby extending eq. (5.10). Suppose that the system of interest is governed by deterministic evolution laws and we dispose of N measurements of the full n-dimensional state vector $\mathbf{X}$: $\mathbf{X}_1, ..., \mathbf{X}_N$ taken in chronological order. In order to predict the value of $\mathbf{X}_{N+1}$ we embed the data available in the full phase space and search along all past states the state $\mathbf{X}_\alpha (\alpha < N)$ closest to $\mathbf{X}_N$ with respect to some appropriately defined distance (e.g. the Euclidean norm). Assuming that the evolution laws are continuous in $\mathbf{X}$ -a property satisfied by a typical physical system- one expects then that the state $\mathbf{X}_{\alpha+1}$ will be the closest to $\mathbf{X}_{N+1}$ and can thus be used as a predictor.

In actual fact, as stressed in Sec. 5.3.1, the number of variables n is a priori unknown and one usually has access to the system via a scalar quantity, say Y, depending on the $\mathbf{X}$'s in a certain (sometimes unknown) way. To cope with this limitation one resorts to the delay reconstruction. For any chosen value of the delay τ and of the embedding dimension m one disposes, on the basis of the N measurements, of $N - m + 1$ state vectors

$$\mathbf{Y}_1 : Y_1, ..., Y_{(m-1)\tau+1}$$

$$- - - - - - -$$

$$\mathbf{Y}_{N-m+1} : Y_{N-(m-1)\tau}, ..., Y_N$$

As before embedding these vectors in the m-dimensional phase space allows one to find the vector $\mathbf{Y}_\alpha$ closest to $\mathbf{Y}_{N-m+1}$ and use $\mathbf{Y}_{\alpha+1}$ as a predictor for $\mathbf{Y}_{N-m+2}$, whose last component will be the desired predictant Y_{N+1}. More generally, one may express the predictant as a combination of more than one predictors, whose different terms are inversely weighted by the aforementioned distances. As a particular case of this latter variant, when the measurements are subject to uncertainties (as it is generally the case) more

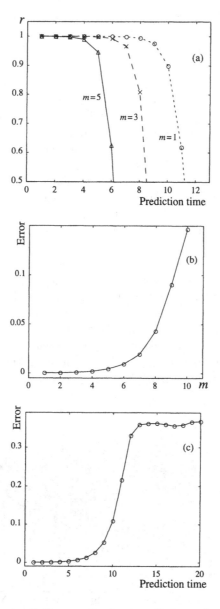

Fig. 5.5. Nonlinear prediction method as applied to the logistic map (eq. (2.18)) using 10 000 data. (a) Dependence of the correlation coefficient r on the prediction time for embedding dimensions $m = 1, 3$ and 5. (b) One step prediction error versus embedding dimension m. (c) Time evolution of the mean prediction error for $m = 1$.

than one predictors responding to the proximity criterion may fall into the uncertainty region, in which case the predictant should be expressed as their arithmetic (non-weighted) mean. The method can likewise be extended to multi-step prediction.

A first measure of the quality of the predictions carried out in this way is provided by the Pearson correlation r (eq. (5.4)) between predicted and observed values, which will also allow one to discriminate between deterministic dynamics, say of the chaotic type, and random noise. In the case of random noise r will display an irregular oscillation around zero and the mean error itself will display a similar kind of oscillation around a certain level whatever the time step considered, since the dynamics is in a sense starting all over again at each step. In contrast, in the presence of chaotic dynamics the correlation will decrease systematically with time and the initial error will tend to amplify (cf. also Sec. 5.5). Distinguishing deterministic chaos from correlated noise proves, on the other hand, to be more tricky. A valuable hint is provided by the switching of the scaling properties of the prediction error as a function of time from exponential behavior (case of deterministic chaos) to power law one (case of correlated noise). It should be emphasized that the behavior of r will also depend strongly on the choices of the lag τ and of the embedding dimension m. A reasonable choice of values of these parameters could be the one maximizing the correlation coefficient.

Figures 5.5a-c summarize the application of the method of nonlinear prediction on the logistic map (eq. (2.18)) at parameter value $\mu = 1$, based on

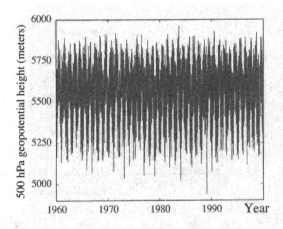

Fig. 5.6. Time evolution of the $500\,hPa$ geopotential height (52°N and 2°E) between 1960 and 2000 as measured daily at 0h UTC (Coordinated Universal Time).

a time series of 10 000 data and $\tau = 1$. In (a) the correlation coefficient r is plotted against the prediction time for different embedding dimensionalities m. As expected, one finds that $m = 1$ is the optimal dimensionality. This is further confirmed in (b), where the one-step mean prediction error is plotted against m. Finally, in (c) the mean prediction error is plotted against the prediction time for $m = 1$.

A more interesting example pertains to the time series depicted in Fig. 5.6, which describes the time development of the height at which the atmospheric pressure is at the 500 hPa level, a quantity introduced already in Sec. 5.2.1 and known as "geopotential height". Figs 5.7a-c summarize, for this quantity, the same information as in Figs 5.5 for the logistic model deduced from the 3 592 data points available. As can be seen, the optimal embedding dimensionality is $m = 2$. Carrying out predictions with this value (Figs 5.7b, c) leads to a behavior similar to Fig. 5.5b, c, except that the initial phase of exponential takeoff is missing. This may be due to the fact that the one-step error is already large enough to bring the system out of the linearized regime (cf. Sec. 2.1). Another explanation could be that the series in Fig. 5.6 is closer to correlated noise than to deterministic chaos. To reach a definitive conclusion one must complement the analysis with information coming from the evaluation of other properties, such as the attractor dimensionality. As it turns out, this information pleads in favor of a deterministic dynamics running on an attractor of dimension $D \approx 8$ and predictability time of about 2 weeks. A further argument in the same direction is the observation (cf. Sec. 5.2.1) that the first four vectors in a principal component analysis account already for half of the variability of the series in Fig. 5.6.

5.4 The monitoring of complex fields

Most of the complex systems observed in nature and in technology, from weather and climate to earthquakes and the combustion chamber of a heat engine, display intricate spatial dependencies of their principal properties. Knowledge on such systems comes from data recorded on various sites by appropriate measuring devices. These data also condition our ability to forecast, since they provide the initial conditions needed to run the various numerical prediction models available in meteorology, oceanography, computational fluid mechanics, and so forth. Now, for reasons of practical accessibility or of financial limitations these observation networks are as a rule sparse, highly inhomogeneous and in some cases fractally distributed. A first question therefore arises as to whether such a network provides an adequate picture of the system of interest and, in particular, whether for

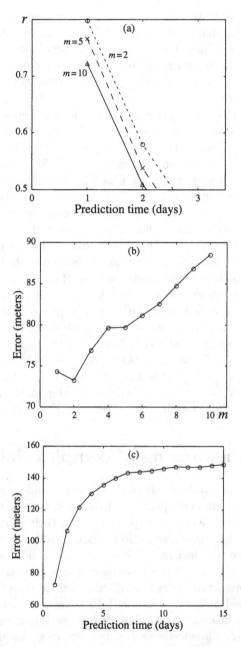

Fig. 5.7. As in Fig. 5.5 but for the time series of Fig. 5.6.

the realistic case of a given finite (and usually rather limited) number of observation sites an optimal distribution can be defined for which errors are minimized. A second question is, how to infer from the data on this limited set of points the full spatial profiles of the corresponding fields, particularly in the perspective of feeding this information into the initial conditions of our predictive model. The present section is devoted to an analysis of these questions.

5.4.1 Optimizing an observational network

If the dynamics were simple, in the sense of reproducibility and stability, the sparsity of the observational network would not introduce any fundamental limitation and a straightforward interpolation would be sufficient to reconstruct the fields of interest. This view must be reassessed when monitoring a complex system where irregularity and instability are the rule, as reflected in the spatial domain by the formation of fractal patterns, such as turbulent eddies or rain fields. Their main characteristic is to display a whole spectrum of coexisting scales. Depending on the sparsity of the network certain scales will be detected reliably, whereas others are likely to be missed altogether (cf. also the preliminary comments made in Sec. 5.2.3). It is therefore important to understand the role of the system size, the number of observation sites and their distribution in space in the quality of the observation of a phenomenon unfolding in certain spatial scales.

Let $X(\mathbf{r}, t)$ be the field to be monitored. We take as the principal quantity of interest the space average of its spatial correlation function (cf. eq. (5.2)),

$$C(t) = \frac{1}{L^2} \int_L d\mathbf{r} \int_L d\mathbf{r}' <\delta X(\mathbf{r}, t) \delta X(\mathbf{r}', t)> \quad (5.16)$$

where L is the system size and δX the deviation from the mean. In what follows we limit ourselves for illustrative purposes to a one-dimensional medium. As usual the brackets denote the average over the probability distribution of the process. If the latter is ergodic this probabilistic average will be equivalent to a time average, and $C(t)$ will be independent of time. In actual fact, since the monitoring is carried out through a discrete set of N observation sites (stations) eq. (5.16) is replaced by its discrete analog

$$C_N(t) = \frac{1}{N^2} \sum_{j=1}^{N} \sum_{k=1}^{N} <\delta X(\mathbf{r}_j, t) \delta X(\mathbf{r}_k, t)> \quad (5.17)$$

where $\mathbf{r}_j$ are the positions of the stations in space.

Turning now to the dynamics, we consider two representative cases for which complexity is reflected by
(i) an exponentially decaying spatial correlation function, expected to arise in the limit of fully developed spatio-temporal chaos,

$$< \delta X(\mathbf{r},t) \delta X(\mathbf{r}',t) > = \frac{g_0}{2\gamma} e^{-\gamma|\mathbf{r}-\mathbf{r}'|} \qquad (5.18a)$$

(ii) a spatial correlation function decaying according to a power law, as it happens in the presence of intermittency,

$$< \delta X(\mathbf{r},t) \delta X(\mathbf{r}',t) > = g_0 |\mathbf{r} - \mathbf{r}'|^{-\mu} \qquad (5.18b)$$

The corresponding integrals and sums in (5.16)-(5.17) can be evaluated exactly in the case of N equidistant stations on a ring of length L (which in a geophysical context would correspond to a latitude belt). Standard manipulations lead then to the following expressions for the relative error $\epsilon = |C_N - C|/C$:

$$\epsilon = \left| \frac{\gamma}{2} \frac{L}{N} \left(\frac{1 + e^{-\gamma L/N}}{1 - e^{-\gamma L/N}} \right) - 1 \right| \qquad (5.19a)$$

in the case of (5.18a), and

$$\epsilon = (1-\mu) \left| \frac{1}{N} \left[1 + 2 \left(\frac{N}{2} \right)^\mu \sum_{k=1}^{\frac{N}{2}-1} \frac{1}{k^\mu} \right] - \frac{1}{1-\mu} \right| \qquad (5.19b)$$

in the case of (5.18b).

Figures 5.8a,b depict the dependence of ϵ on N for different values of γ and μ, keeping the total length L fixed. For each value chosen the data follow a power law in the form of N^{-b}, where b is, respectively, γL and μ dependent and gets smaller as γL and μ are increased. One can define a threshold value of N corresponding to a "tolerance" level of ϵ beyond which it becomes pointless to increase the density of the observational network. For instance, if a tolerable value of error is considered to be 0.2, 50 or so stations would be sufficient to monitor a phenomenon with $\gamma L = 100$ but 200 or so would be needed to monitor a phenomenon with $\mu = 0.7$.

We finally inquire whether there exists an optimal way to position the stations (now taken to be nonequidistant, as it happens in the real world) such that the error ϵ takes a small prescribed value. In its most general formulation this question amounts to minimization of a "cost" function depending on the $N+1$ variables $\mathbf{r}_j$, $j = 1, ..., N$ and N itself. We decompose

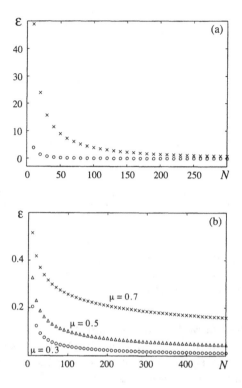

Fig. 5.8. (a) Dependence of the relative error ϵ (eq. (5.19a)) on the number of stations for a correlated process in space according to eq. (5.18a) for correlation lengths $L\gamma = 100$ (circles) and $L\gamma = 1000$ (crosses). (b) Dependence of ϵ on the number of stations N on a ring of length $L = 10\,000$ in the case of a power-law decay of the correlation function given by eq. (5.18b).

this problem into two steps: first, the study of the N dependence of ϵ for prescribed $\mathbf{r}_j$'s; second, the study of the $\mathbf{r}_j$ dependence of ϵ for prescribed N. The station's positions are prescribed

We know already (Fig. 5.8) that for equidistant stations the relative error decreases with N. For nonequidistant stations, to account for large gaps in the network (i.e. oceans, mountain chains, etc. in an environmental context) we adopt the model of the triadic Cantor set (Sec. 2.1) and let the process be of the type of (5.18b) with $\mu = 0.5$. Fig. 5.9 (crosses) depicts the value of ϵ as a function of N for this distribution. We observe a crossover with the N dependence of ϵ for the equidistant case (open circles) for a value of $N \approx 64$. In other words, if a constraint of small N were to be imposed (as it may

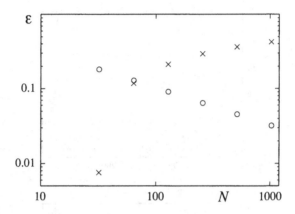

Fig. 5.9. Dependence of ϵ on the number of stations N positioned equidistantly on a ring of length $L = 10\,000$ (open circles) and on the bars of a triadic Cantor set (crosses) in the case of a power-law decay of the correlation function given by eq. (5.18b).

well happen in the real world), the fractal distribution would record reality more faithfully than the homogeneous one. As N increases the homogeneous distribution takes over since the absolute error in the fractal case improves with N less rapidly than in the equidistant case.

The stations' number is prescribed

As classical minimization algorithms fail when the number of variables gets large we resort to a Monte-Carlo type of approach (cf. Sec. 3.6.). The main steps of the procedure are as follows. Starting with a prescribed initial configuration, the positions $\mathbf{r}_j$ are varied randomly and independently in successive steps according to values taken from a uniform distribution centered on the initial positions. The standard deviation of this distribution is chosen to be of the order of the mean distance between stations (L/N), but is further tuned around this basic value according to the parameter values present in the problem. For instance, in the case of correlated processes it is adjusted to the parameter γ proportionally to γ^{-1}, which plays the role of "temperature" in analogy with statistical mechanics. At each step the error value is compared to the preceding one. The configuration is retained for a further step in the process if the value ϵ is less than or equal to the preceding one and rejected otherwise. The simulation is continued until a prescribed (small) value of the error is reached.

In view of determining the dimensionality of the set $\{\mathbf{r}_j\}$ it is necessary to consider rather large values of N. Furthermore, to explore the possibilities

Communicating with a Complex System 157

of optimization as far as possible, we require a very small value of the final error ϵ. To be specific, let us choose $N = 1\,000$ stations in a system of length $L = 10\,000$ units positioned initially as before according to the triadic Cantor set. For a correlation function of the form (5.18b) with $\mu = 0.5$ the initial value of the relative error is $\epsilon = 0.4$. After about 20 000 trials it reaches a value of $\epsilon \approx 2 \times 10^{-8}$. The corresponding spatial distribution of stations is now close to a random Cantor fractal of dimensionality $d \approx 0.80$ (to be compared with $d \approx 0.63$ of the classical Cantor set). For comparison, the relative error for a uniform coverage of L by the stations is in this case $\epsilon \approx 0.03$. We arrive again at the conclusion that a uniform disposition of the stations is far from being optimal. It should be emphasized that the simulation result is both N and initial condition dependent. One cannot exclude the existence of several optima whose accessibility depends on N and/or the initial conditions.

In the literature of multifractals it is known that eq. (5.18b) in one dimension can be generated by a process of dimensionality equal to $1 - \frac{\mu}{2}$ which for $\mu = 0.5$ is, interestingly, close to the dimensionality of 0.80 deduced from the Monte-Carlo simulation. On the other hand, it has been conjectured that a station distribution of dimensionality D is bound to miss phenomena whose dimensionality is $1 - D$ or less. In the context of our example this leads one to identify $\mu/2 = 0.25$ as a minimal network dimensionality for detecting a process like the one generating eq. (5.18b).

5.4.2 Data assimilation

We now address the second question raised in the beginning of the present section, how to infer from sparse data in a spatially extended system the initial conditions needed to run a high resolution predictive model.

Suppose that the space region covered by the model is discretized in the form of a regular lattice of M grid points. Within the same region we also dispose of N irregularly and sparsely distributed observational points (stations) as described in Sec. 5.4.1, with $N \ll M$. As an example in the numerical models used in operational weather forecasting $M \approx 10^7$, but in the current global observational network one hardly reaches a value of $N \approx 10^4 - 10^5$.

The first approach that comes to mind is to interpolate the available observations to the grid points. This can be done, for instance, by adopting a certain analytic representation $X(\mathbf{r}, \lambda)$ of the field of interest in space parameterized by a set of quantities λ (like e.g. a polynomial form), and determine the λ's by minimizing the mean square difference between X and observations Z_j in a prescribed vicinity ΔM of each grid point:

$$\phi(\lambda) = \sum_{j=1}^{N_\Delta} p_j (Z_j - X(\mathbf{r}_j, \lambda))^2 \quad \text{minimum} \tag{5.20}$$

where p_j are empirical weighting coefficients and N_Δ is the total number of observations within the vicinity ΔM.

Now, whenever there is a marked disparity between M and N, spatial interpolation is obviously inadequate. For this reason it is necessary to obtain additional information to prepare initial conditions, based on a first guess of the state of the system at all grid points. The availability of extensive computational facilities provides the background for a solution of this problem, based on the interesting idea that a short-range forecast of the model itself is best suited to provide such a first guess. This leads to the *data assimilation* cycle depicted in Fig. 5.10. Let X be the field generated by the model forecast, H the operator converting X to observed quantities (cf. Sec. 5.2.2). The quantity $H(X)$ will generally differ from the observations Z. The value $\hat{X}$ to be used in the initialization procedure is obtained by adding to

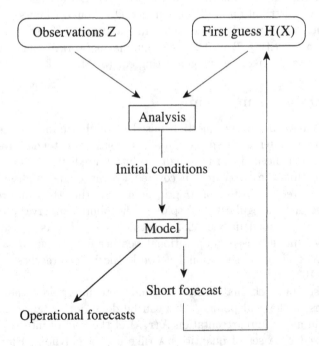

Fig. 5.10. Flow diagram of a typical data assimilation cycle giving access to initial conditions provided by the short-range forecasts of the model itself.

Communicating with a Complex System

X an appropriately weighted increment $Z - H(X)$, usually referred in the specialized literature as "innovation":

$$\hat{X} = X + G(Z - H(X)) \qquad (5.21)$$

where G is the weighting factor (or the weighting matrix in the general case of several coupled fields). Notice the analogy with the formulation of the technique of Kalman filtering outlined in Sec. 5.2.2.

There exist several schemes for implementing the idea of data assimilation, differing in their ways to compute G. Ordinarily, this is carried out by minimizing with respect to $\hat{X}$ a cost function involving weighted quadratic sums of $Z - H(\hat{X})$ and $\hat{X} - X$, measuring the distance of the desired estimate $\hat{X}$ to both the observations and the first guess. In a more sophisticated approach, referred as "four dimensional data assimilation", one also includes within the cost function the distance to observations at each time step within a certain time interval. This method ensures that the model develops in a way to remain as close to reality as possible. An alternative way to incorporate information from the data, referred as "nudging", consists in adding to the evolution equations forcing terms proportional to the distance between the model variable and the observations. Further approaches are based on statistical estimation theory, some of which find their origin in the idea of Kalman filtering introduced in Sec. 5.2.2.

The concept of data assimilation highlights the importance of disposing of a model capturing the principal features of the system under consideration. But real world models are imperfect, and this is a source of fundamental limitations in our ability to harness complex systems. This point is taken up in detail in the following section.

5.5 The predictability horizon and the limits of modeling

As stressed repeatedly in this book, the growth of small errors arising from the finite precision of the experimental data giving access to the initial state or from imperfections inherent in modeling is one of the key manifestations of complexity. It introduces irreducible limitations in the possibility to forecast the future states of the system beyond a predictability horizon generally depending on the nature (scale, etc.) of the phenomenon under consideration, which are especially compelling when the dynamics is unstable as it is the case in the presence of deterministic chaos.

In this section we outline an approach to error growth and predictability associated, successively, to initial and to model errors. The main feature of the approach is to incorporate information on the probabilistic structure of the dynamics through the probability distribution of the underlying attractor. This provides a background for, among others, the technique of ensemble prediction which is increasingly regarded in meteorology as the key for improving traditional weather forecasting, as discussed further in Sec. 6.4.

5.5.1 The dynamics of growth of initial errors

Error growth probes the dynamical (time-dependent) properties of the underlying system. Our starting point are the evolution laws (eqs (2.2)) and their formal solutions, which we write in the form (2.1). To formulate the laws governing error growth we consider a state $\mathbf{x}(t, \mathbf{x}_0 + \epsilon)$ displaced slightly with respect to a reference state $\mathbf{x}(t, x_0)$ on the system's attractor by an "error" ϵ. The instantaneous error $\mathbf{u}_t$ can then be written as

$$\mathbf{u}_t(\epsilon, \mathbf{x}_0) = \mathbf{x}(t, \mathbf{x}_0 + \epsilon) - \mathbf{x}(t, \mathbf{x}_0)$$
$$= \mathbf{F}^t(\mathbf{x}_0 + \epsilon) - \mathbf{F}^t(\mathbf{x}_0) \qquad (5.22)$$

where in writing the second equality we used eq. (2.1). In an unstable system, for any given ϵ, as $\mathbf{x}_0$ runs over the attractor the evolution of $\mathbf{u}_t$ is expected to be highly irregular both in t and in $\mathbf{x}_0$ (see e.g. Fig. 1.2). To identify some reproducible trends we therefore resort to the probabilistic approach, which in the present context amounts to averaging $\mathbf{u}_t$ or more conveniently its norm $u_t = |\mathbf{u}_t|$ over the attractor. Ordinarily the norm chosen is the Euclidean norm, $u_t = (\sum_{i=1}^{n} u_{i,t}^2)^{1/2}$. This leads to the following expression for the mean error,

$$< u_t(\epsilon) > = \int d\mathbf{x}_0 \, \rho_s(\mathbf{x}_0) \, |\mathbf{x}(t, \mathbf{x}_0 + \epsilon) - \mathbf{x}(t, \mathbf{x}_0)| \qquad (5.23)$$

where the invariant probability ρ_s measures the frequency with which different parts of the attractor are visited. Notice that if the system is ergodic in the sense defined in Sec. 3.3 the mean error becomes independent of the choice of initial position on the attractor.

Figure 5.11 depicts the evolution of the mean instantaneous error as given by eq. (5.23) for a typical chaotic system. We observe that error growth follows a logistic-like curve (cf. Sec. 2.7). Three different stages may be distinguished: an initial (short time) "induction" stage during which errors remain small; an intermediate "explosive" stage displaying an inflexion point

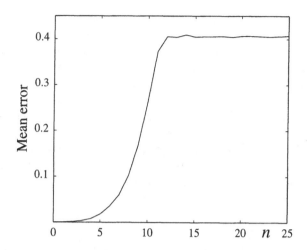

Fig. 5.11. Time dependence of the mean error for the logistic map (eq. (2.18)) at $\mu = 1$ starting from 100 000 initial conditions scattered on the attractor. The mean value of the initial error is $\epsilon \approx 10^{-4}$.

situated at a value t^* of t depending logarithmically on the norm ϵ of the error, $t^* \approx \ln(1/\epsilon)$ where errors suddenly attain appreciable values; and a final stage, where the mean error reaches a saturation level of the order of the size of the attractor and remains constant thereafter. The mechanism ensuring this saturation is the reinjection of the trajectories that would first tend to escape owing to the instability of motion, back to a subset of phase space that is part of the attractor. The very existence of saturation also implies that the dynamical system at hand possesses the property of mixing (Sec. 3. 3) in addition to that of ergodicity.

Let us focus on the initial regime. Since the error remains small, an expansion in which only the lowest order term in ϵ is kept should be meaningful. At the level of eq. (5.22) this leads to

$$\mathbf{u}_t(\epsilon, \mathbf{x}_0) = \mathbf{M}(t, t_0).\epsilon \qquad (5.24)$$

where $\mathbf{M}$ is the fundamental matrix introduced in eq. (2.3). The maximum Lyapunov exponent is then (cf. eq. (2.4))

$$\sigma_{\max} = \frac{1}{t} \ln \frac{u_t(\epsilon, \mathbf{x}_0)}{\epsilon} \qquad (5.25)$$

in the double limit (in the indicated order) of infinitely small initial errors and infinite long times. In this setting error growth in a chaotic system ($\sigma_{\max} > 0$)

is an exponential function of time for short times, with a rate equal to $\sigma_{\max}$. Notice that in an ergodic system $\sigma_{\max}$ may also be computed as a probabilistic average of the logarithm of the local expansion rates. Clearly then, beyond a time horizon of the order of $\sigma_{\max}^{-1}$ and, a fortiori, beyond the time t^* of the inflexion point in Fig. 5.11 errors attain a macroscopic level and predictions become random.

In actual fact, when confronted with the problem of predicting the evolution of a concrete system, the observer is led to follow the growth of a (at best) small but *finite* error over a *transient*, usually limited period of time. In this context the quantity of interest is the finite time version of eq. (5.25) which now depends on both t and $\mathbf{x}_0$, and eq. (5.23) becomes

$$< u_t(\epsilon) > = \epsilon \int d\mathbf{x}_0 \, \rho_s(\mathbf{x}_0) \, e^{t\sigma_{\max}(t,\mathbf{x}_0)} \qquad (5.26)$$

This equation shows that error growth amounts to studying, for finite times, the average over the attractor of an exponential function $< e^{t\sigma_{\max}(t,\mathbf{x}_0)} >$. To recover for such t's the picture of a Lyapunov exponent-driven exponential amplification of the error one needs to identify (5.26) with the exponential of the probabilistic average (or, for an ergodic system, of the long time average) of $\sigma_{\max}(t,\mathbf{x}_0)$, which is the conventional Lyapunov exponent $\sigma_{\max}$. In a typical attractor this is not legitimate since the expansion rates are position-dependent, in which case the average of a nonlinear function like the exponential in eq. (5.26) cannot be reduced to the exponential of an averaged argument. This property is yet another manifestation of the fluctuations of the local Lyapunov exponents referred in Sec. 4.6.

Figure 5.12a depicts the transient behavior of the error for the logistic map (eq. (2.18)) at $\mu = 1$. The open circles describe the error growth curve obtained from the full solution (eq. (5.23)) averaged over 100 000 realizations starting with different initial conditions and an initial error of 10^{-5}. We observe a significant deviation from an exponential behavior corresponding to the same initial error and a rate equal to the conventional (mean) Lyapunov exponent ln 2 (shaded circles), in agreement with the above advanced arguments. Writing, in analogy with (5.26),

$$< u_t(\epsilon) > = \epsilon \, e^{t\sigma_{\text{eff}}} \qquad (5.27)$$

where t is now a discretized time ($t = k\Delta t = k$), we evaluate σ_{eff} from the simulation data. The result, depicted in Fig. 5.12b, shows that σ_{eff} is t-dependent, starting at $t = 0$ with a value significantly larger than ln 2. This entails that error growth is neither driven by the Lyapunov exponent nor follows an exponential law but behaves, actually, in a *superexponential* fash-

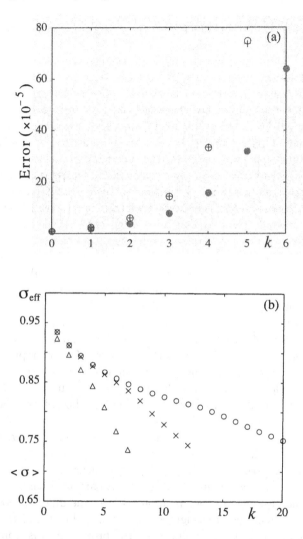

Fig. 5.12. (a) Error growth curve for the logistic map (eq. (2.18)) at $\mu = 1$ with $\epsilon = 10^{-5}$ obtained from numerical simulations averaged over 10^5 initial conditions scattered on the attractor (circles); from the theoretical expression, eq. (5.26) (crosses); and from a purely exponential law whose rate is given by $<\sigma> = \ln 2$ (full circles). (b) Time dependence of the effective rate, σ_{eff}, obtained numerically with $\epsilon = 10^{-3}$ (triangles); $\epsilon = 10^{-5}$ (crosses); and $\epsilon = 10^{-9}$ (circles). For each ϵ the simulation is carried out till a time $n^* \approx \ln(1/\epsilon)$ beyond which the error dynamics leaves the tangent space.

ion. This property further complicates the problem of prediction of complex systems.

In a multivariate system, in addition to the measure u_t of the error vector it is important to have information on the directions along which error is likely to grow most rapidly. In general the directions corresponding to the different expansion and contraction rates are not orthogonal to each other. As it turns out this non-orthogonality provides an additional mechanism of superexponential error growth beyond the one due to the variability of the local Lyapunov exponents, related to the fact that certain linear combinations of perturbations or errors may grow more rapidly than perturbations or errors along a particular direction. In a different vein, a multivariate dynamical system possesses several Lyapunov exponents, some of which are negative. For short times all these exponents are expected to take part in the error dynamics. Since a typical attractor associated to a chaotic system is fractal, a small error displacing the system from an initial state on the attractor may well place it outside the attractor. Error dynamics might then involve a transient prior to the re-establishment of the attractor, during which errors would decay in time.

An important class of multivariate systems are spatially extended systems. Here it is often convenient to expand the quantities of interest in series of appropriate basis functions the members of which represent the different spatial scales along which the phenomenon of interest can develop and, in particular, the different scales along which an initial error can occur. The ideas outlined above imply, then, that the predictability properties of a phenomenon depend in general on its spatial scale.

In summary, error growth dynamics is itself subjected to strong variability since not all initial errors grow at the same rate. As a result, the different predictability indexes such as $\sigma_{\max}$ or σ_{eff}, the saturation level and the time t^* to reach the inflexion point provide only a partial picture, since in reality the detailed evolution depends upon the way the different possible error locations and directions are weighted. This variability is illustrated in Fig. 5.13 depicting the transient evolution of the probability distribution of the error in a model system.

5.5.2 The dynamics of model errors

Natural complex systems are believed to be robust, on the grounds that they are the results of an evolution during which they have adapted to the environmental conditions. One of their specificities is that this robustness often coexists with an unstable motion going on on the attractor, reflected by the increase of small initial errors as seen in the preceding subsection.

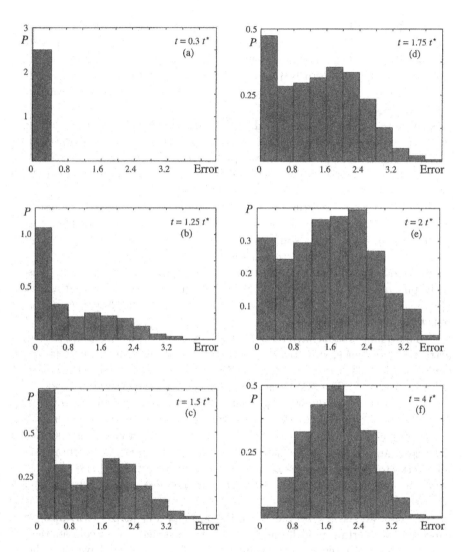

Fig. 5.13. Snapshots of the probability density of the error in a three-variable model system (Chapter 6, eqs (6.22)). Time is normalized to the error corresponding to the inflexion point t^* of the mean error curve ($t^* \approx 20$ time units).

Clearly, robustness is then to be understood in a statistical rather than a pointwise sense.

Closely related to robustness is the concept of structural stability developed in Sec. 2.4. The issue here is how a dynamical system -for our purposes, the model describing a certain process of interest- responds when its evolution laws are slightly perturbed to restitute those descriptive of the real world - the "nature". Such perturbations may be associated to variations of parameters already built in the system (see e.g. eq. (2.2)) reflecting the way it is solicited by the external world. They can also result from the addition of an extra set of evolution equations initially absent in the evolution laws as captured by the approximate description afforded by a model accounting, among others, for processes occurring on fine space and/or time scales that are not resolved in the model under consideration.

In its classical form structural stability addresses the way the attractors of the "approximate world" described by a model are modified by these perturbations. Two types of effects may be expected.

- The nature of the attractors changes qualitatively, as a previously prevailing solution loses its stability and new stable regimes are taking over. This is the phenomenon of bifurcation discussed extensively in this book. It marks the breakdown of structural stability, to which is associated a pronounced sensitivity to the parameters.

- The qualitative behavior remains the same, but the structure of the attractor and, hence, statistical properties such as mean values, correlation times, predictability, etc. are modulated by an amount related to the change of the parameters. The system is then deemed to be structurally stable.

Now from the standpoint of predictability the question, which attractor will eventually be reached following a structural perturbation, is less crucial than the question, how the perturbed system will gradually (that is, as time proceeds) deviate from its initial (unperturbed) regime prior to reaching the final attractor. In this subsection we address this "dynamical" version of the problem of structural stability, to which one may also refer in the spirit of our foregoing discussion as the problem of "model error".

Let $\mathbf{x} = (x_1, ..., x_n)$ be the variables participating in the dynamics as captured by a certain model, and let λ be the parameter (or a combination of parameters) of interest. The evolution laws $\mathbf{f} = (f_1, ..., f_n)$ have then the form of eqs (2.1),

$$\frac{d\mathbf{x}}{dt} = \mathbf{f}(\mathbf{x}, \lambda) \tag{5.28}$$

We suppose that the phenomenon of interest as it occurs in nature is well described by an amended form of (5.28) where λ takes a certain (generally

Communicating with a Complex System 167

unknown) value λ_N close to λ and/or some extra terms associated to physical processes not properly accounted for by the model are incorporated, limiting ourselves for simplicity to the case that the model and "nature" variables span the same phase space. We want to estimate the error between the solutions of eq. (5.28) and the "correct" evolution laws

$$\frac{d\mathbf{x}_N}{dt} = \mathbf{f}_N(\mathbf{x}_N, \lambda_N) \tag{5.29}$$

To this end we set $\lambda = \lambda_N + \delta\lambda$ being understood that $|\frac{\delta\lambda}{\lambda_N}|$ is small and introduce in analogy with the preceding subsection the error $\mathbf{u}$,

$$\mathbf{u} = \mathbf{x} - \mathbf{x}_N$$

Expanding eq. (5.28) in $\mathbf{u}$ and $\delta\lambda$, subtracting (5.29) from the result and keeping the first non-trivial terms one obtains an equation of evolution for the error,

$$\frac{d\mathbf{u}}{dt} = \left(\frac{\partial \mathbf{f}}{\partial \mathbf{x}}\right)_N \cdot \mathbf{u} + \left(\frac{\partial \mathbf{f}}{\partial \lambda}\right)_N \delta\lambda \tag{5.30}$$

where the subscript N implies evaluation of the corresponding quantities at $\mathbf{x} = \mathbf{x}_N$, $\lambda = \lambda_N$. Notice that in a complex system $\mathbf{x}_N$ can be a quite intricate function of time displaying, for instance, chaotic dynamics.

The principal difference of eq. (5.30) with respect to the equation governing error growth due to uncertainty in the initial conditions is the presence of the second term, conferring to it the structure of an inhomogeneous equation. In the limit $\delta\lambda = 0$ only the homogeneous part survives and one obtains a solution in the form of eq. (5.24), where $\mathbf{M}(t, t_0)$ is the fundamental matrix associated with the linearized operator (or Jacobian matrix) $(\partial \mathbf{f}/\partial \mathbf{x})_N$. We are in fact interested here in the opposite limit where $\delta\lambda$ is different from zero but the initial error ϵ is zero. Using the property

$$\frac{d\mathbf{M}(t, t_0)}{dt} = \left(\frac{\partial \mathbf{f}(t)}{\partial \mathbf{x}}\right)_N \mathbf{M}(t, t_0) \tag{5.31}$$

one can check that the solution of (5.30) is given by

$$\mathbf{u}_t = \delta\lambda \int_0^t dt' \, \mathbf{M}(t, t') \cdot (\partial \mathbf{f}(\{\mathbf{x}(t')\}, \lambda)/\partial \lambda)_N \tag{5.32}$$

This relation governs the dynamics of model errors.

As seen in Sec. 2.1 and in the previous subsection, $\mathbf{M}$ is related to the spectrum of the Lyapunov exponents. Its presence in (5.32) suggests that the growth of model errors should, at least in part, be conditioned by the Lyapunov exponents as well. Qualitatively speaking, one may view expression (5.32) as the sum total of partial error growth processes starting at all times t' between 0 and the time of observation t. Each of them displays a time exponential running from t' to t and involving exponents belonging to the full spectrum of Lyapunov exponents multiplied by a term $\partial \mathbf{f}/\partial \lambda$ evaluated at t', playing the role of an "effective" initial error. This can be nicely visualized on the simple example of a linear system

$$\frac{dx}{dt} = \sigma x + \lambda \qquad (5.33)$$

where σ plays the role of the Lyapunov exponent. The corresponding evolution equation for the model error is

$$\frac{du}{dt} = \sigma u + \delta\lambda \qquad (5.34a)$$

and its solution subject to $u(0) = \epsilon = 0$ is

$$\begin{aligned} u_t &= \delta\lambda \int_0^t dt' e^{\sigma(t-t')} \\ &= e^{\sigma t}\delta\lambda \frac{1-e^{-\sigma t}}{\sigma} \end{aligned} \qquad (5.34b)$$

This has the form of an error growth law in the sense of the previous subsection, with the factor multiplying $e^{\sigma t}$ playing the role of the "effective" error. This effective error starts as a linear function of time for short times and reaches a constant value $1/\sigma$ after a time of the order of σ^{-1}. Subsequently, the growth of the model error is conditioned by the Lyapunov exponent and the renormalized "initial" value $\delta\lambda/\sigma$ until nonlinear effects (not included in (5.33)) take over and allow the error to reach a saturation value.

Coming back now to the general result of eq. (5.32) and keeping in mind the preceding remarks one may identify the qualitative features of model error dynamics for different stages of its time evolution. In doing so we will again adopt for reasons explained in the preceding subsection the probabilistic viewpoint and focus on the mean error dynamics $<|\mathbf{u}_t|>$ or, more conveniently, on the mean quadratic error $<\mathbf{u}_t^2>$ and its standard deviation $<\mathbf{u}_t^2>^{1/2}$.

We first focus on the short time behavior. The structure of $\mathbf{u}_t$ depends here crucially on the possibility to approximate eq. (5.32) by power series expansions. This is allowed as long as the "propagator" $\mathbf{M}$ depends smoothly on t as it happens when the parameter λ refers to bulk properties, but breaks down when λ expresses the effect of the boundary conditions in a spatially extended system. Here we limit ourselves to the first case. Utilizing the properties $\mathbf{u}_0 = 0$ and that $\mathbf{M}(t,t)$ is the identity matrix one obtains straightforwardly

$$< u_t^2 > = t^2 \delta \lambda^2 < (\partial \mathbf{f}(\{x(0)\}, \lambda)/\partial \lambda)_N^2 > \quad (5.35)$$

where the average is carried out over the invariant distribution of the attractor of the reference system (the "nature"). This attractor is sampled by the variables $\{\mathbf{x}_N(0)\}$ to which the model variables $\{\mathbf{x}\}$ reduce at $t = 0$ since it was assumed that there are no errors due to the initial conditions. Keeping in mind the comments following eq. (5.32) one expects that the extent of the short time regime described by the above equation will be limited by the inverse of the largest in absolute value Lyapunov exponent. In a dissipative system this exponent is expected to be the most negative one. This introduces a drastic difference between the dynamics of model and initial errors. It also shows that the deficiencies in modeling are revealed more promptly than the limitations due to uncertainties in the initial conditions.

To analyze the later stages of model error growth one needs to augment eq. (5.30) by nonlinear terms in $\mathbf{u}$ and $\delta \lambda$. Simulations on model systems show that error growth follows then, much like initial error, a curve similar to that of Fig. 5.11. Interestingly, the saturation level attained is finite, practically independent of the smallness of $\delta \lambda$, as it reflects the average of typical quadratic distances between any two points of the reference attractor: as time grows the representative points of the reference and approximate systems become increasingly phase shifted, even though the attractors on which they lie may be quite close. We have here a signature of the zero Lyapunov exponent, associated with the borderline between asymptotic stability and instability. Notice that similarly to initial error, the dynamics of individual (non-averaged) model errors are subject to high variability in the form of intermittent bursts interrupted by periods of low error values, giving rise to error probability distributions similar to those of Fig. 5.13.

Much of the above analysis can be extended to the case where the reference system (the "nature") involves an additional set of variables. Modeling is here accompanied by a phase space reduction associated with sheer elimination of the extra variables or, at best, with a procedure expressing them as functions of those retained in the description (see e.g. Sec. 2.7.4). As it turns out the mean quadratic error is again a quadratic function of time in the regime of short times, but the multiplicative coefficient depends now on the additional variables. This dependence becomes especially intricate in the later stages and may even play a determining role in the error evolution.

5.5.3 Can prediction errors be controlled?

Prediction is one of the main objectives of scientific endeavor. As seen in the preceding subsections, the possibility to accomplish this task properly may be compromized by the presence of irreducible sources of errors. A natural question to be raised is, then, to what extent a predictive model can be augmented by an appropriate control algorithm allowing one to keep in check, to the extent of the possible, the development of errors that would tend to reach an unacceptable level.

There exists as yet no comprehensive answer to this question. A growing trend is to model error source terms by stochastic forcings of different kinds, to be added to the model equations. This procedure is especially tempting when error source terms arise from the generally poor accounting of processes not directly expressible in terms of the model variables, as is the case of phenomena evolving on short time and space scales that are not resolved by the model at hand.

One class of stochastic forcings that lends itself to a quantitative analysis are those amenable to a white noise term multiplied by a coupling function generally dependent on the model variables in a non-trivial way (one refers then to a "multiplicative" noise-system coupling, the case of a constant coupling being referred as "additive"). As seen in Sec. 3.2, such a problem can be mapped into a Fokker-Planck equation (eq. (3.8)). By multiplying both sides of this equation by different powers of the model variables and by averaging over the probability distribution one obtains a set of coupled moment equations involving averages, variances, etc. (cf. Sec. 3.5.1). This hierarchy can be decoupled to some extent in the limit where the errors and the strength of the noise are small. By comparing the solutions obtained in this limit with those generated by the reference system (the "nature") one is then in the position to identify some trends regarding the role of the "control" noise in depressing the error and hence in improving the model performance.

As seen earlier there exist two kinds of predictability indices, pertaining to the short time behavior and to the saturation level of the error. Furthermore, in a complex system one should not limit the predictability analysis to the mean error but should address the variability around the mean as well. As a rule, in the short time regime, a control in the form of an additive white noise deteriorates the model performance as far as mean error is concerned. On the other hand it tends to enhance the variability of the processes involved as compared to that predicted by the model and bring it closer to the natural variability. As regards the saturation level, the action of additive white noise control is system dependent. There exists a range of parameters where both mean error and variability can be corrected in the desired sense, but this is only one out of many possibilities. In short, the trends are not only non-universal but are in many cases conflicting about the desired goals. It would undoubtedly be important to pursue the effort by considering a wider class of controls such as multiplicative white noises, more involved Markovian noises or even non-Markovian controls accounting for memory effects.

5.6 Recurrence as a predictor

In the preceding section we identified a number of fundamental limitations associated with the prediction of the future course of a set of initially nearby states of a complex system, as time unfolds. The problem dealt with was in fact the most ambitious version of the idea of prediction since the issue was to produce the full succession of states for an as long as possible time interval, rather than some general characteristics like mean values, correlation functions, and so on.

The question whether, and if so how frequently, under a given set of evolution laws the future states of a system will come close to a particular initial configuration is yet another question having a direct bearing on the issue of prediction. Being more global in nature it is subjected to less compelling restrictions and can for this reason be analyzed in considerable detail. The present section is devoted to this phenomenon of repetition, or *recurrence* of the states of complex dynamical systems, with emphasis on its role as a predictor.

The crossing of the vertical position by a well-serviced pendulum is a recurrent event. Here the event coincides with a particular value of the position coordinate, and the intervening law is Newton's or Hamilton's equations in the presence of gravity. We deal here with the simplest version of recurrence, namely, strict periodicity. But one could also say that in a capitalistic economy, economic growth or slowdown are recurrent events. The event is now an attribute of a whole succession of outcomes interrupted by irregular periods where this attribute is absent, and the intervening laws are no longer directly reducible to nature's fundamental interactions but involve, rather, competing human agents each one of whom wishes to maximize his profit (see Sec. 1.4.4). In a still different vein, the need for survival has made man aware of the repeated appearance (recurrence) of weather patterns associated with different wind and/or precipitation patterns. The Greeks constructed an octagonal structure, the Tower of Winds, which can still be visited in Athens. It depicts the wind from each cardinal point and comprises bas-relief figures representing the typical type of weather associated with such

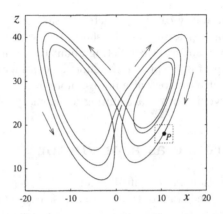

Fig. 5.14. Illustrating recurrence on a two-dimensional projection of the Lorenz chaotic attractor (Chapter 6, eqs (6.20)). Recurrence is here reflected by the fact that the trajectory emanating from point P in the dotted square will re-enter this cell (which plays the role of a coarse-grained state) after some time.

a wind. The event is now a lumped, coarse-grained characterization of the state of the atmosphere that came to be known later on as *Grosswetterlagen* or "atmospheric analogs" (cf. Sec. 5.3.2) and the laws are those of fluid mechanics and thermodynamics as applied to a rotating frame, to which we come in more detail in Secs 6.4 and 6.5.

In the framework of a deterministic description one tends to associate an event with the point that the system occupies in phase space at a given time. But in this view exact recurrence of a state is impossible except for the trivial case of periodic dynamics, since it would violate the basic property of uniqueness of the solutions of the evolution equations (cf. Sec. 2.1). To cope with this limitation, one associates the concept of recurrence with repeated re-entries of the dynamical trajectory in a phase space region possessing finite volume (more technically, "measure"), not necessarily small, as indicated in Fig. 5.14. These are the coarse-grained states referred to above and throughout Chapters 3 and 4, as opposed to the point-like states considered in the traditional deterministic approach. Probabilistic description is a typical instance in which one deals essentially with coarse-grained states.

5.6.1 Formulation

The first quantitative and rigorous statement concerning recurrence goes back to Henri Poincaré, a founding father of nonlinear science. He showed that in a system of point particles under the influence of forces depending only on the spatial coordinates, a typical phase space trajectory will visit infinitely often the neighborhood of any prescribed initial state, provided that the system remains in a bounded region of phase space. The time between two such consecutive visits is referred as the "Poincaré recurrence time". Since then, it has been realized that this major result extends to a very wide class of systems possessing the property of ergodicity, deterministic as well as stochastic, conservative as well as dissipative.

Our starting point is eq. (2.1), linking the instantaneous values of the state variables $\mathbf{x}$ to the initial ones $\mathbf{x}_0$. Consider a finite cell C in the system's phase space and assume that the evolution is started at a point $\mathbf{x}_0$ contained in C, which we denote as $\mathbf{x}_0 \in C$. As the evolution proceeds, the representative point will generally escape from C. We want to evaluate the recurrence probability, that is to say, the probability that it is reinjected into C at some time t being understood that it spent the period prior to t outside C, which we denote as $\mathbf{x}_{t'} \notin C (t' < t)$. It is convenient to introduce a sampling time τ, defining the intervals at which the state of the system is probed. In a continuous time dynamical system τ may be chosen as small as desired, including the limit $\tau \to 0$. On the other hand, in a discrete time system, such as the mapping of a continuous time dynamics on a Poincaré surface of section (Sec. 2.5) or on a Markov partition through the process of coarse-graining (Sec. 3.3), τ has a finite value which can actually be scaled to define the time unit. In both cases the recurrence probability may be expressed as

$$F(C, 0; C, n\tau) = \frac{1}{\text{Prob}(\mathbf{x}_0 \in C)} \text{Prob}(\mathbf{x}_0 \in C, \mathbf{x}_1 \notin C, ... \mathbf{x}_{(n-1)\tau} \notin C, \mathbf{x}_{n\tau} \in C)$$
(5.36)

In principle, the right hand side of this relation is determined from the phase space probability associated to the system. In a deterministic system this quantity obeys to the Liouville equation or its discrete analogue, the Frobenius-Perron equation (Sec. 3.2). This leads to the more explicit form

$$\begin{aligned}F(C, 0; C, n\tau) &= \frac{1}{P(C)} \int_C d\mathbf{x}_0 \rho(\mathbf{x}_0) \int_{\bar{C}} d\mathbf{x}_1 \delta(\mathbf{x}_\tau - \mathbf{F}^\tau(\mathbf{x}_0)) \\ &\quad \cdots \int_{\bar{C}} d\mathbf{x}_{n-1} \delta(\mathbf{x}_{(n-1)\tau} - \mathbf{F}^{(n-1)\tau}(\mathbf{x}_0)) \\ &\quad \int_C d\mathbf{x}_n \delta(\mathbf{x}_{n\tau} - \mathbf{F}^{n\tau}(\mathbf{x}_0))\end{aligned}$$
(5.37)

where $\bar{C}$ denotes the complement of cell C, ρ the invariant probability density and the probability, $P(C)$ to be in cell C is just the integral of $\rho(\mathbf{x})$ over C. The delta functions express the evolution of the initial state $\mathbf{x}_0$ to the states $\mathbf{x}_\tau, \cdots$ up to $\mathbf{x}_{n\tau}$ as stipulated in eq. (2.1).

In practice the explicit evaluation of (5.36)-(5.37) is often an extremely arduous task. Fortunately, for ergodic transformation general results can be obtained for the first moment $<\theta>_\tau$ - the mean recurrence time. Specifically:

- For discrete time systems

$$<\theta>_\tau = \tau/P(C) \qquad (5.38a)$$

- For continuous time systems (allowing for the possibility $\tau \to 0$)

$$<\theta>_\tau = \tau \frac{1 - P(C)}{P(C) - P(C,0;C,\tau)} \qquad (5.38b)$$

where the second term in the denominator is the joint probability to be in C both initially and at time τ. The latter is equal, in turn, to the volume (more technically, the measure) of the intersection of C and of the cell that will be projected into C after the time interval τ, referred as the pre-image, $C_{-\tau}$ of C. One can give an order of magnitude estimate of (5.38b) by noticing that recurrence will be possible only if cell C is including part of the invariant set on which the dynamics is taking place, like e.g. one of the attractors if the system happens to be dissipative. As stressed in Sec. 2.1 and onwards in a complex system this set is generally fractal and occupies a zero phase space volume (in the Euclidean sense). The invariant density ρ is therefore singular, and the probability $P(C)$ can be estimated as ϵ^D where D is the attractor dimension and ϵ a characteristic linear size of the cell reduced by the size of the invariant set. As regards the denominator in eq. (5.38b), a reasonable estimate in the limit of small ϵ and τ is that it mainly depends on the boundary of the cell and on the average phase space speed, $\bar{\nu}$ as the trajectory crosses this boundary. This yields $P(C) - P(C,0;C,\tau) \approx \bar{\nu}\tau\epsilon^{(D-\delta)}$ where δ is a positive number between 0 and 1 accounting for the fractal structure of the boundary. Combining these estimates one thus arrives for small ϵ at the scaling relation

$$<\theta>_\tau \approx \frac{1}{\bar{\nu}}\epsilon^{-(D-\delta)} \qquad 0 \leq \delta \leq 1 \qquad (5.39)$$

The main idea behind this relation is that recurrence is to be sought in a subspace spanned by a limited number of "relevant" variables - those responsible for the structure of the attractor. Inasmuch as in a dissipative system D

is typically much less than the phase space dimensionality recurrence times, while still long for small size cells, are nevertheless substantially less than if the cells where extending over all phase space dimensions. Furthermore, in certain problems one may be interested solely on recurrence of the values of a particular observable rather than of the full state of the system. This will further reduce the corresponding waiting times. In any case, owing to the power law character of the scaling, $<\theta>_\tau$ increases rather slowly for decreasing ϵ's.

In the presence of stochastic dynamics the above formulation is to be partially amended, since eq. (2.1) at the basis of (5.37) no longer holds. The problem of recurrence becomes in many respects simpler as we now briefly summarize on the prototypical example of discrete time Markov processes described by the master equation (3.5). Setting Δt in this latter equation equal to the observation window τ which is taken in turn to be the time unit, we first notice that the (conditional) probability to be in state k at time $n\tau$ starting with state j is

$$P(k, n|j, 0) = (\mathbf{W}^n)_{kj} \qquad (5.40)$$

On the other hand, the waiting times between successive recurrences are mutually independent random variables: once on a certain state, the recurrence process takes place as if this state was the system's initial condition since the memory of the previous transitions is lost owing to the Markov property. This type of situation is described by a relation of central importance in the theory of stochastic processes, known as *renewal equation*:

$$P(j, n|j, 0) = \sum_{n'=1}^{n} F(j, 0; j, n') P(j, n - n'|j, 0) \qquad (5.41)$$

where $F(j, 0; j, n')$ is still defined by eq. (5.36). Using well-established techniques one can then express the recurrence time probability in terms of $\mathbf{W}$. This relation yields an exponential dependence of the recurrence probability on the time n. This gives in turn access to a host of properties including not only the mean recurrence time (for which an expression of the form of eq. (5.38a) holds with $P(C)$ simply replaced by P_j) but also its variance and higher moments. An important feature is that the scatter of recurrence times around their mean is comparable to the mean itself. Notice that under the conditions discussed in Sec. 3.3 all states are recurrent, and the associated mean recurrence times are finite.

5.6.2 Recurrence time statistics and dynamical complexity

The first non-trivial example of recurrence in deterministic dynamical systems pertains to uniform quasi-periodic motion. Such a motion takes place on an invariant set in the form of a two-dimensional torus which is topologically equivalent to a square of side 2π subjected to periodic boundary conditions. It can be parameterized by two angle variables ϕ_1 and ϕ_2 obeying to the evolution equations

$$\phi_1 - \phi_{10} = \omega_1 t$$
$$\phi_2 - \phi_{20} = \omega_2 t \qquad (5.42)$$

ω_1, ω_2 being the two characteristic frequencies. Ergodicity requires these to be incommensurate, that is, to have a ratio which is an irrational number.

We shall take the cell C to be a square whose sides (of length ϵ) are parallel to the ϕ_1 and ϕ_2. The right hand side of eq. (5.38b) can then be easily evaluated, noticing that the invariant density is here uniform. One obtains in this way

$$<\theta>_\tau \approx \frac{4\pi^2}{\epsilon(\omega_1 + \omega_2)} \qquad (5.43)$$

in the limit of small ϵ. On the other hand, recurrence requires that there exists a time t_0 such that the trajectory returns to cell C after an integer number of turns, say k_1 and k_2, around ϕ_1 and ϕ_2, respectively. Keeping in mind eq. (5.42) this yields

$$\omega_1 t_p = 2\pi k_1 + \delta_1$$
$$\omega_2 t_p = 2\pi k_2 + \delta_2 \qquad -\frac{\epsilon}{2} \leq \delta_1, \delta_2 \leq \frac{\epsilon}{2} \qquad (5.44a)$$

Eliminating t_p from these two relations leads to

$$|k_1 - k_2 \alpha| \leq \bar{\epsilon} \qquad (5.44b)$$

where $\alpha = \omega_1/\omega_2$ and $\bar{\epsilon} = \frac{\epsilon}{2\pi}(1 + \frac{\omega_1}{\omega_2})$. We obtain an inequality linking k_1 to k_2 whose solutions must be integers - the type of relation referred as *Diophantine inequality*. A recurrence time is uniquely determined from the values Δk_1, Δk_2 separating two successive solutions of (5.44b) multiplied by the periods corresponding to ω_1, ω_2 plus a small correction associated to the finite size of the cell.

Now relation (5.44b) corresponds to the old-standing problem of number theory of approximating irrationals (here the ratio α) by rationals (here k_1/k_2). A rational number can be written as a finite *continued fraction* whereas for an irrational number an infinite sequence is needed, $\alpha = [\alpha_1, \alpha_2, \cdots] = 1/(\alpha_1 + 1/(\alpha_2 + \cdots))$. Any such sequence truncated to a finite order produces a rational number referred as an approximant to α. Number theory shows that these approximants are the best rational approximations to α, in the sense that any better rational approximation will display a larger denominator. While the absolute values of distances of approximants from α decrease monotonically as higher order approximants are taken, their signs are alternating. These properties can be nicely visualized and classified exhaustively thanks to a mathematical construct known as *Farey tree*, each branch of which is an ordered sequence $k_{1,n}/k_{2,n}$, such that $k_{1,n} < k_{1,n+1}$ and $k_{2,n} < k_{2,n+1}$, converging to an irrational number.

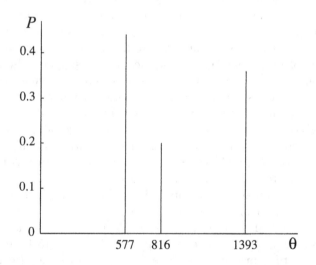

Fig. 5.15. Probability histogram of return times (measured by the number of cycles along the ϕ_2 direction) for the dynamical system of eq. (5.42), with $\omega_1 = 1$, $\omega_2 = \sqrt{2}$ and a cell size $\epsilon = 0.002$.

The study of inequality (5.44b) using these concepts shows that, as expected, there exists an infinity of solutions. More surprising is the fact that the corresponding differences Δk_1, Δk_2 between two successive solutions may take only three allowable values, two of which are "fundamental" and the third one is their sum. Recurrence time probability reduces therefore to three discrete peaks, as illustrated in Fig. 5.15. Such a strict selection, forbidding most of the possible recurrence time values, reflects the deterministic nature of the dynamics. In this respect it facilitates prediction. On the other hand the structure of the probability distribution implies a high scatter around the mean, illustrating the inadequacy of averages (eq. (5.43)) in the prediction of recurrent events.

The specific location of the probability peaks in Fig. 5.15 depends on the choice of cell size and on the underlying dynamics. In particular:

- For fixed ω_1, ω_2, the Δk's (and thus the recurrence times) are shifted to larger values in a stepwise fashion as the size is decreased.

- For fixed (and sufficiently small) ϵ the dependence of the recurrence times on α is highly fluctuating. This reflects the fact that certain irrational numbers (the "Diophantine numbers") are less well approximated by rationals than others as in the case, for instance, of the golden mean $\alpha = (\sqrt{5} - 1)/2$. In such cases the fractions $\Delta k_1/\Delta k_2$ lying in a small neighborhood of α determined by ϵ will be formed by large values of Δk_1 and Δk_2, implying both high and little spread values of the recurrence times.

Part of the above analysis can be extended to quasi-periodic systems displaying more than two incommensurate frequencies. The main result on the existence of a finite number of recurrence times is supported by numerical evidence, but several fundamental aspects of the problem remain open.

We turn next to systems giving rise to chaotic dynamics. Since recurrence can only be formulated (except for the trivial case of periodic dynamics) in terms of cells of finite extent in phase space, a natural approach is to project the full phase space dynamics into a phase space partition. In fully developed chaos one can construct Markov partitions associated with this process of coarse graining. In simple cases such as in eq. (2.18) for $\mu = 1$ these are formed by cells separated by points either belonging to unstable periodic points or being pre-images of periodic points. An immediate consequence is that having identified the transition matrix **W** mediating the passage of the trajectories between the cells one can apply eqs (5.40) and (5.41) to compute the recurrence time probabilities. These turn out to have an exponential structure. They also present, typically, gaps for certain values of times or intervals thereof, reflecting the presence of selection rules imposed by the deterministic nature of the underlying dynamics.

The situation is quite different in systems that display intermittent chaos for which a phase space partitioning not only typically leads to intricate non-Markovian properties, but also violates the renewal equation (5.41): the recurrence times are no longer mutually independent random variables, owing to the memory implied by the deterministic dynamics. The computation of the recurrence time probabilities may nevertheless be achieved in some representative cases. The main result is that there exist partition cells j

in which this probability follows a power law structure in the range of long recurrence times, e.g. $F(j,0;j,n) \approx n^{-3}$. This entails, in particular, that although the mean recurrence time is finite the second moment (in the above example) and the corresponding variance diverge logarithmically. Clearly, this wide scatter will have a direct bearing on the capability of predicting a reliable value for this event.

As in the quasi-periodic case, the specific characteristics of the probability distributions and of their moments depend in a significant way on the principal indicators of the complexity of the underlying system. Specifically:

- Typically, there is a strong variability of the mean recurrence times as one moves across the attractor or, more generally, the invariant set where the dynamics is taking place. This is due (cf. eqs (5.38a,b)) to the non-uniformity of the invariant probability of most chaotic systems. Furthermore, because of the divergence of nearby trajectories, close initial conditions may recur at very different times since the trajectories may visit quite different regions of phase space prior to reinjection to their cell of origin. Actually, even when the invariant probability happens to be uniform, the variances and higher moments of the recurrence times turn out to depend strongly on the position of the recurrence cell on the attractor.

- Most (but not all) chaotic systems are organized in phase space on a skeleton of unstable periodic orbits densely distributed in phase space. A chaotic trajectory is, typically, "trapped" in a neighborhood of a periodic orbit for some time, leaves subsequently this neighborhood to get trapped in one associated to another periodic orbit, and so forth. Because of this, the return of the chaotic orbit to a certain neighborhood in phase space is conditioned by the period(s) of the periodic orbit(s) passing through this neighborhood, around which it remained locked during its journey outside this neighborhood.

- More quantitatively, it can be shown that in chaotic attractors the invariant probability $P(C)$ to be in cell C is the limit of $P_T(C)$ as T tends to infinity, where

$$P_T(C) = \sum_j \Lambda_{jT}^{-1} \qquad (5.45)$$

Here Λ_{jT} is the expansion rate of a periodic orbit of period T passing through the cell C (supposed to be sufficiently small), and the summation is taken over all periodic orbits of this type. In view of eq. (5.38a) this relation establishes an explicit connection between mean recurrence times and dynamical instability. A more comprehensive connection becomes possible for piecewise linear maps possessing a Markov partition mapping the dynamics on a stochastic process. In this case the recurrence time distribution in a partition cell can be expressed in terms of the expansion rates of the subclass of periodic orbits which cross this particular cell only once, whose period corresponds to the particular recurrence time of interest. These expansion rates are, in turn, products of local expansion rates of the map evaluated at the points visited by the periodic orbit.

5.7 Extreme events

A typical record generated by a natural, technological or societal system consists of periods where a relevant variable undergoes small scale fluctuations around a well-defined level provided by the long term average of the values available, interrupted by abrupt excursions to values that differ significantly from this level. Such *extreme events* are of great importance in a variety of contexts since they can signal phenomena like the breakdown of a mechanical structure, an earthquake, a severe thunderstorm, flooding, or a financial crisis. Information on the probability of their occurrence and the capability to predict the time and place at which this occurrence may be expected are thus of great value in, among others, the construction industry or the assessment of risks. While the probability of such events decreases with their magnitude, the damage that they may bring increases rapidly with the magnitude as does the cost of protection against them. These opposing trends make the task of prediction extremely challenging.

Extreme events are of special relevance in complexity research. In many respects the ability of complex systems to adapt and to evolve is conditioned by the occurrence of extreme events in, for instance, the form of transitions between states or the exceedance of thresholds. In the present section some salient features of extreme events are summarized from the perspective of complex systems with emphasis on, once again, the issue of prediction.

5.7.1 Formulation

The basic question asked in connection with extreme events is, given a sequence $X_0, \cdots, X_{n-1}$ constituted by the successive values of an observable monitored at regularly space time intervals $0, \tau, \cdots, (n-1)\tau$, what is the probability distribution of the largest value x found in the sequence, $M_n = \max(X_0, \cdots, X_{n-1})$:

$$F_n(x) = \text{Prob } (X_0 \leq x, \cdots, X_{n-1} \leq x) \qquad (5.46)$$

Obviously, $F_n(a) = 0$ and $F_n(b) = 1$, a and b being the upper and lower ends (not necessarily finite) of the domain of variation of x. In actual fact $F_n(x)$

is therefore a *cumulative probability*, related to the probability density $\rho_n(x)$ (in so far as the latter exists) by

$$F_n(x) = \int_a^x dx' \rho_n(x') \tag{5.47}$$

As n increases $F_n(x)$ will shift toward increasing values of x, being practically zero in the remaining part of the domain of variation. The possibility to elaborate a systematic theory of extremes depends on the cumulative probability $F(x)$ of the underlying process ($F(x) = \text{Prob}(X \leq x)$). If the latter satisfies certain conditions, then one can zoom the vicinity of the upper limit of x through appropriate scaling transformations and explore the possibility of universal behaviors.

In many instances a criterion for qualifying an event as extreme is that some of the X_i's in a sequence $X_0 \cdots, X_{n-1}$ exceed a threshold u. Denoting an arbitrary term in the X_i sequence by X, then the excess probability distribution of X over u is given by the conditional probability

$$F_u(y) = \text{Prob}\,(X - u \leq y | X > u), \quad y \geq 0 \tag{5.48}$$

Another statistic relevant to extreme events is the distribution of local maxima. Focusing on any three successive members X_0, X_1, X_2 of a time sequence, and given that X_0 is governed by the invariant probability density $\rho(x)$, the quantity to be evaluated is

$$\psi(x) = \text{Prob}\,(X_1 \leq x | X_0, X_2 < X_1) \tag{5.49}$$

Contrary to $F_n(X)$ in the limit of n going to infinity, this quantity involves the short time behavior of the underlying dynamical system. Notice that in the limit where the time interval τ between successive observations is taken to tend to zero the probabilities in (5.48) and (5.49) are amenable to the joint probabilities of X and its first time derivative, and of X and its first and second time derivatives, respectively.

Finally as we saw in Sec. 1.4.2, a most important class of extreme events are those associated with the Hurst phenomenon. Our starting point is again the sequence $X_0 \cdots, X_{n-1}$. The mean, $\bar{X}_n$ and the standard deviation, C_n of this record are (we assume that the X's have a finite variance):

$$\bar{X}_n = \frac{1}{n}(X_0 + \cdots + X_{n-1}), \quad C_n = [\frac{1}{n}((X_0 - \bar{X}_n)^2 + \cdots + (X_{n-1} - \bar{X}_n)^2)]^{1/2} \tag{5.50}$$

Subtracting $\bar{X}_n$ from each of the values of the record leads to a new sequence of variables that have zero mean :

$$x_0 = X_0 - \bar{X}_n, \cdots, x_{n-1} = X_{n-1} - \bar{X}_n \tag{5.51}$$

Next we form partial sums of these variables, each of them being the cumulative sum of all values up to a particular value x_k,

$$S_1 = x_0, \quad S_2 = x_0 + x_1, \cdots, \quad S_n = x_0 + \cdots x_{n-1} \tag{5.52}$$

The set of these sums will have a maximum and a minimum value, $M_n = \max S_k$, $m_n = \min S_k$ when the k's run up to n. The *range*, r of the phenomenon described by the sequence is then quite naturally defined as

$$r_n = M_n - m_n \tag{5.53a}$$

or, in a rescaled form,

$$r_n^* = r_n/C_n \tag{5.53b}$$

In a complex system one expects that r should display a pronounced variability. To sort out systematic trends one should thus compute its average value or perhaps a higher moment thereof. The basic quantity involved in such an averaging is the probability distribution of the event

$$F(u, v, n) = \text{Prob}\,(M_n \le u,\ m_n \ge -v, n) \tag{5.54}$$

where u and v are taken to be positive, $M_n > 0$ and $m_n < 0$. Since $r = M - m$, this probability should be integrated over all values of u in the interval from 0 to r prior to the calculation of the moments of r. It is this sort of average, $<r>$ that displays the power law behavior n^H for a large body of data, as mentioned in Sec. 1.4.2.

Throughout the present subsection extremes have been associated to sequences of states rather than to single states. On the other hand, in the literature some authors simply identify extremes to states located in the tail of a statistical distribution. In this logic, then, a probability distribution in the form of a power law (cf. Sec. 1.4.4) would be a mechanism for generating extremes with a non-negligible probability. This potentially confusing view should be differentiated clearly from that underlying the generally accepted interpretation of the concept of extreme.

5.7.2 Statistical theory of extremes

There exists a powerful statistical theory of extremes. In its classical version to which we limit ourselves in this subsection, it is concerned with the case where the variables in the sequence $X_0, \cdots, X_{n-1}$ are independent and identically distributed random variables (iidrv's). The distribution $F_n(x)$ of M_n in eq. (5.46) is then evidently given by

$$F_n(x) = F^n(x) \tag{5.55}$$

and we are interested in the limit of n tending to infinity. But as stressed in the preceding subsection, for any x less than the upper end point b of $F(x)$ $F^n(x)$ will tend to zero, so that the distribution of M_n would degenerate to a peak on b. To cope with this difficulty one introduces a linear renormalization of the variable M_n,

$$z_n = \frac{M_n - b_n}{c_n} \tag{5.56}$$

Appropriate choices of the constants b_n and c_n allow then one to stabilize the location and scale of z_n as n increases, essentially by zooming the neighborhood of this limit. Consequently, one now seeks for limiting distributions of the form $\text{Prob}(z_n \leq x)$ or, with (5.56),

$$\text{Prob}\,(M_n \leq c_n x + b_n) \tag{5.57}$$

Actually such limiting distributions exist only when certain continuity conditions on $F(x)$ at its right endpoint b are satisfied. These rule out a number of distributions, some of them quite common like the Poisson distribution, for which there are no non-degenerate limit distributions. But when they are satisfied, it turns out that all limiting distributions fall into just three universality classes,

$$\text{Prob}\left(\frac{M_n - b_n}{c_n} \leq x\right) \to H(x) \tag{5.58}$$

with

$$H_I = \exp(-e^{-x}) \quad -\infty < x < \infty \tag{5.59a}$$

(Gumbel distribution)

$$H_{II} = \begin{matrix} 0 & x \leq 0 \\ \exp(-x^{-\alpha}) & x > 0, \ \alpha > 0 \end{matrix} \tag{5.59b}$$

(Fréchet distribution)

$$H_{III} = \begin{matrix} \exp(-(-x)^\alpha) & x \leq 0, \ \alpha > 0 \\ 1 & x > 0 \end{matrix} \quad (5.59c)$$

(Weibull distribution)

Notice that all three distributions can be written in the generic form, referred as the generalized extreme value distribution (GEV)

$$H_\xi(x) = \exp(-(1 + \xi x)^{-1/\xi}) \quad (5.60)$$

with $\xi = 0$, $\xi > 0$ and $\xi < 0$ successively for the types I, II and III.

The main idea behind the proof of this remarkable *Fisher-Tippett theorem* is to convert (5.58) in the limit of n tending to infinity, to a functional equation for H by observing that the limit of the left hand side when instead of n one chooses windows nt for any fixed t should tend to $H^t(x)$. The three types of solutions of this functional equation can be viewed as *attractors*, the domain of attraction of a given type being the class of distributions $F(x)$ converging to it in the sense of eq. (5.58). For instance, the Gumbel distribution attracts the distributions $F(x)$ for which $\bar{F}(x) = 1 - F(x)$ is exponential like $Ke^{-\lambda x}$, with $c_n = \lambda^{-1}$ and $b_n = \lambda^{-1}\ln(Kn)$; the Fréchet distribution attracts distributions for which $\bar{F}(x)$ is of the Pareto type (encountered in Sec. 1.4.4) $Kx^{-\alpha}$, with $c_n = (Kn)^{1/\alpha}$; and the Weibull distribution attracts distributions for which $\bar{F}(x) = K(b-x)^\alpha$, with $c_n = (Kn)^{-1/\alpha}$, $b_n = b$. The three distributions (5.59a-c) themselves are *maximally stable* in the sense that for all $n \geq 2$ all maxima $M_n = c_n x + b_n$ are described by the same law, up to a (trivial) affine transformation $x \to \gamma x + \delta$ where γ and δ are constant.

In addition to the foregoing, a key question for the purposes of prediction is what is the mean waiting time between specific extreme events. This question can be answered straightforwardly for iidrv's since the "success" probability of the event exceeding a level x is $p = \bar{F}(x) = 1 - F(x)$. Consequently, the probability distribution of the time of first exceedance is given by the distribution

$$r_k = \underbrace{p}_{1\ success} \underbrace{(1-p)^{k-1}}_{k-1\ failures} \quad k = 1, 2, ... \quad (5.61a)$$

and the desired return time is just the mean value of k with this distribution,

$$<k> = p^{-1} = \frac{1}{1 - F(x)} \quad (5.61b)$$

Eq. (5.61a) allows one to compute higher moments as well. The conclusion is in line with the one drawn previously on other predictability related issues, namely, that the fluctuations around $<k>$ are comparable to the mean itself. This makes the task of prediction especially hard given that, in addition, one deals here with events occurring with low probability.

Throughout this subsection the assumption of iidrv's was adopted. Intuitively, it would seem that events that are sufficiently rare -as is the case for extreme events- can indeed be reasonably treated as independent of each other. Nonetheless, a more elaborate study shows that eqs (5.59) still apply to correlated sequences, essentially as long as the time autocorrelation function falls to zero faster than $(\ln n)^{-1}$.

5.7.3 Signatures of a deterministic dynamics in extreme events

As stressed in this book, complex deterministic dynamics in the form of abrupt transitions, a multiplicity of states or spatio-temporal chaos is at the basis of a wide variety of phenomena encountered in our everyday experience. It is therefore natural to inquire whether a theory of extremes can be built for such systems and, if so, to what extent it reduces for practical purposes to the classical statistical theory summarized in the preceding subsection or, in contrast, bears the signature of the deterministic character of the underlying dynamics.

Our starting point is similar to that in Sec. 5.6, eq. (5.36). It amounts to realizing that in a deterministic dynamical system (eqs (2.1)-(2.2)) the multivariate probability density to realize the sequence $X_0, \cdots, X_{n-1}$ (not to be confused with $\rho_n(x)$ in eq. (5.47)) is expressed as

$$\rho_n(X_0, ..., X_{n-1}) = (\text{Prob to be in } X_0 \text{ in the first place}) \times$$

$$\times \prod_{k=1}^{n-1} (\text{Prob to be in } X_{k-1} \text{ given one was in } X_0 \text{ } k \text{ time units before}) \quad (5.62)$$

In both stochastic and deterministic systems the first factor in eq.(5.62) is given by the invariant probability density, $\rho(X_0)$. In contrast, the two classes differ by the nature of the conditional probabilities inside the n-fold product. While in stochastic systems these quantities are typically smooth, in deterministic dynamical systems they are Dirac delta functions, $\delta(X_{k\tau} - f^{k\tau}(X_0))$, where the formal solution of the evolution equations is denoted by $f(X_0)$ rather than by the notation adopted so far (eq. (2.1) and Sec. 5.6), to avoid confusion with the cumulative distribution. The upper index denotes as before the order in which the corresponding operator needs to be iterated.

By definition the cumulative probability distribution $F_n(x)$ in eq. (5.46) -the relevant quantity in a theory of extremes- is the n-fold integral of eq. (5.62) over $X_0, ..., X_{n-1}$ from the lower bound a up to the level x of interest. This converts the delta functions into Heaviside theta functions, yielding

$$F_n(x) = \int_a^x dX_0 \, \rho(X_0) \, \theta(x - f^\tau(X_0))...\theta(x - f^{(n-1)\tau}(X_0)) \qquad (5.63)$$

In other words, $F_n(x)$ is obtained by integrating $\rho(X_0)$ over those ranges of X_0 in which $x \geq \{f^\tau(X_0), ..., f^{(n-1)\tau}(X_0)\}$. As x is moved upwards new integration ranges will thus be added, since the slopes of the successive iterates $f^{k\tau}$ with respect to X_0 are, typically, both different from each other and X_0 dependent. Each of these ranges will open up past a threshold value where either the values of two different iterates will cross, or an iterate will cross the manifold $x = X_0$. This latter type of crossing will occur at x values belonging to the set of periodic orbits of all periods up to $n-1$ of the dynamical system.

At the level of $F_n(x)$ and of its associated probability density $\rho_n(x)$ these properties entail the following consequences. (i) Since a new integration range can only open up by increasing x and the resulting contribution is necessarily non-negative $F_n(x)$ is a monotonically increasing function of x, as indeed expected. (ii) More unexpectedly, the slope of $F_n(x)$ with respect to x will be subjected to abrupt changes at the discrete set of x values corresponding to the successive crossing thresholds. At these values it may increase or decrease, depending on the structure of the branches $f^{k\tau}(X_0)$ involved in the particular crossing configuration considered. (iii) Being the derivative of $F_n(x)$ with respect to x, the probability density $\rho_n(x)$ will possess discontinuities at the points of non-differentiability of $F_n(x)$ and will in general be non-monotonic.

Properties (ii) and (iii) are fundamentally different from those familiar from the statistical theory of extremes, where the corresponding distributions are smooth functions of x. In particular, the discontinuous non-monotonic character of $\rho_n(x)$ complicates considerably the already delicate issue of prediction of extreme events.

Having identified some universal signatures of the deterministic character of the dynamics on the properties of extremes we now turn to the derivation of the analytic properties of $F_n(x)$ and $\rho_n(x)$ for some prototypical classes of dynamical systems.

Fully developed chaotic maps in the interval. A dynamical system exhibiting this kind of behavior possesses a mean expansion rate larger than one and an exponentially large number of points belonging to unstable periodic trajectories. In view of the comments following eq. (5.63), these properties will show up through the presence of an exponentially large number of points in which $F_n(x)$ will change slope and an exponentially large number of plateaus of the associated probability density $\rho_n(x)$. One may refer to this latter peculiar property as a "generalized devil's staircase". It follows that the first smooth segment of $F_n(x)$ will have a support of $O(1)$ and the last one an exponentially small support, delimited by the rightmost fixed point of the iterate $f^{(n-1)}$ and the right boundary b of the interval. Since $F_n(x)$ is monotonic and $F_n(b) = 1$, the slopes will be exponentially small in the first segments and will gradually increase as x approaches b. Notice that the bulk of the probability mass is *not* borne entirely by the last (exponentially small) layer. These properties differ markedly from the structure found in the classical statistical theory of extremes. Figs 5.16a,b depict (full lines) the functions $\rho_{20}(x)$ and $F_{20}(x)$ as deduced by direct numerical simulation of the tent map, a variant of the logistic map (eq. (2.18)) for $\mu = 1$ defined by the iterative function $f(X) = 1 - |1 - 2X|$, $0 \leq X \leq 1$. The results confirm entirely the theoretical predictions. The dashed lines in the figure indicate the results of the statistical approach for the same system for which $\rho(x)$ happens to be a constant, leading to $F(x) = x$. One would think that this approach should be applicable since this system's correlation function happens to vanish from the very first time step. Yet we see substantial differences, both qualitative and quantitative, associated to considerable fluctuations, discontinuities and a non-monotonic behavior. Notice that since $F(x)^n = x^n$, upon performing the scaling in eq. (5.56) with $c_n = n^{-1}, b_n = 1 - n^{-1}$, one should fall for this system, according to the statistical approach, into the domain of attraction of (5.60) with $\xi = -1$. However, the piecewise analytic structure of the actual $F_n(x)$ precludes the possibility of existence of such simple limiting distributions, at least under linear scaling. These differences underlie the fact that in a deterministic system the condition of statistical independence is much more stringent than the rapid vanishing of the correlation function. The simplicity of this dynamical system allows one to establish a further result of interest, namely, that the probability mass in the internal $(1 - 1/n, 1)$ is of $O(1)$. This also happens to be the case in the classical statistical theory, but despite this common property the structure of the corresponding cumulative distributions is completely different.

Intermittent chaotic maps in the interval. As seen in Sec. 4.3, these are weakly chaotic systems in which the expansion rate in certain regions is close to unity. We take, without loss of generality, this region to be near the

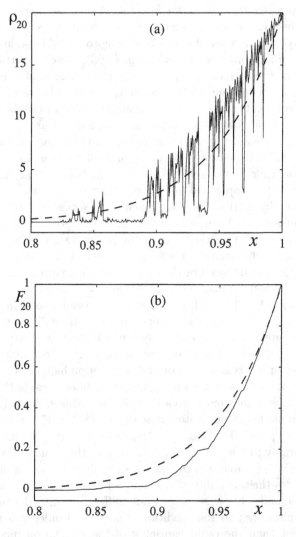

Fig. 5.16. Probability density (a) and cumulative probability distribution (b) of extremes for the tent map as obtained numerically using 10^6 realizations. Dashed curves represent the prediction of the classical statistical theory of extremes. The irregular succession of plateaus in $\rho_{20}(x)$ and the increase of the slope of $F_{20}(x)$ in the final part of the interval are in full agreement with the theory. The irregularity increases rapidly with the window and there is no saturation and convergence to a smooth behavior in the limit of infinite window.

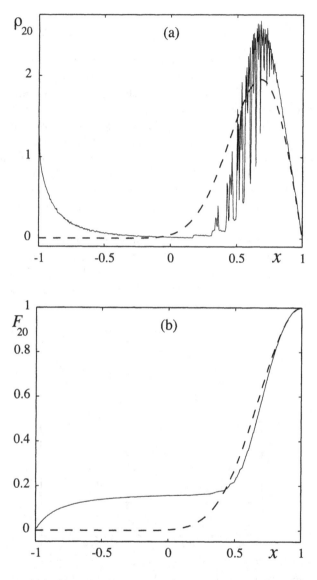

Fig. 5.17. As in Fig. 5.16 but for the cusp map. The irregularities pointed out in connection with Fig. 5.16 subsist, the main new point being the presence of a more appreciable probability mass in the left part of the interval.

leftmost boundary a

$$f(X) \approx (X - a) + u|X - a|^z + \epsilon \qquad a \leq X \leq b \qquad (5.64)$$

where $z > 1$ and ϵ measures the distance from strict tangency of $f(X)$ with the $f(X) = X$ axis. As $\epsilon \to 0$ successive iterates $f^k(X)$ will follow the $f(X) = X$ axis closer and closer, and will thus become increasingly steep at their respective reinjection points where $f^k(X) = b$. As a result the positions of these points (and hence of the (unstable) fixed points other from $X = a$ too, whose number is still exponentially large) will move much more slowly towards a and b compared to the fully chaotic case. Two new qualitative properties of $F_n(x)$ can be expected on these grounds, reflecting the presence of long tails of the correlation function in the system: the probability mass borne in the first smooth segment of this function near $X = a$ and the length of the last smooth segment near $X = b$ will no longer be exponentially small. This is fully confirmed by direct numerical simulation of eq. (5.64) for the symmetric cusp map, $f(X) = 1 - 2|X|^{1/2}, -1 \leq X \leq 1$, as seen in Fig. 5.17. Using the explicit form of $f(X)$ one can check straightforwardly that $F_n(x) \approx 1 + x$ as $x \to -1$, $F_n(0) \approx n^{-1}$, a final segment of $F_n(x)$ of width $O(n^{-1})$ and $F_n(x) \approx 1 - n(1-x)^2/4$ as $x \to 1$.

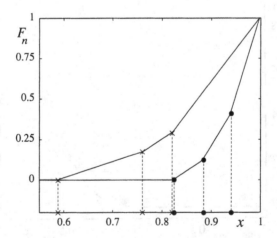

Fig. 5.18. Cumulative probability distribution $F_n(x)$ of extremes for uniform quasi-periodic motion, eq. (5.42), with $\frac{\omega_1}{\omega_2} = (\sqrt{5} - 1)/3$. Upper curve corresponds to $n = 4$ and lower to $n = 10$. Notice that in both cases the number of slope changes in $F_n(x)$, whose positions are indicated in the figure by vertical dashed lines, is equal to 3.

Uniform quasi-periodic behavior. The canonical form for this behavior was already considered earlier in our discussion on recurrence times, eq. (5.42). The invariant density is uniform, but the fundamental difference with cases (i) and (ii) above is that there is no dynamical instability (all expansion rates are constant and equal to unity), and there are neither fixed points nor periodic orbits. Yet the structure of $F_n(x)$, depicted in Fig. 5.18, is similar to that of Figs 5.16b, 5.17b, displaying again slope breaks. These arise from the second mechanism invoked in the beginning of this subsection, namely, intersections between different iterates of the mapping functions. Remarkably, these intersections give rise to integration domains in the computation of $F_n(x)$ that can only have three sizes, for reasons similar to those giving rise to the three possible values of recurrence times as seen in Sec. 5.6.2.

5.7.4 Statistical and dynamical aspects of the Hurst phenomenon

In 1951, William Feller calculated the probability distribution of the range (cf. eq. (5.54)) for the reference case where X_n are independent random variables with a common distribution of finite variance. The partial sums S_n converge then, in the limit of continuous time, to a process $S(t)$ of independent increments described by a stochastic differential equation of the form (3.7) in which the right hand side contains only the noise term. For this process, known as Wiener process $W(t)$, the probability density satisfies a Fokker-Planck equation (eq. (3.8)) in which only the diffusion term is present and turns out to be a Gaussian whose variance increases linearly with t. Feller computed the associated distribution $F(u, v, t)$ in eq. (5.54) by linking it to the solution of the diffusion equation for the variable $S(t)$ in the interval $-v < S < u$ subject to zero boundary conditions when $S = -v$ or $S = u$. He deduced the probability distribution of the range r as an infinite sum of terms involving Gaussian distributions of arguments $kr/t^{1/2}$, with k running from 1 to ∞. This led to explicit expressions of $<r>$ and its variance,

$$<r^*> \approx \sqrt{\frac{\pi}{2}} T^{1/2}$$

$$<\delta r^{*2}>^{1/2} \approx \left(\frac{\pi^2}{6} - \frac{\pi}{2}\right)^{1/2} T^{1/2} \qquad (5.65)$$

where $T = n\tau$ and τ is the sampling time. The increase of the range with the observation window thus follows a power law with exponent $H = 1/2$. But as seen in Sec. 1.4.2, in many real world data the rescaled range is governed by a power law of the form T^H where the Hurst exponent H is greater than 1/2 (typically around 0.7). This implies in turn that the successive X's are not independent: S_k has somehow a persistent effect on X_k, i.e. highs tend to be followed by highs, lows by lows. The question arises, then, what are the mechanisms that could generate such persistencies and the associated Hurst exponents.

To get a feeling about the question let us first consider that the process X itself, rather than the cumulative sums S that it generates, satisfies a stochastic differential equation of the form (3.7) containing only the noise term (which, we note in passing, is not a stationary process !). S becomes then, essentially, the integral of a Wiener process,

$$S(t) = \int_0^t dt' W(t')$$
$$= \int_0^t dt' (t - t') R(t') \qquad (5.66)$$

where R is a white noise process. Under these conditions one can establish straightforwardly that the normalized range r^* follows a power law dependence on T, with a Hurst exponent $H = 1$. As it turns out this is a particular case of a more general setting, in which $S(t)$ is related to the white noise process by the relation

$$S(t) = \int_{-\infty}^t dt' (t - t')^{H' - \frac{1}{2}} R(t') \qquad (5.67)$$

referred as fractional Brownian motion, which produces Hurst exponents different from 1/2 as long as $H' > 1/2$. When $H' \leq 1, X$ is a stationary process and $H = H'$. But when $H' > 1$ (it is 3/2 in eq. (5.66)) the X process is non-stationary and $H = 1$. Although very successful this model cannot be inverted easily, in the sense of identifying a class of dynamical systems generating a process X compatible with these properties. The case $H = 1, H' = 3/2$ is an exception, since it is the limit of the generic form of eq. (3.7) in the absence of a deterministic contribution. This limit is realized in critical situations around a steady state, where the dominant eigenvalue of the Jacobian matrix of f vanishes. But as soon as a deterministic contribution is added, the situation changes qualitatively. For instance, considering a single variable and setting $f = -\lambda X$ in eq. (3.7) generates a stochastic process known as Ornstein-Uhlenbeck process, for which $H = 1/2$. It is worth noting, however, that the convergence to this result tends to be very slow and one needs very long observation times $T = n\tau$ before approaching the correct Hurst exponent.

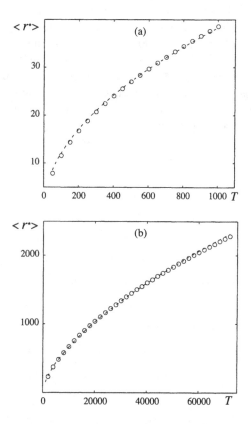

Fig. 5.19. Dependence of the scaled range on observation time T for a model system generating fully developed chaos (a) and weak chaos (b). Empty circles stand for the numerical result averaged over 10^4 realizations and dashed line for the best fit Hurst exponent $H = 1/2$ for (a) and $H = 0.63$ for (b).

As in the previous subsections, we finally inquire whether deterministic dynamical systems may generate behaviors similar to the Hurst phenomenon. Fig. 5.19a depicts the dependence of the scaled range on T for a model system generating fully developed chaos. The corresponding Hurst exponent is $1/2$, compatible with the presence of weak correlations in this system. In a different vein, evidence has been provided that the fluctuations of the phase of a chaotic trajectory in continuous time fully chaotic systems around its local averages constitutes a fractional Brownian motion process with $H \approx 0.74$. As regards weak chaos like e.g. in the presence of intermittency, the situation is more puzzling. Fig. 5.19b summarizes some results for a system of this kind. In all cases a power law linking the normalized range to the observation time holds. But while for windows up to 1 000 the Hurst exponent is 0.77, it falls to 0.63 for windows up to 72 000. One might argue that the conclusion to be drawn is that for all practical purposes the Hurst exponent is non-classical, a result that seems to be compatible with the presence of long range correlations in the system. Still, one cannot exclude a (very slow) convergence to a classical exponent $H = 1/2$. If so one would be tempted to speculate that this conclusion should extend to all systems whose variables evolve according to a bounded and stationary process. Aside from these rather partial results, the dynamical origin of the Hurst phenomenon remains an open question.

Chapter 6
Selected topics

6.1 The arrow of time

The differentiation of the past and the future is an essential ingredient of complexity in connection with such issues as evolution, adaptation, or robustness. And yet it raises an apparent paradox since, as stressed in Sec. 2.2.1, the elementary laws describing microscopic level dynamics are invariant under time reversal, a property which at the level of eq. (2.2) can be expressed as

$$f'_i \equiv f_i(\{X'_j\}) = -f_i \qquad (6.1)$$

with

$$X'_j(t) = X_j(-t) \qquad (6.2)$$

Faced with the above at first sight fundamental opposition, the majority of the scientific community adopted for a long time the attitude that irreversibility -the breakdown of time reversal symmetry- cannot be understood solely from first principles, and that to accommodate it into physics one needs to resort to heuristic approximations. This rather awkward situation whereby one of the most universal phenomena of nature was "ostracized" from being qualified as fundamental, was reversed in the 1980's and onwards. This radical change of perspective, at the origin of a renaissance in the field of nonequilibrium phenomena and basic complexity research, was made possible by the cross-fertilization between nonlinear science, statistical mechanics, thermodynamics, and numerical simulation techniques. In this section we outline the developments that led to the basic laws of irreversibility and comment on their status and role in fundamental and applied research.

According to the view expressed in this book irreversibility finds its most natural manifestation in the behavior of systems composed of a large number of subunits and evolving on macroscopic space and time scales. Such systems may exhibit, when appropriate conditions are met, properties that can in no way be reduced to those of their constituent parts. Emergent properties of this kind -and one may legitimately think that irreversibility is one of them- are best captured at a level of description involving macroscopic observables accounting for collective properties (cf. subsection 2.2.2). Here the breakdown of time reversal is concomitant with the occurrence of dissipative processes like e.g. friction or diffusion, where part of the energy available becomes trapped within the system and cannot be converted to work. Contrary to conservative phenomena these processes live on a part of phase space -the attractor- whose dimensionality is strictly less than that of the total number of variables involved, and enjoy the property of asymptotic stability.

6.1.1 The Maxwell-Boltzmann revolution, kinetic theory, Boltzmann's equation

A cubic centimeter of a gas at 1 atmosphere and ambient temperature is known to contain about 10^{20} particles, moving at speeds of the order of the speed of sound and giving rise to a total of something like 10^{34} collisions per second. These numbers, which are hard to visualize in their enormity, force one to realize that a conventional description involving the positions and velocities of the individual particles is not operational, since these variables are expected to vary in time in a very irregular fashion. Rather, one is tempted to focus on a single representative particle's position $\mathbf{r}$ and velocity $\mathbf{v}$ and inquire about the probability distribution $f(\mathbf{r}, \mathbf{v}, t)$ of presence of particular $\mathbf{r}$ and $\mathbf{v}$ values at any given time t. Familiar as it may sound in view of the developments outlined in Chapter 3 of this book this systematic introduction of probabilistic ideas into physics was in fact immensely bold and startling when it was formulated by James Clerk Maxwell and Ludwig Boltzmann in the 1860's, a time when the atomistic view of matter still met an almost universal hostility.

Building on this idea Boltzmann derived in 1872 by what appeared to be completely mechanical arguments an evolution equation for $f(\mathbf{r}, \mathbf{v}, t)$ based on the decomposition of its rate of change into a free flow and a collisional part (Fig. 6.1a, b):

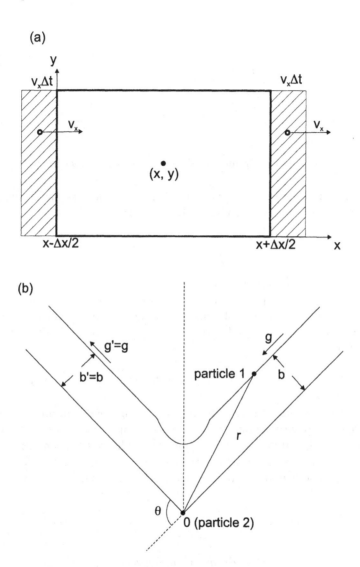

Fig. 6.1. Illustration of the decomposition into free flow (a) and collisions (b) in the Boltzmann equation (6.3)-(6.6). The hatched areas in (a) represent the entrance and exit regions of particles during a time interval Δt. In (b), the trajectories are drawn schematically in a coordinate system attached to one of the particles. θ measures the angle between initial and final relative velocity, and b the distance of closest approach of the two particles that would be observed in the absence of interactions.

$$\frac{\partial f_1}{\partial t} = \left(\frac{\partial f_1}{\partial t}\right)_{free\ flow} + \left(\frac{\partial f_1}{\partial t}\right)_{coll}$$

$$= \left(\frac{\partial f_1}{\partial t}\right)_{free\ flow} + \underbrace{J_+}_{\substack{transitions \\ leading \\ to\ the\ state \\ of\ interest}} - \underbrace{J_-}_{\substack{transitions \\ out\ of \\ the\ state \\ of\ interest}} \quad (6.3)$$

where the subscript of f refers to the position and velocities of a particle numbered "1".

A moment's inspection of Fig. 6.1a visualizing the free flow in the space spanned by $\mathbf{r}$ and $\mathbf{v}$- the so-called μ space- leads to

$$\left(\frac{\partial f_1}{\partial t}\right)_{free\ flow} = -\mathbf{v}_1 \cdot \frac{\partial f_1}{\partial \mathbf{r}_1} \quad (6.4)$$

The situation concerning $J_\pm$ is far more involved since these quantities depend, in principle, on particle 1 as well as on the joint distribution of *all* other particles in the gas. Boltzmann's stroke of genius was to propose, in the limit of a dilute gas, a reduced description involving only single particle distributions,

$$J_\pm = \text{(prefactors determined from the mechanics of the collision process)} \cdot f_1 f_2 \quad (6.5)$$

This is the famous *Stosszahlansatz*: a seemingly natural, innocent statement which in reality contains a heuristic assumption of a probabilistic nature since it stipulates that at the moment of the collision particles 1 and 2 are uncorrelated and hence that their joint probability can be factorized. Combining these ingredients one arrives at the Boltzmann equation

$$\frac{\partial f_1}{\partial t} + \mathbf{v}_1 \cdot \frac{\partial f_1}{\partial \mathbf{r}_1}$$
$$= \int d\mathbf{v}_2 d\mathbf{v}'_1 d\mathbf{v}'_2 w(\mathbf{v}_1, \mathbf{v}_2 | \mathbf{v}'_1, \mathbf{v}'_2) f'_1 f'_2$$
$$- f_1 \int d\mathbf{v}_2 d\mathbf{v}'_1 d\mathbf{v}'_2 w(\mathbf{v}'_1, \mathbf{v}'_2 | \mathbf{v}_1, \mathbf{v}_2) f_2 \quad (6.6)$$

Here the primes denote evaluation of the corresponding quantities after the collision process has taken place and w accounts for the mechanics of the collision and for momentum and energy conservation of the pair of colliding particles.

Eq. (6.6) provides an autonomous, closed description for a reduced, "relevant" quantity (the distribution f). As such it has set a prototype for further developments that would have been impossible had one followed the standard view that information on the evolution of a physical system requires information on the full details concerning its structure and its initial preparation. More than 130 after Boltzmann we cannot but be deeply impressed by realizing that, at his time, the idea of a mesoscopic description based on Markovian type equations following a somewhat similar logic (Sec. 3.2) was still to come! From the mathematical standpoint the Boltzmann equation constitutes a nonlinear partial integro-differential equation - a very complex object indeed. There exists a rich literature devoted to the questions of existence, uniqueness and nature of its solutions, as well as to the search or approximation schemes for describing the long time regime. We do not dwell on these aspects here but, rather, summarize the principal results drawn from eq. (6.6) by Boltzmann and his followers.

(i) <u>The H-theorem</u>. In a homogeneous system, the quantity

$$H = k_B \int d\mathbf{v}\, f\, (\ln f - 1) \qquad (6.7)$$

k_B being a universal constant (known as Boltzmann's constant) decreases monotonously, $dH/dt \leq 0$, until f reaches its equilibrium form given by the Maxwell-Boltzmann distribution, $f_{eq} = \frac{1}{V}(m/2\pi k T)^{3/2} \exp(-\frac{mv^2}{2kT})$, m being the mass of the particles, T the temperature and V the volume. Substitution of f_{eq} into (6.7) reproduces the value of entropy in the state of equilibrium, $S_{eq} = -H_{eq}$. The monotonic behavior of H in time, known as the H-theorem, should provide, then, the microscopic origin of the second law of thermodynamics.

(ii) <u>The laws of macroscopic physics</u>. Multiplying eq. (6.6) by the mass, velocity and energy of the representative particle and integrating over $\mathbf{v}_1$ one derives, in the long time limit, the laws of hydrodynamics and chemical kinetics (see e.g. eq. (2.6)) as applied to dilute systems. Furthermore, one obtains the values of the transport coefficients such as the diffusion, viscosity and heat or electrical conductivity coefficients in good agreement with experiment.

These results constitute a major scientific breakthrough and yet, almost as soon as they were enunciated, gave rise to a series of negative reactions all of them stemming from the fact that because of the Stosszahlansatz, eq. (6.6) contains in fact a mixture of exact mechanical and approximate probabilistic considerations. The objections crystallized in the famous reversibility (Loschmidt's) and recurrence (Zermelo's) paradoxes. The former states that if there is an evolution leading to a monotonic decrease of H in time, then there is bound to be another one, obtained by time reversal, in which H must behave "anti-thermodynamically", i.e. increase. The second one states that since by the Poincaré recurrence theorem (cf. Sec. 5.6) states recur, the H-function is bound to reverse its course at some stage. Boltzmann made a valiant attempt to respond, based essentially on the enormity of recurrence times in a molecular system (of the order of ten to the power of the Avogadro number) and on the practical impossibility to reverse the velocities of all particles. He also pointed out that the states satisfying the H-theorem and showing thermodynamic behavior are in fact to be regarded as "most probable states" in the sense of constituting a set of states on which the system spends the overwhelming part of its time, the chance of falling into any other state being exceedingly small. But that was not good enough for his detractors. In fact, in advancing these arguments Boltzmann tacitly admitted that the second law of thermodynamics is to be interpreted as a statistical property. This reinforced further the impression that irreversibility is an empirical concept outside the realm of fundamental physics.

6.1.2 First resolution of the paradoxes: Markov processes, master equation

A first resolution of the paradoxes was provided by the, by now famous, "dog flea" model of Paul ad Tatiana Ehrenfest. $2N$ numbered balls are distributed in boxes A and B. An integer from 1 to $2N$ is drawn at random and the ball corresponding to that number is moved from the box in which it is found to the other one. The process is then repeated to any desired number of times. One can show that the transitions between the states are fully reversible and thus in agreement with microscopic physics, in the sense that the probability of realizing a state, say ℓ, at time $t + \Delta t$ given that the system was in state k at time t, is equal to the probability of having realized this state at time $t - \Delta t$ subject to the same conditions at time t. Furthermore, every initial state recurs with probability one. Yet the system tends irreversibly toward a configuration in which the balls are distributed among the two boxes according to the binomial law

$$P_s(k) = \frac{2N!}{k!(2N-k)!} 2^{-2N} \tag{6.8a}$$

which in the limit of large N tends in turn to the Gaussian distribution

$$P_s(k) \approx \exp\left(-\frac{(k-N)^2}{2N}\right) \tag{6.8b}$$

Notice that the mean recurrence time of state k (eq. (5.38a)) turns out to be here equal to $1/P_s(k)$. This is a reasonable number for k close to N, which corresponds in view of eq. (6.8) to the most probable state. On the contrary, for $2N = 40$ and state $k = 1$ which from the standpoint of (6.8) is an extremely improbable state, one obtains the incredibly large value of 2^{80} time units! This provides, a posteriori, a rationale for Boltzmann's response to the Zermelo paradox.

An important insight afforded by this realization is that while microscopic reversibility and recurrence are properties of individual states (albeit coarse-grained ones in the present context!) and of trajectories joining them, the approach to an asymptotic regime (the analog of thermodynamic equilibrium) is a property of probability distributions. Again, this clarifies further the ideas underlying Boltzmann's original response to Loschmidt and Zermelo.

The reader will have realized by now that the Ehrenfest model belongs to the general class of stochastic processes of the Markov type introduced in Sec. 3.2. In fact, the very form of the Boltzmann equation is reminiscent of the master equation (eq. (3.5)) -or more appropriately of its continuous time limit, $\Delta t \to 0$- where one also writes out a balance equation for the transitions between states. As seen in Sec. 3.3 once a process is described by eq. (3.5), the set of available states consists entirely of ergodic states, and the transition probability matrix is a stochastic matrix, then any arbitrary initial probability distribution will approach irreversibly the unique time-independent solution, P_s of the equation. This property can be expressed in a form very similar to Boltzmann's H-theorem by introducing an H-function related to the excess of information entropy (eq. (4.1)) with respect to its steady-state value,

$$H = \sum_k P(k,t) \ln \frac{P(k,t)}{P_s(k)} \tag{6.9}$$

(in fact, one could use for this purpose instead of the logarithm any convex function $F(x)$ of $x = P(k,t)/P_s(k)$). Using eqs (3.5) - (3.6) as well as the property valid for any convex function F,

$$\sum_k m_k F(\bar{x}) \leq \sum_k m_k F(x_k)$$

where $\bar{x} = \sum_i m_i x_i / \sum_i m_i$ and $m_i \geq 0$, one shows then straightforwardly that

$$H(t + \Delta t) \leq H(t) \tag{6.10}$$

The equality holds only when $P_k(t)$ reaches the time-independent distribution $P_s(k)$. In other words, the problem of irreversibility is solved once the premises underlying the master equation are secured. The derivation of (6.10) highlights on the other hand the *statistical character* of the H-theorem obtained under these conditions which, as seen earlier, goes back in its original formulation to Boltzmann himself. In plain terms, $P_s(k)$ is maintained in its (time-independent) form thanks to ongoing transitions between states, but each of the individual realizations of the stochastic process is subjected to fluctuations and is in fact coming back every now and then to its original configuration. The question to be raised is then first, whether this view is an acceptable one and second, under what conditions can the full scale microscopic dynamics be amenable to a Markov process. The objective of Sec. 6.1.4. will be to address this problem from the standpoint of nonlinear dynamics and chaos theory.

6.1.3 Generalized kinetic theories

The need to clarify the assumptions underlying the Boltzmann equation and to extend it to a wider class of systems than dilute gases became a major preoccupation since the 1940's. Key advances along this line have been realized by Nikolai Bogolubov, Léon Van Hove and Ilya Prigogine. Their starting point is the Liouville equation (eq. (3.3)) and its quantum counterpart, from which evolution equations for reduced probability densities $\bar{\rho}$ are derived by projecting the full phase space density ρ into a subspace of phase space. As a first step this procedure leads to exact, formal equations of the form

$$\frac{\partial \bar{\rho}}{\partial t} + K\bar{\rho} = \int_0^t d\tau G(t-\tau)\bar{\rho}(\tau) + D(t) \quad (6.11)$$

Here $K\bar{\rho}$ stands for the contribution of the mean field, $D(t)$ depends on the initial correlations in the subspace of phase space complementary to the one of $\bar{\rho}$, and G describes the memory effects accompanying the reduction of the description from ρ to $\bar{\rho}$. The important point is that, contrary to (3.5), eq. (6.11) is both non-Markovian and non-closed. A great deal of effort has been devoted to identifying conditions linked to first principles in a more clearcut way than Boltzmann's Stosszahlansatz, under which closed Markovian kinetic equations involving only one or two particle probability densities could be derived. Generally speaking these conditions require the limit of long times to be taken. They involve ideas such as the many-particle probabilities becoming functionals of the one-particle density f_1 (Bogolubov), the short range character of the initial correlations allowing one to neglect the term $D(t)$ in eq. (6.11) in the long time limit (Prigogine), or in the quantum mechanical context the progressive randomization of the phases of the wave functions describing the individual states (Van Hove). Their universality cannot be assessed, as they rely on perturbative expansions whose convergence is hard to establish. Furthermore, even if one accepts them, it has so far proved impossible to establish for the resulting *generalized kinetic equations* an H-theorem in the most general case. Still, an important advance is that in all cases where the approach to equilibrium can be established a common mechanism turns out to be at work, namely, the abundance of *resonances*. The main point is that in a many-particle system the dynamics can be viewed as a free-flight motion interrupted by collisions. The interplay between these two processes can be seen most clearly in a weakly coupled system. A perturbative solution of the Liouville equation in which only the dominant terms are retained leads here to an equation of the form (6.11) which, in the above mentioned double limit of long times and of short ranged initial correlations reduces to a Markovian evolution equation for $\bar{\rho}$. The right hand side of this equation displays a *collision operator* containing a multiplicative contribution of the form

$$(2\pi/L)^3 \sum_\ell |V_\ell|^2 \delta[\mathbf{l}\cdot(\mathbf{p}_j - \mathbf{p}_k)]$$

in which L is the system size, $\mathbf{p}_j$, $\mathbf{p}_k$ are the momenta of particles j and k, V_ℓ the Fourier transform of the interaction potential between j and k, δ the Dirac delta function and $\mathbf{l}$ a vector whose components are integers multiplied by $2\pi/L$. The evolution equation for $\bar{\rho}$ would be non-trivial provided that the above expression exists and is finite. This implies, in turn, that the argument of the delta function should vanish, $\mathbf{l}\cdot(\mathbf{p}_j - \mathbf{p}_k) = 0$. Now, in the framework of the perturbative approach adopted $\mathbf{p}_j, \mathbf{p}_k$ are referring to free-particle motion and are therefore related to the inverse of the time needed by the particles to cross the characteristic dimension L of the system. The vanishing of the argument of the delta function expresses, then, the fact that the characteristic frequencies of the free motion are commensurate (cf. Sec. 5.6.2). This is, precisely, the phenomenon of resonance. Dynamical systems theory shows that resonances are "dangerous" in the sense that a small perturbation around the reference system (here the free particle motion) can give rise to unstable motions. As long as $\mathbf{l}$ takes only discrete values, the resonance conditions will be satisfied only for certain $\mathbf{p}$'s and the collision operator will vanish almost everywhere in phase space. But in the limit of a large system $\mathbf{l}$ becomes a continuous variable. As a result, it is expected that resonances will appear almost everywhere and that the collision operator will exist and be finite, driving initial nonequilibrium states toward a final equilibrium distribution. It should be pointed out that resonances are also closely related to the property of non-integrability (Sec. 2.2.1) and the nonexistence of invariants of the motion -other than the total energy- depending smoothly on the interparticle interactions which, if present, would hinder the approach to equilibrium.

Generalized kinetic equations have been at the origin of spectacular progress for the study of dense fluids and plasmas and for the microscopic evaluation of transport coefficients in the linear range of irreversible phenomena close to equilibrium. Of special interest is the hydrodynamic mode approach to transport in dense fluids, where a remarkable synthesis is achieved between a perturbative analysis of the Liouville equation incorporating information on the microscopic level dynamics of the individual particles and the collective behavior described at the macroscopic level by the equations of fluid dynamics.

6.1.4 Microscopic chaos and nonequilibrium statistical mechanics

Generalized kinetic theories rest on approximations whose validity is difficult to assess. Furthermore, there is no explicit link between the structure of the kinetic equations and the indicators of the microscopic dynamics in phase space. In view of the fundamental importance and ubiquity of irreversibility in the natural world it would certainly be desirable to arrive at a description free of both these limitations. This has been achieved in the last two decades by the cross-fertilization between dynamical systems, nonequilibrium statistical mechanics and microscopic simulation techniques.

Mapping to a Markov process

A first series of attempts pertains to the class of strongly unstable chaotic systems, in which each phase space point lies at the intersection of stable and unstable manifolds. It takes advantage of the existence in such systems of special phase space partitions -the Markov partitions- introduced in Sec. 3.3 whose boundaries remain invariant under the dynamics, each partition cell being mapped at successive time steps into a union of partition cells. If the projection of the full phase space dynamics onto such a partition commutes with the Liouville operator, then the Liouville equation can be mapped into an exact Markovian equation exhibiting an H-theorem. This provides a rigorous microscopic basis of coarse-graining. The mapping is generally not one-to-one as the projection operator is not invertible, reflecting the loss of information associated with this process. Ilya Prigogine and coworkers succeeded in constructing an invertible transformation operator Λ which transforms the dynamical evolution operator U_t (essentially the exponential of the

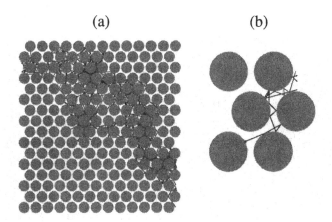

Fig. 6.2. Complex dynamics (a) and sensitivity to the initial conditions (b) arising from the collisions between a light elastic sphere and a set of heavy particles arranged on a regular lattice. In (a), the trajectory fills eventually the entire space available. In (b), neighboring trajectories emanating from the low left part first remain close to each other, but are subsequently clearly differentiated after some six collisions (see Color Plates).

Liouville operator) into a Markov evolution operator W_t

$$W_t = \Lambda U_t \Lambda^{-1} \qquad (6.12)$$

thereby mapping in a one-to-one way the Liouville equation to an irreversible evolution.

Escape rate formalism and deterministic thermostats

A second line of approach draws on the prevalence of microscopic chaos at the level of individual particle trajectories (in a classical setting) in a system of colliding particles. The origin of this property is the universal phenomenon of collision as illustrated in Fig. 6.2 on a model known as the Lorentz gas, where a beam of light particles is injected against a lattice of heavy immobile scatterers. Consider two collisional events involving a particular heavy scatterer, in which the initial positions of the light particles are identical but their velocities deviate by a small angle $\delta\theta_0$ (Fig. 6.3). The distance between the points on which the light particles will touch the heavy sphere just before collision will be then $\delta s_0 = \ell \delta\theta_0$, where ℓ is the mean length of the trajectory of a light particle between collisions, referred as "mean free path". Once on these contact points the particles will undergo specular reflection, implying that the resulting velocities will deviate by an angle $\delta\theta_1 \approx \frac{2\delta s_0}{r} = \left(\frac{2\ell}{r}\right)\delta\theta_0$, where r is the radius of the heavy sphere. Since

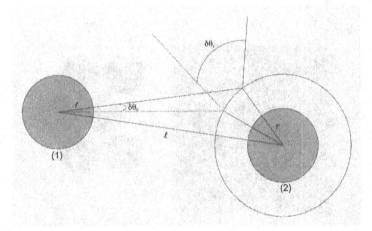

Fig. 6.3. The defocusing character of the process of collision, here illustrated on the collision between two elastic disks (1) and (2). In the coordinate system where (2) is at rest, a small change $\delta\theta_0$ in the impact angle of two nearby trajectories entails a deviation between these trajectories after collision by an angle $\delta\theta_1$ larger than $\delta\theta_0$.

ℓ/r is greater than unity (which allows one to argue as if the two incident trajectories were parallel) this deviation is, clearly, larger than the initial one. We refer to this as the "defocusing character" of collisions. Continuing the reasoning, the angular deviation after the next collision will be $\delta\theta_2 \approx \frac{2\ell}{r}\delta\theta_1 = \left(\frac{2\ell}{r}\right)^2 \delta\theta_0$, etc. In other words there is exponential amplification of the initial uncertainty - a typical signature of deterministic chaos. The corresponding Lyapunov exponent σ can be estimated by dividing $\ln |\delta\theta_n/\delta\theta_0|$ by the time needed to undergo these collisions, which in turn can be approximated by $n\tau$ where τ is the mean time between collisions. This yields, finally,

$$\sigma \approx \frac{1}{\tau} \ln(2\ell/r) \qquad (6.13)$$

A typical intercollisional time is given by the ratio of a typical interparticle distance in a fluid, $\ell \approx 10^{-8}m$ and a typical particle velocity, $10^2 ms^{-1}$. This leads us to $\tau \approx 10^{-10}s$, indicating that the predictability horizon in this phenomenon of microscopic level chaos is quite short compared to the times associated to the chaotic dynamics exhibited by macroscopic observables like e.g. velocity of bulk motion in a fluid, temperature, concentration, etc. One can convince oneself that on a long time scale this will generate, for each particle, a diffusional motion provided that ℓ/r is larger than a threshold value. At the same time two initially nearby particles will undergo quite different sequences of collisions, but each of them will eventually fill the entire space as expected in a diffusion process.

Two different methodologies have been developed to analyze the repercussions of this microscopic level chaos in the foundations of irreversibility.
The escape rate formalism. This formalism is built on the observation that in a Lorentz gas type system there exists a set of "exceptional" trajectories that remain trapped in a limited portion of phase space and hence cannot participate in any transport process. This set forms a fractal object which repels the trajectories not belonging to it. Accordingly, transport can be viewed as an escape from this fractal repellor. The corresponding transport coefficients (diffusion, etc.) can then be expressed in terms of the Lyapunov exponents of the repellor and its Kolmogorov-Sinai entropy or its fractal dimension. No boundary conditions need to be imposed. Furthermore, the (transient) nonequilibrium states associated with the transport process can be computed as generalized eigenfunctions of the Liouville operator and turn out to be fractal distributions. A remarkable fact is that these states decay to equilibrium in spite of the fact that the underlying system is conservative and time reversible. Mathematically, this is reflected by the existence of (generalized) eigenvalues of the Liouville operator possessing a negative real part. These properties are reminiscent of Boltzmann's H-theorem, but to this date there is no satisfactory definition of entropy for an arbitrary nonequilibrium state that would increase monotonically with time in this type of system. This, and many other questions touching on the foundations of irreversibility remain open.
Deterministic thermostats. The question here is how to express in purely mechanical terms a constraint maintaining a system away from equilibrium. One of the most popular ways to achieve this is to augment (in a classical system) the Hamiltonian dynamics by dynamical friction terms preserving the time reversal symmetry. One obtains in this way a dissipative dynamical system where the rate of phase space volume contraction and the entropy production at the nonequilibrium steady state are shown to be related to the sum of the Lyapunov exponents. The nonequilibrium steady states generated in this formalism are fractal attractors, contrary to the equilibrium states which extent over the entire phase space available. Furthermore, they satisfy an interesting fluctuation theorem (in the sense of Sec. 4.5, see also Sec. 6.2.1) involving the entropy production and the probability of fluctuations associated with particles moving in a direction opposite to the applied constraint. A more fundamental approach is to respect the conservative character of the local dynamics and express the nonequilibrium constraints through appropriate boundary conditions. This allows one to derive from first principles

the formalism or irreversible thermodynamics and, in particular, the bilinear expression of entropy production as a product of macroscopic fluxes and associated generalized forces (cf. eq. (2.8)). In this setting, entropy production is no longer universally related to the phase space volume contraction.

Statistical mechanics of dynamical systems

Yakov Sinai, David Ruelle and Rufus Bowen have shown that the long time behavior of a large class of chaotic dynamical systems is determined by an invariant probability measure characterized by a variational principle, thereby establishing a link with ergodic theory and equilibrium statistical mechanics. A major difference is that contrary to equilibrium measures, Sinai-Ruelle-Bowen (SRB) measures are smooth only along the expanding directions while they possess a fractal structure along the contracting ones. This provides a rationale for the fractal character of the nonequilibrium steady states generated by deterministic thermostats. Furthermore, it has been the starting point of a series of developments aiming to characterize low-order deterministic dynamical systems showing complex behavior through probability densities and the spectral properties of their evolution operators such as the Liouville or the Frobenius-Perron operators. These approaches have provided insights to questions motivated by nonequilibrium statistical mechanics which had remained unsolved for a long time, owing to the formidable difficulties arising from the presence of a large number of interacting particles.

6.2 Thriving on fluctuations: the challenge of being small

As stressed from the very first sections of this book complexity is closely related to the emergence of new properties in a system of coupled subunits, reflecting a state of collective behavior that can in no way be reduced to individual level processes. Ordinarily, this observation is considered to be tantamount to taking the "macroscopic" limit where the number N of subunits tends to infinity. This allows one to sort out some universal trends, and it is certainly not an accident that some of the most clearcut manifestations of complexity from pattern forming instabilities and chaos to irreversibility itself are, indeed, encountered in macroscopic level systems. In this view, then, the limit $N \to \infty$ plays the role of a "complexity threshold" much like the role played by the thermodynamic limit in equilibrium statistical mechanics in building a universal theory of phase transitions. True, as stressed from Chapter 3 onwards, macroscopic behavior may be blurred -or even in some cases be superseded- by the fluctuations. But even for such finite size effects, the best way to understand what is going on is to take the asymptotic limit where N, although now finite, is still considered to be large.

On the other hand, many natural and artificial systems of paramount importance turn out, simply, to be small. Examples of such systems, whose size is usually in the nanometer range, include biological regulatory and energy transduction machines at the subcellular or the macromolecular level; and domains of homologous chemical species formed on a surface catalyst, which mix rather poorly with domains including other chemical species with which they eventually react (Sec. 3.5.1). As one reaches such small scales fluctuations begin to play a substantial role in the behavior. They can lead, in particular, to observable deviations from the predictions of the mean field analysis that are not described adequately by any known macroscopic theory, including classical irreversible thermodynamics. A similar failure may arise in systems like small animal or human societies as encountered, for instance, in isolated communities or in higher mammals, even though the "physical" size of the individual subunit is here well beyond the nanometer range.

To what extent can systems of the above kind exhibit non-trivial behavior sharing common universal features? The basic idea is that in the absence of asymmetries such as those provided by nonequilibrium constraints, the behavior of a small system is bound to be blurred by the fluctuations. In particular, owing to the property of detailed balance (cf. Sec. 4.4) there will be no systematic long term reproducible trends such as, for instance, a net flux of energy or matter traversing the system or any form of mechanical work performed against the environment. The situation changes radically in the presence of asymmetries provided, in particular, by externally imposed nonequilibrium constraints or by internally driven irreversible processes converting for instance energy rich initial reactants to energy deficient products. Here one would be entitled to expect the appearance of clearcut and reproducible trends, but since the fluctuations will still be a major driving force behind the evolution these trends will necessarily bear the signature of their presence. Now, as seen in Secs 4.4, 4.5 and 5.7 fluctuations in small systems possess some universal properties, provided that exceptional events -in the sense of the "large deviations" of Sec. 4.5 and of the "extremes" of Sec. 5.7- are accounted for in the description. The conjunction of the two ingredients, asymmetry and (strong) fluctuations, may thus be expected to provide mechanisms allowing a small system to cope with -and sometimes take advantage of- the fluctuations and perform non-trivial tasks. Below, we show on some prototypical examples how this can happen. What is more, it will turn out that some of the resulting trends are so robust that one could legitimately speak of a second "complexity threshold", this time in the range of descending values of N, below which new horizons that are inaccessible to a macroscopic system are opened.

6.2.1 Fluctuation dynamics in nonequilibrium steady states revisited

In an equilibrium system, owing to time reversal invariance, the effect of a process proceeding in a certain "direct" way is counteracted by the effect of a process proceeding in the "reverse" way. For instance, if equilibrium is secured by the contact between the system and an environment at constant temperature T (a "heat bath"), the probability of absorbing a certain amount of heat Q from the environment is identical to that of releasing it, i.e. $\text{Prob}(Q)/\text{Prob}(-Q) = 1$. A similar relation holds true for the transfer of heat -or of any other property for that matter- per unit surface between two regions within the system during a certain time interval τ.

Consider now a system maintained in a nonequilibrium steady state. The simplest such setting is a system in which different parts of its surface are maintained at different temperatures, like in the thermal convection experiment described in Sec. 1.4.1. A second typical situation to which we focus here pertains to systems submitted to external forces, like e.g. a system of charged particles in an external electric field or a fluid under shear. As the presence of the forces tends to pump energy into the system, to maintain a time-independent state one must evacuate the excess energy to the environment. This is ensured by the presence of an adequate "thermostat" like, for instance, a heat bath at fixed temperature T and of size much larger than the system size. The question one now asks is, how the nonequilibrium constraints and the thermostating will be manifested at the level of a microscopic trajectory embedded in the full phase space in which the dynamics is taking place. Let this trajectory be partitioned into segments on which the system spends equal amounts of time τ. We denote by a the value of net flux J_τ (of heat, of charge, ...) detected during this interval and convene to count a as positive if J_τ is in the direction of the field. Giovanni Gallavotti and Eddie Cohen proved that under these conditions, and provided that the underlying system possesses sufficiently strong chaotic properties, a *fluctuation theorem* similar to those introduced in Sec. 4.5 determines the relative probabilities of J_τ and its reverse:

$$\frac{\text{Prob } (J_\tau = a)}{\text{Prob } (J_\tau = -a)} = e^{dN\tau a} \qquad (6.14)$$

in the limit of long time windows τ, where d stands for the dimensionality of the embedding space. The important point to be stressed is that this result is not restricted to small fluctuations nor to near-equilibrium states. Furthermore it is valid for large as well as small systems, provided that the strong chaoticity condition is satisfied.

Eq. (6.14) is often formulated in terms of the entropy production σ, since one may argue that for a given value of the constraint σ is just proportional to J (cf. eq. (2.8)). Actually this supposes that the local equilibrium assumption remains valid at the microscopic level, for small systems arbitrarily far from equilibrium, which is not necessarily the case. On the other hand, if accepted, this view leads to some interesting conclusions. It indicates that it is more likely to deliver a certain amount of heat to the environment (σ positive) during the interval τ than it is to absorb an equal amount of heat from the environment (σ negative). In other words, the system in the nonequilibrium steady state will on the average dissipate heat, a property that may be regarded as an expression of the second law of thermodynamics. In a macroscopic system the ratio of the two probabilities, $\text{Prob}(\sigma)/\text{Prob}(-\sigma)$ grows exponentially with the system size (as it also does with the window size τ), entailing that the probability to absorb heat is negligible. But for small systems this probability can be significant. Although, on average, such systems will produce heat, there will be stages of their evolution during which they will absorb heat from the environment. This will allow them, in particular, to use fluctuations in a constructive way - a process often referred as "rectifying thermal fluctuations".

Results of this kind shed light in the foundations of irreversibility in connection, in particular, with Loschmidt's paradox introduced in Sec. 6.1.1. They show how macroscopic level irreversibility can arise from time-reversible dynamics at the microscopic level: time-reversed steps are present during certain stages of the evolution but they become increasingly rare as the system size and/or the observation window increase, provided the dynamics displays sufficiently strong instability properties. They also have their limitations. In particular, the strong chaoticity assumption may be compromised in such situations as intermittent systems or Hamiltonian systems of few variables, where anomalous transport is known to occur.

6.2.2 The peculiar energetics of irreversible paths joining equilibrium states

A second typical setting where nonequilibrium and irreversibility are manifested is when a system undergoes a transformation between an initial state of equilibrium with respect to its environment, to a final one that is eventually also coming to equilibrium with this environment. This can be achieved by varying in time a control parameter λ affecting the system's overall state such as, for instance, the end-to-end length of a macromolecule either by external manipulation or through the coupling between simultaneously on-going processes. To each stage of the transformation there will be associated a variation of the system's total energy E, say δE_λ, which is nothing but the work, $-\delta W$ performed on the system. But even if we fix once and for all the way the parameter λ varies in time between its initial, λ_1 and final, λ_2 values the transformation could still occur in many possible ways, as there will be an infinite number of microscopic pathways joining the associated initial and final equilibrium states. These pathways will be characterized by different time dependencies of the system's microscopic degrees of freedom, $X_i(t)$ (in a classical setting, the coordinates and the momenta of the constituting particles) and hence, by different amounts by which the value of the *internal energy*, $\delta U \equiv \delta E_{\{X_i\}}$ will vary. Energy conservation -essentially equivalent to the first law of thermodynamics- requires that $\delta E_{tot} = \delta E_\lambda + \delta E_{int}$ or alternatively $\delta E_{tot} = \delta U - \delta W$ be equal to the energy exchanged in the form of heat, $\delta_e Q$ between the system and its environment. Summing all the elementary steps between states 1 and 2 leads then, to

$$\Delta_e Q = \Delta U - W \tag{6.15}$$

In a small system all terms in the above relation will fluctuate in magnitude and presumably even in sign. The extent of these fluctuations will be described by the probability distributions $P(W)$ and $P(Q)$ which appear therefore, once again, as the central quantities of interest.

If the transformation joining the initial and final states is reversible, then $\Delta_e Q = T \Delta S$ where T is the environmental temperature and ΔS the change of the system's entropy due to the transformation. Eq. (6.15) entails, then, that $W = \Delta U - T \Delta S$ which is nothing but the difference of the values of Helmholtz's free energy, ΔF (another "thermodynamic potential" in the terminology of Sec. 2.2) between the initial and final equilibria. But if the transformation is irreversible as it will be typically the case in a real world situation, $\Delta_e Q/T$ represents only the entropy flux $\Delta_e S$. To recover the full entropy variation ΔS one has then to write $\Delta_e Q = T(\Delta S - \Delta_i S)$ where $\Delta_i S$ is the entropy *produced* during the transformation. As seen in Sec. 2.2.3 this is precisely the quantity whose the second law of thermodynamics asserts the non-negativity in a macroscopic system, $\Delta_i S \geq 0$. Introducing these different elements into eq. (6.15) yields

$$\begin{aligned} W &= \Delta U - T \Delta S + T \Delta_i S \\ &= \Delta F + T \Delta_i S \end{aligned}$$

This implies that in an irreversible transformation W comprises, in addition to the part ΔF (still related to the free-energy difference between the initial and final equilibria), a contribution associated to the entropy production to which one may refer as the "irreversible work".

$$W_{irr} = T\Delta_i S \qquad (6.16)$$

In a macroscopic system this part is bound to be non-negative. For such macroscopic systems the total work W is thus bounded from below by ΔF.

To what extent are fluctuations likely to change this picture? Using the formalism of equilibrium statistical mechanics along with some general properties of conservative, time reversible dynamical systems, Christopher Jarzynski showed that in actual fact

$$< \exp\left(-\frac{W_{irr}}{k_B T}\right) >= 1 \qquad (6.17)$$

Here k_B is Boltzmann's universal constant and the brackets indicate an average taken over all possible microscopic trajectories joining the initial and final states. Now the validity of this equality requires some of the individual W_{irr}'s to be negative although on the average, $< W_{irr} > \geq 0$. Clearly, these properties are manifestations of, respectively, microscopic level reversibility and macroscopic level irreversibility both of which conspire to produce result (6.17) and are otherwise mutually perfectly compatible. Interestingly, for those trajectories for which $W_{irr} < 0$ the system can perform work by exploiting ("rectifying" in the terminology of Sec. 6.2.1) the presence of fluctuations.

From the probabilistic point of view, the bulk of Prob(W_{irr}) is in the region of positive values of W_{irr}. But to comply with (6.17) it must comprise a tail in the range of negative W_{irr}'s. In a macroscopic system this tail can be shown to be associated with probability values that decrease exponentially with the system size. But in a small system its effects will be of importance. In particular, they will induce deviations of Prob(W_{irr}) from a Gaussian form -which will emerge as the limiting distribution for small fluctuations- and, more generally, they will be responsible for its asymmetric structure.

At first sight eq. (6.17) appears to be more general than (6.15), since it does not require the strong instability properties needed to establish this latter relation. On the other hand eq. (6.17) relies on the formalism of equilibrium statistical mechanics which also imposes conditions, albeit milder ones, on the system at hand and its environment. Furthermore, it overlooks what is happening between the stage where the system has just reached the value λ_2 of the control parameter through a nonequilibrium pathway and the final stage of equilibration with the environment.

6.2.3 Transport in a fluctuating environment far from equilibrium

In this subsection we show how the peculiar thermodynamic properties of small systems surveyed above can give rise to a number of generic phenomena providing the explanation of the functioning of biological energy transduction devices and opening the way to some tantalizing perspectives.

The level of description we adopt is, quite naturally, the mesoscopic one (Sec. 2.2). We consider the minimal model of a system described by a single variable $x(t)$, say the position coordinate of a particle embedded in a medium and subjected to an external potential $V(x)$. An example of $V(x)$ is provided by the action of the charged parts of the subunits of a biological macromolecule on an effector moving along the molecule. The potential is assumed to be periodic in x with period L, $V(x+L) = V(x)$, a restriction aimed to account for the closed, finite geometry of the kind of media arising in many problems of interest, including biological ones at the subcellular level. The time evolution of x is then given by Newton's equation

$$m\frac{d^2x}{dt^2} = -V'(x) + \text{friction terms}$$
$$+ \text{ Langevin (fluctuating) forces}$$

where $-V' = -dV/dx$ represents the force acting on the particle and m its mass. Actually in most of the cases of interest -and especially in small systems where m is a small quantity- one may restrict oneself to the "overdamped" case where the friction terms are much larger than the "inertial" terms (the left hand side). In the linear limit where friction is just proportional to the velocity dx/dt this equation reduces then to the familiar Langevin form (eq. (3.7))

$$\frac{dx}{dt} = \frac{-1}{\zeta}V'(x) + R(t) \qquad (6.18)$$

Here ζ is the friction coefficient and $R(t)$ is modeled as a Gaussian white noise as in Sec. 3.2. Notice that in the state of thermodynamic equilibrium the variance of this noise is related to ζ by the fluctuation-dissipation theorem expressing, at our mesoscopic level of description, the time reversible character of the microscopic dynamics.

As seen in Sec. 3.2, eq. (6.18) can be mapped into a Fokker-Planck equation (eq. (3.8)) for the probability distribution $P(x,t)$ of x. This equation can be solved exactly in the steady state ($dP_s/dt = 0$). A very general result of this analysis is that whatever the detailed structure of $V(x)$, in this state there is no net current J (essentially given by the average value of dx/dt) traversing the system: $J_s = <dx/dt>_s = 0$. This result is in many respects natural since, owing to the periodicity of $V(x)$ and the white noise character of $R(t)$, there is no overall asymmetry built on the system. In particular, in writing eq. (6.18) we assumed tacitly that there are no systematic nonequilibrium constraints acting on the system: nonequilibrium states do exist, but only as transients as the system evolves in time to eventually reach its unique stationary state of equilibrium. On the other hand, as it turns out the vanishing of J_s holds equally well for highly asymmetric potentials $V(x)$, for instance in the form of a saw-tooth dependence on x. This may appear counter-intuitive, since for such potentials the positive and negative x directions need a priori not be equivalent. It is customary to call such $V(x)$'s *ratchet potentials*, in analogy with the mechanical rectifying devices that suppress motion in an undesired direction. On the grounds of this analogy one might thus surmise that a ratchet like $V(x)$ could take advantage of the fluctuations favoring a displacement in a certain direction over those favoring the opposite direction. The vanishing of the steady state flux J_s shows that this mechanism does actually not work in equilibrium since in this regime there is no preference between the "forward" and the "reverse" directions on the grounds of microscopic reversibility.

We now augment eq. (6.18) to account for nonequilibrium constraints. This can be achieved in many ways depending on the particular setting and system under consideration, but the simplest one is to add to the equation a systematic force F (arising for instance from an external constant electric field acting on the (charged) particle of interest):

$$\frac{dx}{dt} = -\frac{1}{\zeta}(V'(x) - F) + R(t) \qquad (6.19a)$$

This relation keeps the structure of eq. (6.18) by incorporating the original potential and the constant external force F into an effective potential

$$V_{eff} = V(x) - xF \qquad (6.19b)$$

It can be analyzed along the same lines as (6.18). As one could expect from the foregoing discussion the presence of F (the nonequilibrium constraint) breaks the $x \to -x$ symmetry and this gives rise now to a non-vanishing flux in the steady state, $J_s = <dx/dt> \neq 0$. In the limit of a weak force the flux-force relation is a straight line passing through the origin ($J_s = F = 0$), but this symmetry is broken in the case of large F's (nonlinear response). Whatever the regime, however, the flux is aligned to the direction on which F is acting.

A more surprising observation is that keeping F constant and monitoring the flux as a function of certain other parameters, may lead to a crossover regime in which there is an inversion of the flux. A fairly general mechanism for this to happen is to allow for the possibility that the variance of the random force R in eq. (6.19a) is a function of time, say a periodic one of period τ. In physical terms this means that the medium in which the system of interest is embedded is not in a state of equilibrium of constant temperature T, but is externally driven in a way that the system is unable to equilibrate with it. Under these conditions one can show that there exists a threshold value τ_c where J_s becomes inverted. This property, to which one may refer as "fluctuation induced transport", is not potential independent: it can be materialized only for certain classes of asymmetric (ratchet) potentials. A similar inversion effect arises when nonequilibrium is induced by modulating periodically or randomly the potential V or the force F ("flashing" and "rocking" ratchets, respectively).

In summary, two ratchet like devices that have quite similar properties may operate in opposite directions. This possibility is of special interest in connection with the classes of small systems referred as *molecular motors*, to which we now briefly turn.

A molecular motor is a system where fluctuation-induced transport is achieved by coupling the dynamics of the "mechanical" variable x introduced above to a chemical reaction proceeding far from equilibrium. What is happening, basically, is that the coupling induces an effective modulation of a reference (ratchet) potential $V(x)$, which now becomes dependent on the particular state in which the reaction is found at a given moment. This confers to the overall system the character of a flashing ratchet.

More technically, the setting is similar to eq. (6.19) but the form of the potential $V(x)$ is now changing depending on the chemical state m ($m = 1, ..., M$) of the motor. In parallel with the evolution of x the motor performs transitions between the different m's, whose rates $k_{m \to m'}(x)$ are depending on the mechanical variable x. It is this double dependence of V on m and of k on x that is at the origin of the above mentioned coupling.

A classic example of molecular motor is the kinesin enzyme. The shape of this macromolecule is rather elongated (about 100 nm in its longest direction and 10 nm in the two other ones). One of its ends consists of two heads each of which can bind an adenosine triphosphate (ATP) molecule and contains at the same time a site connecting kinesin to a network of intracellular polymer filaments known as microtubules. As for its other end, it has a fork like shape capable of seizing a particular lead to be carried out along the microtubules for such purposes as replication, transcription and repair of DNA and the translation of RNA. While the head is attached to the microtubule, a chemical reaction can occur in which ATP is hydrolyzed to produce adenosine diphosphate (ADP) and inorganic phosphate (P_i). The energetics of a living cell is such that the level of intracellular ATP (from which the ATP bound to the kinesin is drawn) and that of ADP (to which the ADP produced by the hydrolysis is released) differ by several orders of magnitude from the values they would have if the reaction occurred in equilibrium. There is, therefore, practically no chance that the inverse process of ADP to ATP conversion could take place. Because of this there is a net energetic gain of about 20 times the energy $k_B T$ of the random thermal motion per reaction step, which enables the motor to proceed along the microtubules - typically, by steps of 10 nm or so. In the previously introduced terminology the ATP hydrolysis is associated with the transition between the different chemical states m, and the mechanical variable x with the position of the motor along the microtubules. Finally, the reference potential $V(x)$ arises from the interactions between the head and the microtubules and is locally modulated by the presence of the products of the ATP hydrolysis.

It should be stressed that in addition to their minute size, molecular motors differ from conventional ones used in everyday life and in industrial processing in other significant respects as well. They can operate isothermally, whereas most of the familiar energy producing devices involve significant temperature variations arising from combustion. In addition their yield is strictly vanishing unless the crucial transitions between the different chemical states take place in an irreversible way, contrary to thermal engines where, on the contrary, the efficiency is evaluated in the idealized limit where the successive steps occur slowly and can thus be assimilated to reversible transformations (Carnot cycles, etc.).

6.3 Atmospheric dynamics

In Sec. 1.4.2 we summarized a number of facts highlighting the variability of atmospheric dynamics over a wide range of space and time scales, and the difficulty to make long-term predictions in view of the growth of small errors. In the present section these phenomena are revisited in the light of the concepts and tools developed in the preceding chapters many of which were in fact originally inspired from studies on the physics of the atmosphere, one of the most authentic prototypes of complexity.

There exists a whole hierarchy of atmospheric models, each tailored to respond to specific needs such as desired resolution, running time limitation, and space or time scales of the phenomenon to be described. These models can be traced back, in one way or another, to the classical laws of mass, moisture, momentum and energy balance complemented by a set of "diagnostic" relations specifying how the pressure and the moisture, momentum and heat sources and sinks depend on the variables retained in the description. In addition, other variables may be introduced for particular applications such as the concentrations of ozone, carbon dioxide and other trace constituents.

The equations describing the above processes are nonlinear partial differential equations of the first order in time. They therefore correspond to a dynamical system possessing an infinite number of degrees of freedom (cf. Sec. 2.2.2). In all practical applications they are, however, reduced to a set of ordinary differential equations either by discretizing space on a finite grid or by expressing the various fields in terms of truncated expansions in a function basis appropriate to the geometry of the problem under consideration (e.g. Fourier series or spherical harmonics).

The first step in the process of forecasting consists in identifying the phase space point that represents most adequately the initial condition available from observation (initialization). The next step is to calculate numerically, by additional discretization in time, the trajectory of the dynamical system in phase space (model integration). To reach a high spatial resolution one includes the maximum number of degrees of freedom compatible with the computing power available. This brings the complication that the essential traits may be masked by a multitude of secondary fluctuations. Furthermore the cost of a given integration limits considerably the number of trajectories that one may compute. This precludes reliable statistical analysis or a systematic exploration of the behavior in terms of the parameters.

6.3.1 Low order models

The difficulties accompanying the integration of large numerical models of the atmosphere have prompted some investigators to explore another approach, in which one tries to reduce systematically the number of degrees of freedom while preserving at the same time the structure of the underlying equations. The merit of such simplified models is to concentrate on the most important qualitative phenomena; to allow for systematic exploration of the behavior in terms of the parameters; and to enable statistical analysis by generating, within reasonable time limits, a large number of phase space trajectories.

One of the most famous low-order models, which actually has been at the origin of modern ideas on deterministic chaos, is a simplified model of

Selected Topics 219

thermal convection developed by Edward Lorenz. The phenomenology of thermal convection under laboratory conditions was surveyed in Sec. 1.4.1. But thermal convection is ubiquitous in the atmosphere as well, where it constitutes one of the major sources of variability on space scales of the order of a few kilometers and plays an instrumental role in the formation and further evolution of clouds. Lorenz's model is a limiting form of the mass, momentum and energy balance equations appropriate to a shallow fluid layer. Furthermore, the velocity and temperature fields are expanded in Fourier series, keeping one Fourier mode for the vertical component of the velocity and two Fourier modes for the temperature variation. One arrives then at the equations:

$$\frac{dx}{dt} = \sigma(y - x)$$
$$\frac{dy}{dt} = rx - y - xz$$
$$\frac{dz}{dt} = xy - bz \qquad (6.20)$$

The variable x measures the rate of convective (vertical) turnover, y the horizontal temperature variation and z the vertical temperature variation. The parameters σ and r are proportional, respectively, to the Prandtl number (depending entirely on the intrinsic properties of the fluid) and to the Rayleigh number, a dimensionless parameter proportional to the thermal constraint ΔT introduced in Sec. 1.4.1. Finally, the parameter b accounts for the geometry of the convective pattern. In addition to reproducing the thermal convection instability for $r \geq 1$, eqs (6.20) reveal a surprisingly rich variety of behaviors and have been used extensively to illustrate the onset of deterministic chaos in systems involving few variables.

Another interesting family of low-order models stems from a limiting form of the full balance equations in which the different fields are averaged vertically. The reduced variables, which now depend only on the horizontal coordinates, are expanded in orthogonal functions. Furthermore the momentum balance is approximated by the balance between the Coriolis force, accounting for the earth's rotation around its axis, and the atmospheric pressure gradient. The equations assume the general form

$$\frac{dx_i}{dt} = \sum_{j,k} a_{ijk} x_j x_k + \sum_j b_{ij} x_j + C_i \qquad (6.21)$$

A highly truncated version which has received much attention is

$$\frac{dx}{dt} = -y^2 - z^2 - ax + aF$$

$$\frac{dy}{dt} = xy - bzx - y + G$$

$$\frac{dz}{dt} = bxy + xz - z \qquad (6.22)$$

in which x denotes the strength of the globally averaged westerly current, y and z are the strength of the cosine and sine phases of a chain of superimposed waves. The unit of the variable t is equal to the damping time scale of the waves, estimated to be five days. The terms in F and G represent thermal forcings: F stands for the cross-latitude heating contrast, whereas G accounts for the heating contrast between oceans and continents. Finally, parameter b stands for the strength of the advection of the waves by the westerly current.

A study of system (6.22) along the lines of the previous chapters reveals the existence of complex dynamics for a wide range of parameter values. Figs 6.4 depict the time series of the variable x, and a 2-dimensional projection of the phase space portrait of the attractor. A finer analysis confirming the chaotic character of the process leads to a maximum Lyapunov exponent $\sigma_{max} \approx 0.22$ time units^{-1} for the parameter values of the figure. One can show that chaos arises in this model from a complex sequence of bifurcations, in particular, bifurcations of homoclinic loops, that is to say, trajectories that are biasymptotic to a fixed point as t tends to $+\infty$ and $-\infty$.

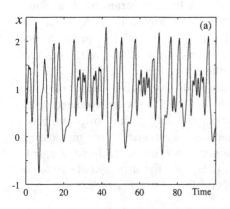
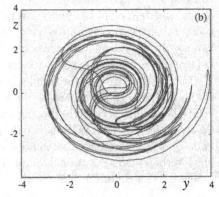

Fig. 6.4. Time evolution of variable x, (a) and 2-dimensional projection of the attractor on the yz plane, (b) of the dynamical system of eqs (6.22). Parameter values are $a = 0.25$, $b = 4$, $F = 8$ and $G = 1.25$.

In summary, models (6.21) and (6.22) provide some generic mechanisms for the existence of deterministic chaos in the atmosphere, thereby constituting a basis for understanding the origin of the phenomenon of sensitivity to the initial conditions and the well-known difficulties (see also Sec. 5.5) to issue reliable forecasts beyond a certain time period. For model (6.22) this period is of the order of the inverse Lyapunov exponent which, as we saw, is 0.22 time units^{-1}. Since the time unit chosen corresponds to about 5 days this leads us to a predictability horizon of about 3 weeks. This number is somewhat higher than -but of the same order as- the predictability horizon admitted by the practitioners in the field.

Low order models can also account for large scale phenomena related to the global circulation, when one incorporates in the balance equations information pertaining to topography (mountains and valleys). When thermal effects are discarded and, as before, an average over the vertical direction is performed one arrives at a model of the form

$$\frac{dv_r}{dt} = kv_i(u_0 - \beta/2k^2) - Cv_r$$
$$\frac{dv_i}{dt} = -kv_r(u_0 - \beta/2k^2) + f_0\frac{\tilde{h}}{2H}u_0 - Cv_i$$
$$\frac{du_0}{dt} = -f_0\frac{\tilde{h}}{H}v_i - C(u_0 - u_0^*) \qquad (6.23)$$

that goes back to the pioneering work of Julius Charney. Here v_r, v_i are, respectively, real and imaginary parts of the first Fourier mode of the gradient of a quantity ψ known as *stream function*, related to the horizontal wind components u and v by $u = \partial\psi/\partial y, v = -\partial\psi/\partial x$; u_0 is the velocity of the zonal mean flow that does not depend on y; f_0 is a mid-latitude value of the Coriolis parameter; β is a constant arising from the assumption that the Coriolis parameter depends linearly on y; C is a damping parameter; H is a length scale ($10^4 m$); $\tilde{h}$ is the first Fourier mode of the orographic profile, u_0^* stands for the forcing of the mean flow, and k is the wavenumber of the flow.

Equations (6.23) predict multiple steady-state solutions in the range of intermediate wavelength scales $L(\sim 3 \times 10^6 m)$. The great interest of this discovery is to provide a qualitative explanation of the phenomenon of persistent flow regimes in mid-latitudes, also referred as "blocking", which constitute one of the principal elements of low-frequency atmospheric variability (time scale of weeks or so). Their main characteristics are a preferred geographic localization (in contrast to the familiar "zonal" flows), a certain persistence, a tendency to recur (see also Sec. 5.6), a variable lifetime, and a rapid transition toward the zonal regime.

6.3.2 More detailed models

In view of the huge societal and economic impact of the weather, atmospheric forecasts cannot be limited to the qualitative features provided by low order models. Currently, forecasts are issued on the basis of very detailed models involving fine scale variables. Their number (of the order of tens of millions) is only limited by the present day computer technology and the availability of data that need to be fed into the model integration scheme when the spatial mesh becomes finer and when short time processes are incorporated. A tempting idea is that as both of these limitations are gradually relaxed one will eventually be in the position to deliver more and more reliable forecasts valid over increasingly long times.

One of the best known and most widely used operational forecasting models of this kind is the model developed at the European Centre for Medium Range Weather Forecasts (ECMWF), designed to produce weather forecasts in the range extending from a few days to a few weeks. This involves daily preparation of a N-day forecast of the global atmospheric state (typically, $N=10$ days), using the present day's state as the initial condition. Since the equations are solved by stepwise integration, intermediate range (1, 2, ...) day forecasts are routinely achieved as well. Capitalizing on the fact that the model produces rather good 1-day forecasts, one expects that the state predicted for a given day, 1 day in advance, may be regarded as equal to the state subsequently observed on that day, plus a relatively small error. By comparing the 1- and 2- day forecasts for the following day, the 2- and 3- day forecasts for the day after, and so on, one can then determine upon averaging over several consecutive days how the mean error evolves in time.

The result for the period 1982-2002 (in steps of 5 years), summarized in Fig. 6.5, establishes beyond any doubt the presence of error growth and of sensitive dependence on the initial conditions in the atmosphere. The solid line in the figure depicts the *doubling time* of the initial error. In 1982 this value was about 2 days, as determined by Edward Lorenz in a seminal work in which the ECMWF data were first analyzed in this perspective. As can be seen, this value has dropped to about 1.2 days in 2002. This decrease of predictability time appears at first sight as paradoxical, since during this 20-year period one has witnessed significant technological advances such as an increase of spatial resolution, a substantial improvement of parameterization schemes and an almost three-fold decrease of initial (1-day forecast) error (dashed line in Fig. 6.5). It reflects the fact that although the accuracy of forecasts for a few days ahead is considerably increasing we may gradually be reaching the limits of predictability in a wider sense: *detailed* forecasting of weather states at sufficiently long range with models of increas-

Selected Topics

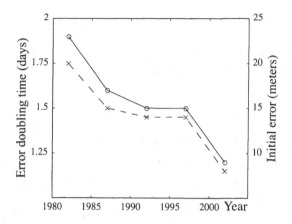

Fig. 6.5. Evolution (in steps of 5 years) of the one-day forecast error in meters (dashed line) and doubling time of the initial error in days (full line) of the 500 hPa Northern Hemisphere winter geopotential height as obtained from the ECMWF operational weather forecasting model.

ing sophistication may prove impractical, owing to the complexity inherent in the dynamics of the atmosphere. Sensitivity to the initial conditions -the principal signature of deterministic chaos- is thus not an artifact arising when low order models are used but is, rather, deeply rooted in the physics of the atmosphere.

6.3.3 Data analysis

A model necessarily provides only a schematic view of reality. In view of the uncertainties inherent in the modeling processes (see also Sec. 5.5) it is, therefore, desirable to dispose of methodologies providing information on the nature of atmospheric dynamics on the sole basis of experimental data. Once available these can also be applied, if appropriate, to the analysis of the output of a mathematical model.

An array of methodologies for tackling this issue was presented in Chapter 5. They all have their limitations, notably in connection with the limited number of data usually available when monitoring a real-world complex system like the atmosphere. Nevertheless, when applied in conjunction with theoretical ideas such algorithms do provide valuable information as they give hints on the minimum number of key variables involved in the dynamics, and on a pragmatic modeling in which irrelevant modes are filtered out. In particular, using the technique of dynamical reconstruction (Sec. 5.3.1) one is led to the conclusion that certain large scale aspects of atmospheric variability may be captured by chaotic attractors of dimensions (depending on the variable involved and the time and space scale considered) varying from 6 to 11, the corresponding predictability times being in the range of up to a few weeks. This latter conclusion adds further credence to the presence of deterministic chaos in the atmosphere.

6.3.4 Modeling and predicting with probabilities

We have stressed repeatedly that the probabilistic approach constitutes the natural tool for tackling complex systems. When implemented on a mathematical model supposed to represent a concrete system, this approach amounts to choosing a set of initial conditions compatible with the available data; to integrate the model equations for each of these initial conditions; and to evaluate the averages (or higher moments) of the quantities of interest over these individual "realizations". In the context of atmospheric dynamics this procedure is known as *ensemble forecasts*. Its principal merit is to temper the strong fluctuations associated with a single realization and to sort out quantitative trends in relation with the indicators of the intrinsic dynamics of the system at hand.

Figure 6.6 illustrates schematically the nature of ensemble forecasts. The dark circle in the initial phase space region $\delta\Gamma_0$ stands for the "best" initial value available. Its evolution in phase space, first after a short lapse of time (region $\delta\Gamma_1$) and next at the time of the final forecast projection (region $\delta\Gamma_2$) is represented by the heavy solid line. Now, the initial position is only one of several plausible initial states of the atmosphere, in view of the errors inherent in the analysis. There exist other plausible states clustered around it, represented in the figure by open circles. As can be seen, the trajectories emanating from these ensemble members differ only slightly at first. But between the intermediate and the final time the trajectories diverge markedly, presumably because the predictability time as defined in Sec. 5.4 has been exceeded: there is a subset of the initial ensemble including the "best" guess that produces similar forecasts, but the remaining ones predict a rather different atmospheric state. This dispersion is indicative of the uncertainty of the forecast. It constitutes an important source of information that would not be available if only the "best" initial condition had been integrated, especially when extreme situations are suspected to take place in a very near future. It should be pointed out that such uncertainties frequently reflect *local* properties of the dynamics such as local expansion rates and the orientation of the associated phase space directions.

The probabilistic approach can also be applied for developing predictive

Selected Topics

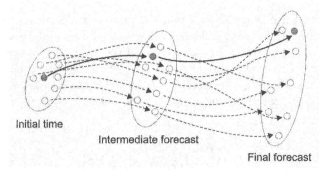

Initial time
Intermediate forecast
Final forecast

Fig. 6.6. Illustrating the nature of ensemble forecasts. The ellipsoids represent three successive snapshots of an ensemble of nearby initial conditions (left) as the forecasting time increases. The full line represents the traditional deterministic single trajectory forecast, using the best initial state as obtained by advanced techniques such as data assimilation. The dotted lines represent the trajectories of other ensemble members, which remain close to each other for intermediate times (middle ellipsoid) but subsequently split into two subensembles (right ellipsoid), suggesting that the deterministic forecast might be unrepresentative (after D. S. Wilks).

models on the sole basis of data. An example of how this is achieved pertains to the transition between atmospheric regimes such as the onset of drought described in Sec. 1.4.2, see Fig. 1.6. The main idea is that to be compatible with such data, the underlying system should possess (as far as its hydrological properties are concerned) two coexisting attractors corresponding, respectively, to a regime of quasi-normal precipitation and a regime of drought. In a traditional modeling the system would choose one or the other of these attractors depending on the initial conditions and would subsequently remain trapped therein. In reality under the influence of the fluctuations generated spontaneously by the local transport and radiative mechanisms, or of the perturbations of external origin such as, for instance, surface temperature anomalies, the system can switch between attractors and change its climatic regime by the mechanism of transitions outlined in Sec. 3.5.2. This is at the origin of an intermittent evolution in the form of a small scale variability around a well-defined state followed by a jump toward a new state, which reproduces the essential features of Fig. 1.6. The idea can of course apply to a host of other problems, including the transition between the zonal and blocked atmospheric flows described in Sec. 6.3.1 (eqs (6.23)).

In their quantitative form the models belonging to this family appear in the form of Langevin equations and the associated Fokker-Planck equations, if there is evidence from the data that the process is Markovian. The model allows then to calculate the residence time of the system in the neighborhood of one of the attractors, and hence to predict the duration of a drought. The main tool for this is the property that the mean residence time τ_1 and its successive moment $\tau_2, \tau_3, \cdots$ satisfy a relation known as the backward Fokker-Planck equation involving the adjoint of the Fokker-Planck operator in eq. (3.8),

$$f(x)\frac{d\tau_n}{dx} + \frac{q^2}{2}\frac{d^2\tau_n}{dx^2} = -n\tau_{n-1} \qquad (6.24)$$

complemented by appropriate boundary conditions. Applied to the data of Fig. 1.6 this method leads to a mean drought lifetime much longer than the regime of quasi-normal rainfall. A very important point is that the dispersion around these average values is quite large, a property that is actually generic for all transitions requiring the system to overcome a "barrier" separating two stable states. This complicates further the task of making reliable predictions.

In certain instances the Markovian assumption is contradicted by the data. Usually this is manifested by a significant deviation of the distribution of residence times from the exponential form, which is characteristic of a Markov process. Such an information may be available if the data cover a large number of transitions as is often the case for transitions between zonal and blocked flows. Modeling is subtler in these non-Markovian cases, since there exists no universal equation describing such processes. On the other hand one may be in the position to simulate the process directly by a Monte Carlo type approach, on the basis of the probability of occurrence of each of the states available and of the residence time distributions (cf. Sec. 5.3.2). This can constitute, again, the starting point of statistical predictions replacing traditional methods of prediction whose validity is questionable in the presence of abrupt transitions.

6.4 Climate dynamics

The data surveyed in Sec. 1.4.2 suggest that a reasonable model of the climatic system should be able to account for the possibility of large-scale transitions between glacial and interglacial regimes and of a pronounced variability over intermediate time scales.

Ordinarily, predictions on future climatic trends are carried out on the basis of detailed numerical models, referred as *general circulation models* (GCM's). The latter derive from weather forecasting models in which the spatial grid is coarser and additional couplings and parameterizations are incorporated to account for the role of the cryosphere (sea ice and continental ice sheets), biosphere, etc. For reasons similar to those invoked in Sec. 6.3.1 one is tempted to inquire whether in the context of climate dynamics one can also set up models capturing the essence of the physics involved, while amenable at the same time to a systematic qualitative and quantitative analysis.

6.4.1 Low order climate models

Climatic change essentially reflects the energy and mass balance of the planet. Energy balance describes how the incoming solar radiation is processed by the system: what parts of that spectrum are absorbed and what are re-emitted, how they are distributed over different latitudes and longitudes as well as over continental and ocean surfaces. The most significant parts of mass balance as far as long term climate is concerned are the balance of continental and sea ice and of some key atmospheric constituents like water vapor and carbon dioxide.

As a case study, let us argue in very global terms, considering the earth as a zero-dimensional object in space receiving solar radiation and emitting infrared radiation back to space. In such a view there is only one important state variable, the mean temperature T, which evolves in time according to the heat balance equation

$$\frac{dT}{dt} = \frac{1}{C}\left[Q\left(1-a(T)\right) - \epsilon_B \sigma T^4\right] \qquad (6.25)$$

Here C is the heat capacity of the system, Q the solar constant, and a the albedo, which expresses the part of the solar radiation emitted back to space. The temperature dependence of this quantity accounts for the "surface-albedo feedback", namely, the enhancement of reflectivity at low temperatures owing to the presence of continental and/or sea ice. The last term in eq. (6.25) describes the emitted thermal radiation as a modification of the well-known Stefan-Boltzmann law of black body radiation. σ is the Stefan constant and ϵ_B is an emissivity factor accounting for the fact that the earth does not quite radiate as a black body.

A study of eq. (6.25), summarized in Fig. 6.7, shows that the system can admit up to three steady states. Two of them, T_+ and T_-, are stable and correspond, respectively, to the present-day climate and a cold climate reminiscent of a global glaciation. The third state, T_0 is unstable and separates the above two stable regimes.

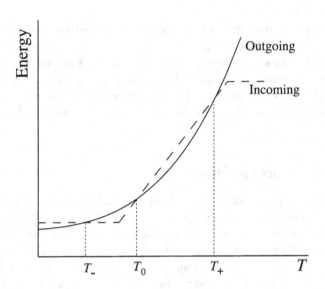

Fig. 6.7. Incoming $[Q(1-a(T))]$ and outgoing $[\epsilon\sigma T^4]$ radiative energy curves as functions of T (global averaged temperature). For plausible parameter values their intersection gives rise to three steady states: T_-, T_0 and T_+.

The evolution of a system involving only one variable can always be cast in a form involving a potential U (cf. Sec. 2.4) which in the present case is defined by

$$U(T) = -\frac{1}{C}\int dT\,[Q(1-a(T)) - \epsilon_B\sigma T^4]$$

We call $U(T)$ the *climatic potential*. Under the conditions of Fig. 6.7 this potential has two wells separated by a maximum.

Now the climatic system is continuously subjected to statistical fluctuations, the random deviations from deterministic behavior inherent in any physical system, which essentially account for the error committed in replacing the full dynamics by an equation involving averages only. As we do not have sufficient information enabling us to write an evolution equation for the underlying probability distribution, we assimilate the effect of the fluctuations to a random force, $F(t)$ descriptive of a Gaussian white noise. The energy balance equation, eq. (6.25), now becomes a stochastic differential equation of the form

$$\frac{dT}{dt} = -\frac{\partial U(T)}{\partial T} + F(t) \qquad (6.26)$$

The important new element introduced by this enlarged description is that different states become connected through the fluctuations. Stated differently, starting from some initial state the system will sooner or later reach any other state. This holds true as well for the stable states T_+ and T_- taken as initial states, whose deterministic stability is then transformed to some sort of metastability. As seen in Sec. 3.5.2, the time scale of this phenomenon is determined by two factors: the potential barrier $\Delta U_\pm = U(T_0) - U(T_\pm)$ and the strength of the fluctuations as measured by the variance q^2, of $F(t)$. The order of magnitude estimate is given by essentially the same arguments as those underlying Kramer's theory of chemical reactions in liquid phase. The mean transition time form state T_+, or T_- via the unstable state T_0, is then

$$\tau_\pm \approx e^{\frac{\Delta U_\pm}{q^2}} \qquad (6.27)$$

It is clear that if fluctuations are reasonably small this time will be exceedingly large for any reasonable magnitude of the potential barrier. Typical values can be 10^4 to 10^5 years, bringing us the range of the characteristic times of Quaternary glaciations. Still, we cannot claim that eq. (6.27) provides an explanation of glaciation cycles, as the passage between the two climatic states T_+ and T_- remains a random process, without a preferred periodicity.

Let us now repeat the above analysis by adding to the solar constant Q a small term of the form $Q\epsilon \sin \omega t$, whose period $\Delta = 2\pi/\omega$ accounts for phenomena affecting the mean annual amount of solar energy received by the earth, such as the change of the eccentricity of the earth's orbit around the sun because of the perturbing action of the other bodies of the solar system. The effect of these variations on the energy budget is known to be very small, of the order of 0.1 percent. How this small effect may be amplified to leave a clearcut signature in the glaciation record is a major problem of climatology. As it turns out, if Δ is much smaller than the transition time $\tau_\pm$ the external forcing has, indeed, practically no effect. But when Δ is comparable to $\tau_\pm$ the response of the system to the forcing is tremendously enhanced. Essentially, the presence of the forcing lowers the potential barrier and in this way facilitates the passage between climatic states. Moreover, because of the periodicity of the signal, the system is entrained and locked into a passage between two kinds of climate across the forcing-depressed barrier characterized by some average periodicity. We see here a qualitative mechanism of glaciations in which both internally generated processes and external factors play an important role. The simple model of *stochastic resonance* discussed above allows us to identify the parameters that are likely to be the most influential in the system's behavior.

Naturally, the schematic character of the model calls for the development of more sophisticated descriptions which, while still amenable to an analytical study, incorporate additional key variables in the description. A particularly interesting class of such models comprises those describing the interaction between average surface temperature and continental or sea ice. For plausible values of the parameters they predict time-periodic solutions of the limit cycle type with periods comparable to the characteristic times of glaciations. However, by incorporating the effect of the fluctuations in the description, one finds that coherence in the oscillating behavior will disappear after a sufficient lapse of time, owing to the weak stability properties of the phase of the oscillator. The coupling with a periodic external forcing allows under certain conditions for the stabilization of the phase and thus, once again, for a regime of entrainment and phase locking between the climatic system and the external forcing. Furthermore, there exists a range of values of the amplitude and period of the forcing for which the response is chaotic, exhibiting a variability much like the one shown in Fig. 1.5.

6.4.2 Predictability of meteorological versus climatic fields

Climate is, by definition, the set of long-term statistical properties of the atmosphere. Our experience shows that the quantities associated to this average description display new properties as compared to those of the fine scale variables. In this subsection we outline an approach for deriving these properties which in its conceptual basis goes actually beyond the atmosphere-climate context and applies to a wide class of problems, in which different levels of description prove to be necessary to capture the complexity of the observed behaviors.

Let $\{X_i\}$ be a set of atmospheric short scale variables evolving according to the basic balance equations as alluded in Sec. 6.3.1. We define the averaged variables associated to climate as running means over a time interval ϵ whose merit as compared to ordinary (infinite time) averaging is to keep track of the evolution that the averaged variables $\{\bar{X}_i\}$ might still undergo in time and space. As an example, the running mean of X_i over time will be given by

$$\bar{X}_i = \frac{1}{\epsilon} \int_t^{t+\epsilon} dt' X_i(t') \qquad (6.28)$$

A typical value of ϵ in climate theory is of the order of $10\,yrs$. For simplicity we consider here asymmetric averaging, to avoid the appearance of spurious negative time values in the integrand. We now apply the averaging operator on both sides of a typical evolution equation (eq. (2.2)),

$$\frac{d\bar{\mathbf{X}}}{dt} = \overline{f(\mathbf{X}, \lambda)} \qquad (6.29)$$

Clearly, in the limit in which the right hand side of (6.29) could be linearized for all practical purposes around a well-defined time independent state, the properties of the averaged observables would follow straightforwardly from those of the initial variables. The situation is far more complex in the presence of nonlinearities since now $\overline{f(\mathbf{X})} \neq f(\bar{\mathbf{X}})$: in other words, to determine the evolution of $\bar{\mathbf{X}}$ one needs to know not only $\bar{\mathbf{X}}$ but also the entire hierarchy of moments $\overline{\delta \mathbf{X}^n} = \overline{(\mathbf{X} - \bar{\mathbf{X}})^n}$, $n > 1$. It follows that, in principle, there should be a strong feedback between local, short scale properties and global averages, unless the moments $\overline{\delta \mathbf{X}^n}$ could be neglected for $n \geq 2$. This is definitely not the case for a typical atmospheric variable where one witnesses large deviations around the mean, comparable to the value of the mean itself. More generally, this should hold true in any nonlinear dynamical system in which the evolution equations admit multiple solutions or give rise to chaotic dynamics. The lack of commutation of the averaging operator with the evolution law constitutes therefore a fundamental issue at the basis of the complex relation between short scale "weather" variability and global climate.

The traditional way to cope with this difficulty is to perform a truncation of the hierarchy to the first few orders or, more generally, to appeal to a *closure* relation linking high order averages to lower order ones. Such closures amount essentially to introducing phenomenological, "diagnostic" type of relations between averaged momentum, energy, etc. fluxes and averaged observables such as temperature, and/or their spatial derivatives. They are difficult to justify from first principles, even in very simple cases. Nevertheless, they are widely used in climate studies where they are responsible for the great number of parameters appearing in numerical models of climate prediction, which need to be fitted from observational data. The sensitivity of the model's response to the values chosen becomes, then, one of the properties to be checked before assessing the predictive value of the model at hand.

One of the merits of the low order atmospheric models introduced in Sec. 6.3.1 is to allow for a systematic analysis of this question. The principal results are as follows.

(i) The size and the effective dimensionality of the attractor associated to the averaged observables are reduced (Figs 6.8). This is in agreement with the idea that averaging erases the fine structure of the original dynamics.

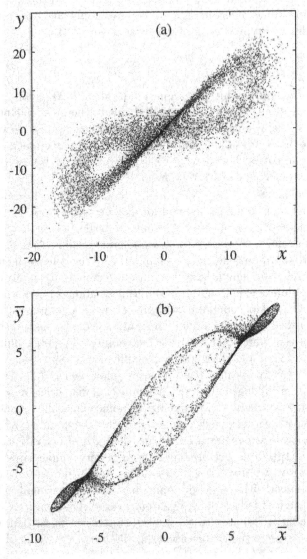

Fig. 6.8. Two-dimensional projection of the attractor associated to the Lorenz thermal convection model, eq. (6.20), on the xy plane. Parameter values $\sigma = 10$, $b = 8/3$ and $r = 28$. (a) attractor in the original phase space; (b) attractor in the averaged variables $\bar{x}$, $\bar{y}$ with averaging window $\epsilon = 0.8$ time units.

(ii) The correlation time of the averaged observables is enhanced and the associated power spectrum is more organized.

(iii) The Lyapunov exponents of the original and averaged systems are formally identical, since the associated variables are related by a linear transformation. Nevertheless, the averaged observables are practically more predictable in the sense that initial errors limited to the subspace of these observables grow more slowly and their saturation levels are depressed. They still undergo, however, an unstable dynamics unless of course the (unrealistic) infinite time averaging is performed. This fact is often overlooked in climate studies where it is tacitly assumed that in the absence of anthropogenic forcing climate is in a unique, essentially stable configuration and that in the presence of a forcing a comprehensive knowledge of past history and of the forcing suffices to deduce reliably the system's response for all subsequent times.

Paleoclimatic time series similar to those of Fig. 1.5 have also been analyzed by the methods of dynamical reconstruction (Secs 5.3.1 and 6.3.3). The investigations suggest the existence of low-dimensional chaotic attractors over the long time scales underlying the global (averaged) information contained in the data of Fig. 1.5. The instability of motion on such a chaotic attractor may well provide the natural mechanism for the amplification of the weak external signals arising from the earth's orbital variability. Furthermore, as we saw in Chapters 3 and 4, when viewed through an appropriate window, a chaotic attractor can be mapped into a stochastic process. In this respect, one may regard the fluctuations responsible for the transitions in the model of Sec. 6.4.1 as a shorthand description of the variability of an underlying chaotic attractor.

6.4.3 Climatic change

In the long run and as long as the principal parameters of the climatic system remain invariant, the earth radiates as much energy back into space as it receives from the sun. Even so, as seen in this section and in Sec. 6.3, the atmosphere and the climate undergo complex dynamics giving rise to chaotic behavior and/or to a multiplicity of states. Because of this the forecasting of atmospheric and climate dynamics is subjected to irreducible limitations.

This realization comes in a time of considerable technological, societal and political pressure to improve the period over which the future states of the atmosphere and climate can be reliably predicted. This pressure, which would already be understandable under normal conditions, is further increased by the suspicion to which we already alluded in Sec. 1.4.2, namely, that some of the parameters of the climatic system are actually not remaining invariant.

As a result of these variations our climate could change dramatically over the next decades instead of fluctuating over a range determined by the nature of its attractor. It is generally admitted that the primary responsible for effects of this kind are the "greenhouse gases" like carbon dioxide, which regulate the transfer of energy within the atmosphere and control the release of heat back to space. An increase in the atmospheric CO_2 arising from man's industrial activities -a well documented fact over the last decades- may be expected to disturb the overall radiative equilibrium and bring about an increase in the global average surface temperature. Whence the major question, are we heading toward a global warming?

Complex systems research can contribute to this major debate in many respects. On the modeling side it must be realized that sophisticated, million variable numerical models run on the most powerful supercomputers acquire their full interest and significance only when a qualitative insight of the nature of the system at hand has been reached. Long-term climatic forecasts, reported routinely, based on a static formulation in which the system adjusts to a "quasi steady-state" level corresponding to the instantaneous value of the relevant parameters are, clearly, inadequate. They must be supplemented by studies on the system's dynamical response, capable of taking into account the entire repertoire of possible behaviors and formulated, whenever necessary, in probabilistic terms. Furthermore, the very possibility of intrinsically generated complexity makes the distinction between "normal state" and "fluctuation" deduced from models much subtler than usually thought, thereby placing the distinction between natural variability and externally induced systematic change on a completely new basis. For instance, at the peak of the last glaciation, CO_2 levels were considerably lower than the pre industrial level, whereas during the Cretaceous period (about 100 million years ago) they were substantially above those existing today. On a shorter time scale, while during the first half of the 20th century there was a warming trend, between the 1940's and the mid 1970's relatively stable (and perhaps even slightly cooler) conditions have been reported. Most importantly, it makes no sense to push a large numerical model, with all its inherent uncertainties arising from heavy parameterizations, beyond its own natural (and as yet largely unexplored) predictability limits.

Complex systems research can also contribute to the physics of climatic change through the elaboration of generic evolutionary scenarios based on low-order models, in which the control parameters are now taken to depend on time in a way to fit the available data. The main conclusions of the studies conducted so far can be summarized as follows.

(i) At the level of a purely macroscopic, mean field analysis a systematic slow variation of a control parameter can enhance the stability of a state that would otherwise become unstable beyond a certain threshold, thereby postponing the occurrence of a bifurcation.

(ii) Let now the above description be augmented to account for the fluctuations. The transition, if any, is then "anticipated" by an increase of the standard deviation of the fluctuations of the monitored variable around the reference state that would otherwise prevail.

(iii) Under certain conditions the transition may be quenched altogether as the system becomes "frozen" for all practical purposes in a state that would otherwise not be sustained. This is due to the difficulty to overcome through the action of the fluctuations the increasingly large barrier separating this state from what should normally be the "dominant" state. This complicates further the possibility of reliable long term predictions.

On the monitoring side, the idea that the underlying system is a complex, highly unstable dynamical system, should have repercussions on the nature of data to be collected and their sampling frequency. Furthermore, it is legitimate to inquire whether certain essential features of atmospheric dynamics, notably those associated with an irregular distribution of an observable in space and time, may be adequately reconstituted from the current global observational network. It is well known that this network has been extremely sparse in the past. The disparity between the past situation and the present-day one imposes severe constraints on our ability to reliably detect small-amplitude climatic changes, such as temperature increases since the preindustrial period. Even today, despite impressive effort toward improving this state of affairs, the global observational network remains highly inhomogeneous and fractally distributed, a feature that may be responsible for missing events that are sparsely distributed in space, as, for instance, extreme events. For a given (finite) number of stations it is therefore important to seek for an *optimal distribution*, if any, for which errors are minimized. This brings us to the analysis of Sec. 5.4.

All in all, the new insights afforded by viewing atmospheric and climate dynamics from the standpoint of complex systems research is likely to generate new ideas and new tools that could revitalize and reorient the traditional approach to earth and environmental sciences in a healthy manner.

6.5 Networks

As discussed in Sec. 2.6, much of the specific phenomenology of complex systems can be traced in one way or the other to the coupling between their constituent elements. This coupling reflects the fact that two or more elements participate jointly in certain processes going on in the system, and can be materialized in different ways: direct interactions, e.g. electrostatic interactions between charged groups along two macromolecular structures (cf. Sec. 6.2.3), or cell-to-cell and individual-to-individual contacts in a population; spatial couplings on a scale longer than the range of intermolecular interactions, e.g. transport of mass, energy or information between different spatial regions in a spatially inhomogeneous system; finally, through purely functional relationships subsisting in the limit of a spatially homogeneous system, e.g. activation or inhibition processes at the molecular, genetic or population level.

In each of the above cases, upon performing a discretization of space into discrete cells -a procedure that is inherent, in fact, in any method of numerical resolution of the evolution equations or of simulation of the underlying processes- one arrives at a lumped variable representation of the system. As seen throughout Chapter 2, an advantage of such a representation is to embed the dynamics into a finite dimensional phase space, which constitutes the starting point of a qualitative as well as quantitative characterization of the system's complexity. In this section we focus on another aspect of the lumped variable representation, namely, the mapping of the system into a *network*. An early example of such a view is provided by electrical circuits - the standard engineering approach to electromagnetic phenomena. Currently one witnesses a renewed interest in networks following the realization that they are ubiquitous in a wide spectrum of other disciplines as well, from information science to biology, linguistics and sociology.

6.5.1 Geometric and statistical properties of networks

Networks as introduced above are graphs, whose nodes represent the elementary units constituting the system of interest and the links connecting the nodes stand for the effect of the interactions between the corresponding units on their subsequent evolution (Fig. 6.9).

There exists a powerful mathematical theory of graphs, dealing with different ways to characterize their structure. A most important indicator of the network structure is the *degree distribution* $P(k)$, defined as the probability that k links are merging at a randomly selected node among the N nodes of the graph. A second one is the *clustering coefficient*, the ratio between the number of links L_i existing between a set of k_i nodes connected to a given node i and the maximum number $k_i(k_i - 1)/2$ of possible links: $C_i = 2L_i/[k_i(k_i - 1)]$. In many problems it is also of interest to estimate the "diameter" ℓ of the network as a function of its size N, measured by the maximal distance along pairs of connected nodes. A more algebraic approach to graphs is based on the *adjacency matrix* $\mathbf{W}$, an $N \times N$ matrix

Selected Topics

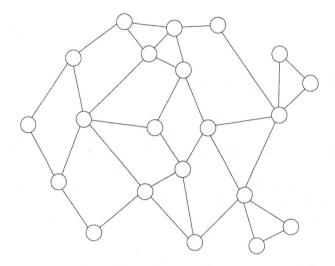

Fig. 6.9. Graph representation of a network.

whose elements W_{ij} are equal to unity if nodes i and j are connected and zero otherwise. The eigenvalues of this matrix provide useful information on the response of the underlying system to external disturbances and can also be linked to its topological properties.

All of the above quantities can be determined for the subclass of graphs known as random graphs whose theory, pioneered by Paul Erdös and Alfrèd Rényi, serves as a reference for the analysis of more general graph families. These are graphs whose N nodes are connected by n links chosen randomly from the $N(N-1)/2$ possible links. The mean value of the random variable n is, then, $<n> = pN(N-1)/2$, where p is the probability that a pair of nodes is connected. This turns out to be a key variable, since many properties of interest change suddenly (for large size graphs) when p crosses a critical (N-dependent) threshold p_c. For instance, there exists a p_c below which the network is composed of isolated clusters and above which a cluster spanning the entire network is emerging.

Using simple combinatorial arguments one can convince oneself that the degree distribution of a random graph is given by the binomial distribution. In the limit of large N this tends in turn to the Poisson distribution,

$$P(k) \approx e^{-pN}(pN)^k/k! \qquad (6.30a)$$

Since this function falls off rapidly with k, the vast majority of nodes will have a number of links close to the average value $<k> = pN$. In a similar vein, the clustering coefficient is merely given by

$$C_i = \frac{<k>}{N} = p \tag{6.30b}$$

and the graph diameter behaves as

$$\ell \approx \ln N / \ln <k> \tag{6.30c}$$

This latter result shows that in a random graph of given size paths between connected points tend to be short, a property also referred as the *small-world* property.

The adjacency matrix of a random graph is an object of special interest. Indeed, the distribution of its eigenvalues turns out to fit surprisingly well those encountered in many problems of quantum spectroscopy of large atoms and nuclei and of solid state physics. For this reason, random matrix theory constitutes a useful model for understanding how non-trivial statistical behavior can arise by gradually tuning the coupling (here represented by the connectivity p) between the different subunits involved.

Regular lattices, in which nodes are connected to a limited number of nearest neighbors in exactly the same way throughout, lie on the other end of the spectrum of network topologies as compared to random graphs. Their clustering coefficient is much higher, the degree distribution is just a sum of isolated peaks and their diameter increases much faster with N, $\ell \approx N^{1/d}$ where d is the dimension of embedding space.

To what extent do the networks encountered in nature and in technology fit either of the above two models? To address this question one needs first to treat all connections in such networks on an equal footing. This is not necessarily appropriate: a "passive" relation between nodes i and j is, clearly, fundamentally different from a positive or negative feedback; and likewise for linear versus nonlinear dependencies. But if we accept this idea despite its limitations, the answer turns out to be far from clearcut. There is no doubt that nearly regular networks do exist in the real world, for instance, in materials science (crystalline lattices, etc.) and in situations where connections are ensured by transport processes of physico-chemical origin like diffusion (reaction-diffusion systems, etc.). Random, or nearly random, networks have also been used successfully to account for key features of disordered materials or of situations in which environmental noise plays a dominant role in establishing connections between units having nearly identical affinities for each other. Still, there is growing evidence that in a host of other situations, from the worldwide web to metabolic, ecological and social networks, one lies between these two extremes. Two different manifestations of this diversity are:

- High clustering and, at the same time, small diameter;
- Degree distributions $P(k)$ that fall off with k much more slowly, in particular, like a power law $P(k) \approx k^{-\gamma}$ where γ lies between 1 and 3.

A major development since the late 1990's has been to identify universal mechanisms leading to these properties. To satisfy both the high clustering and small world properties a one-parameter model interpolating between an ordered finite-dimensional lattice and a random graph has been developed. The key is to randomize with probability p (keeping the size N fixed) connections between initially close neighbors, thereby introducing long range connections between nodes that would otherwise not communicate. As for the power law dependence of $P(k)$, the strategy for achieving such *scale free* (in the sense of Sec. 1.4.4) networks proposed by Albert-Laszlo Barabasi and coworkers is completely different. Rather than keep N fixed one considers a growing network where a new node is added at each time step connected preferentially to nodes whose degree is the highest. This algorithm can be implemented to create, in particular, hierarchical networks like in Fig. 6.10, exhibiting in the limit $N \to \infty$ the scale free property along with the small world one and a high clustering coefficient. Whatever the implementation, the most ubiquitous feature of scale free networks is the existence of few highly connected nodes, to which one refers as "hubs", along with less and less connected ones.

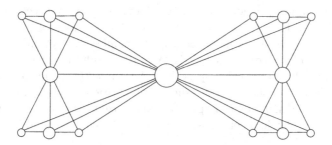

Fig. 6.10. A hierarchical network possessing highly connected nodes ("hubs") along with less connected ones obtained iteratively, starting from a single node. In each step two replicas of the preceding cluster are created and each of their nodes is connected to those of the preceding generations.

6.5.2 Dynamical origin of networks

Throughout the developments summarized above the geometric structure of the network is taken for granted in the sense that (a), the connections are not weighted by the nature of dynamical interactions they are supposed to represent (linear or nonlinear, passive, positive or negative feedback, etc.); and (b), dynamical processes that may be at the origin of the network itself are addressed in an ad hoc way. In this subsection we outline some ideas on the dynamical origin of networks and provide a complementary classification of interaction networks based on nonlinear dynamics.

The point is that before becoming "frozen" in the form of a graph, a naturally generated network (we exclude from the very start networks arising entirely from design) was part of a certain invariant subset Γ of the phase space spanned by variables related in one way or the other to the entities involved. This space is partitioned into K non-overlapping cells each of which represents a coarse grained state of the underlying dynamical system (see Sec. 3.3), and a time window τ for monitoring is chosen. As the dynamics induced by the evolution laws of these variables unfolds in phase space, the trajectory performs transitions between these cells thereby creating connections between the associated states, now viewed as the set of nodes of a "network in the making". A connection of this sort involving a node/state will be *outgoing* if the transition leads from i to a node $j \neq i$, or *incoming* if it leads to i starting from $j \neq i$. If the underlying dynamics is deterministic but sufficiently complex the partition in conjunction with the observation window will induce in phase space a probabilistic process, not necessarily Markovian. This may entail the presence of several connections involving a given network node, in addition to single loops connecting directly i to itself. This will be, a fortiori, the case if the underlying dynamics is a stochastic process to start with. In either case the dynamics will determine, for free, the structure of the adjacency matrix $\mathbf{W}$ of the graph. The question is, then, to determine the statistical properties of the connections and to relate them to the indicators of the dynamics such as correlation times, Lyapunov exponents and so forth. As an example, we summarize the principal properties of the adjacency matrices generated by fully developed chaos and intermittent chaos.

Fully developed chaos

We choose as representative illustration a variant of the logistic map (eq. (2.18)) at parameter value $\mu = 1$ known as tent map, considered already in Sec. 5.6. For this dynamical system a partitioning of the unit interval (the phase space Γ) into K equal segments (cells) constitutes a Markov partition. The corresponding adjacency matrix, in which the cells play the role of network nodes, can be constructed analytically and has the following structure:
- the leftmost node and the node containing the non-trivial fixed point $x_s = 2/3$ have one outgoing and one incoming connection;
- the remaining ones have two outgoing and two incoming connections.

Clearly, this structure is reminiscent of a regular network.

In the above procedure, the connections were established as the system was run with a time window of one unit. Suppose now that the window is increased to n time units. Owing to the Markovian character of the partition the adjacency matrix will now be $\mathbf{W}^n$. As n increases -for the tent map this happens already for $n = 2$- it becomes more and more filled until all its elements are nonzero and equal to $1/K$, implying that any given node is connected to all other nodes. This corresponds to a small world network (albeit a rather trivial one), since any two nodes are one connection apart.

Intermittent chaos

We focus on the symmetric cusp map (cf. also Sec. 5.6), a prototypical model displaying this kind of behavior:

$$X_{n+1} = f(X_n) = 1 - 2|X_n|^{1/2}, \quad -1 \leq X \leq 1 \quad (6.31)$$

The map is tangent to the bissectrix at the marginally stable point $X = -1$ and possesses an unstable fixed point at $X_s = 3 - 2\sqrt{2}$. The local expansion rate varies from one at $X = -1$ to infinity at $X = 0$.

We first consider a partition whose rightmost cell is $C_K = (X_s, 1)$ and the remaining $K - 1$ are delimited by the successive $K - 1$ preimages of X_s. This partition keeps the Markov property in a topological sense. By construction, the transitions allowed for a window 1 are $C_i \to C_{i+1}$ for $i \leq K - 1$, $C_K \to \bigcup_{i=1}^{K-1} C_i$, the last being associated with the fast "turbulent" reinjection process following the slow "laminar" stage induced by the presence of the marginally stable fixed point. In terms of connectivity, nodes 2 to $K - 2$ possess one outgoing and two incoming connections; node 1 possesses one outgoing and one incoming connection; and node K possesses $K - 1$ outgoing and one incoming ones, thereby playing the role of a hub. Notice the absence of intermediate values of links k between 3 and K. This is due to the existence of selection rules arising from the deterministic nature of the dynamics.

To obtain a richer structure we next consider networks generated by a phase space partitioning into K equal size cells - definitely not a Markov partition in either the metric or the topological sense. To get an idea of what one might expect for a window 1 as far as connectivity is concerned we evaluate the images of the leftmost and rightmost points of a partition cell $(X_0, X_0 + \frac{1}{K})$, taking to fix ideas $X_0 \geq 0$. Using eq. (6.31) we obtain

$$X'_\ell = 1 - 2\sqrt{X_0}, \quad X'_r = 1 - 2\sqrt{X_0 + \frac{1}{K}} \quad (6.32a)$$

The image of the cell spans therefore an interval of size

$$\Delta X' = |X'_\ell - X'_r| = 2(\sqrt{X_0 + \frac{1}{K}} - \sqrt{X_0}) \qquad (6.32b)$$

For $X_0 = 0$ this yields $\Delta X' = 2K^{-1/2}$, which is much larger than K^{-1} if K is large. It follows that those cells near $X = 0$ will have several outgoing connections since they will be projected into a union of several partition cells, thereby playing the role of hubs. If on the contrary X_0 is at finite distance from 0, expanding (6.32b) in K^{-1} yields $\Delta X' \approx \frac{1}{K\sqrt{X_0}}$. This implies that $\Delta X'$ will be substantially larger than, or close to K^{-1} according to whether X_0 (in fact $|X_0|$, if the argument is extended to $X_0 < 0$) is smaller than or close to unity, respectively. Now, this factor is nothing but the local expansion rate of the map, whose logarithm is the local Lyapunov exponent: cells with small expansion rates ($|X_0|$ close to one) will thus tend to have a small number of outgoing connections, the opposite being true for cells close to 0.

Figure 6.11a depicts the result of the numerical evaluation of the connections emanating from the different partition cells for $K = 10^3$. The agreement with the above advanced analytic arguments is complete, thereby substantiating the link between connectivity and the indicators of the underlying dynamics (Lyapunov exponents or expansion rates). Clearly, the network displays a hierarchical structure with a dominant hub around $X = 0$, a few nearby less important satellite hubs and a majority of nodes with few connections. Also detectable in the figure is a fine structure, in the sense of connectivity jumps between nearby nodes arising from the conjunction between the arguments underlying eq. (6.32) and the constraint that the number of connections must be integer. An alternative view of the above is provided in Fig. 6.11b, where the degree distribution $P(k)$ for the outgoing (full line) and total (dashed line) connections is drawn.

We finally consider the connectivity properties of the network generated by the above type of partition for increasing windows 2, 3, etc. Fig. 6.12a (to be compared with Fig. 6.11a) depicts the main result. As can be seen the number of both principal and secondary hubs increases with the window. These hubs are interrupted by dips within which the connectivity is small, as it also happens when one approaches the boundaries. Notice that there is always a local maximum at $X = 0$ albeit not always the dominant one. These effects can be understood qualitatively by the following argument, here presented for window 2. Considering a given window larger than 1 amounts to dealing with a higher iterate of $f(X)$ in eq. (6.31), here $f^2(X) = 1 - 2|1 - 2\sqrt{|X|}|^{1/2}$. This function is equal to -1 at $X = \pm 1$ like $f(X)$, as

Selected Topics 243

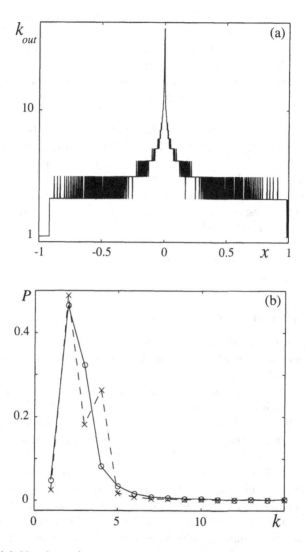

Fig. 6.11. (a) Number of outgoing connections versus X for the cusp map (eq. (6.31)), as obtained numerically from a partition of the phase space into 1 000 equal intervals during a time period covering 2×10^6 transitions and a time window $\tau = 1$. (b) Probability of the number of outgoing (full line) and total (incoming and outgoing) connections (dashed line) for $1 \leq k \leq 15$. Notice that $P(k_{\text{in}})$ vanishes for $k_{\text{in}} > 4$ whereas the cut-off of $P(k_{\text{out}})$ is $k_{\text{out}} = 45$.

well as at $X = 0$ (unlike $f(X)$), where it possesses an infinite derivative (like $f(X)$). Furthermore, it takes its highest value at $X = \pm 1/4$ where the derivative is also infinite. In short, the highest expansion rates (and thus the maximum local Lyapunov exponents) are around $X = 0, \pm 1/4$. By the preceding argument these should be then the locations of the principal hubs. On inspecting Fig. 6.12a we see that this is in full agreement with the results of the numerical evaluation.

Figure 6.12b provides the degree distribution $P(k)$ of the network. The structure is more intricate than in Fig. 6.11b. Limiting attention to the tail, one can fit the data with a power law $P(k) \approx k^{-\gamma}$ and with increasing confidence as the window increases ($\gamma \approx 1.8$ for window 4). The network generated by the dynamics shares, in this respect, properties of a scale-free network. In other respects, however, it keeps track of its deterministic origin manifested in particular in the form of "selection rules" forbidding certain types of connections.

Further comments on the dynamical origin of networks will me made in Sec. 6.6 in connection with biological systems.

6.5.3 Dynamics on networks

We now turn briefly to the role of the geometry of a network, assumed to be prescribed once and for all, on the properties of dynamical processes going on on the network.

A typical setting is to associate the network nodes to elements undergoing a certain dynamics -multi stable, periodic, chaotic, etc.- and the links to the coupling between these elements in space. As seen in Secs 2.6 and 2.7 nearest neighbor couplings can then lead to synchronization when appropriate conditions are met, and global couplings tend to facilitate the onset of synchronization. More in line with Sec. 6.5.1. one may compare, for a given finite number N of elements, regular arrays, small world networks with an underlying regular structure, and random graphs. The natural quantity to monitor is the network connectivity p, the ratio of the actual number of connections to the maximum possible one. Numerical experiments on coupled chaotic oscillators show in each case the existence of an N-dependent critical value p_c of p above which the system becomes synchronized. For a given N it turns out that the largest such value corresponds to a regular array and decreases as one switches, successively, to random graphs and to small world networks with nearest and next to nearest neighbor interactions. More quantitatively, for regular arrays p_c is size-independent and is determined by the local properties and the ratio of the largest to the first non-trivial eigenvalue of the adjacency matrix; for random graphs p_c behaves as $\ln N/N$ for large

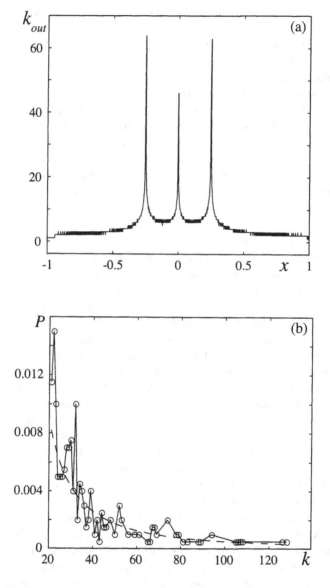

Fig. 6.12. (a) As in Fig. 6.11a but with a time window $\tau = 2$. (b) As in Fig. 6.11b but for the total number of connections k. Dashed line stands for a best bit with a power law $P \sim k^{-1.8}$.

N; and for small world ones as $1/N$. We stress that prior to synchronization different degrees of order can be achieved, as already mentioned in Secs 2.6 and 2.7. An interesting perspective is to extend this analysis to stochastic dynamics, where intrinsic or environmental fluctuations play a role. Some interesting results have been reported recently in this direction, in connection with random walks on complex lattices.

A second setting is to associate the network nodes to a set of variables $X_1, \cdots, X_N$ determining the state of the system and the links to the couplings between these variables arising from their joint participation in processes contributing to the time evolution of the $X's$. Here the links in such a "functional" network are, as a rule, state and time-dependent unless one linearizes around a reference steady state. The resulting network's adjacency matrix has then a structure determined entirely by the Jacobian matrix of the underlying dynamical system (cf. eq. (2.10)). The network geometry is thus intimately related to the dynamics, and a natural question in line with the ones raised earlier is how this geometry can affect, e.g., the loss of stability of the reference steady state signaling the onset of non-trivial behavior around this state. This raises the "stability versus complexity" problem: how does the size N and the connectivity of the network affect the existence of unstable eigenvalues of the Jacobian matrix? It can be argued legitimately that the larger the N the easier will be to destabilize. On the other hand natural systems are the results of evolution and many artificial systems are the results of design, both of which may filter those links that would tend to threaten stability as N increases. Yet the very ability to evolve and to adapt requires some degree of instability. How a network geometry may allow the system to strike the right balance between these tendencies remains an open problem, owing to the difficulty to determine the full structure of the eigenvalues of the adjacency matrix for complex geometries. For small networks an ingenious method to assess stability on the basis of the *logical relations* between the elements involved (positive or negative feedbacks, etc.) has been developed by René Thomas and applied successfully to a wide spectrum of problems of interest in biological regulatory processes.

Closely related to the foregoing is the concept of robustness, namely, how the network responds to perturbations. A first aspect of the problem pertains to the way the relevant variables respond to the perturbations, keeping the network geometry fixed. Clearly, this problem cannot be addressed unless the dynamics going on on the network is specified. But there is a second aspect, where the principal statistical properties of the network summarized in Sec. 6.5.1 are tested against changes of the network geometry, such as the removal of links or nodes. The question is of obvious importance in, for instance, the drug or mutation resistance of metabolic and genetic networks, the viability of an ecosystem threatened by environmental stresses, or the reliability of communication systems in the presence of failures of different sorts. Here one deals with an interesting version of the problem of structural stability (cf. Sec. 2.4) in which emphasis is placed on the statistical properties rather than on the dynamical response, that can be addressed (at least in part) on the sole basis of the network structure.

Until recently, the dominant idea was that robustness is secured when a sufficient amount of redundancy is present. The developments outlined in this section open the way to a new approach to this important problem. We briefly summarize the main results.

- Generally speaking, scale free networks are more robust than random networks against random node removals, but less robust against specific perturbations affecting hubs.

- Let a connected random graph of size N be subjected to the removal of a fraction p of its links. Then, there is a critical probability $p_c(N)$ beyond which the graph becomes disconnected. Likewise, if the graph is subjected to the removal of a fraction r of its nodes, there is a threshold value r_c at which the network breaks into small isolated clusters.

- Under analogous conditions, a scale-free network breaks at small clusters at a higher threshold value r_c. As a matter of fact, in the limit of N tending to infinity and as long as the power γ in the k dependence of $P(k)$ is less than 3, r_c tends to 1: scale free networks do not collapse under random failures. As regards targeted removals of nodes bearing a large number of connections fragmentation occurs already for small r's as long as $\gamma > 2$, but the critical value r_c presents a maximum around a γ value between 2 and 3.

These results constitute a valuable framework for reassessing and revisiting from a new angle a number of questions of concern in complex systems research in connection with evolution and adaptive behavior or, more generally, the different ways complex systems may cope with the environment in which they are embedded.

6.6 Perspectives on biological complexity

Since the middle of the 20th century, we are being exposed to the message that the properties of living organisms are recorded on the structure of the deoxyribonucleic acid -the DNA- and on the structure of the proteins coded by it, that eventually associate in a stereospecific fashion to produce all known macroscopic level biological structures and functions.

This view dominated biology until the 1990's. It was adopted enthusiastically by the vast majority of specialists in the field and led to a number of impressive breakthroughs. It thus came to be regarded as the prototype of predictive science and, on a conceptual level, contributed significantly by its reductionist character to the reinforcement of the Newtonian paradigm as the sole sensible approach to the understanding of nature. But during the last decade there has been increasing awareness that this attitude needs to be tempered by a more global view in which emphasis is placed on systems rather than molecules, on dynamics rather than structure. An organism is not just made up of many parts, but the parts are different in form and function, in distinctive, often complementary ways and communicate to each other in a highly intricate manner. Biological systems are hierarchical or, to use a term that became popular lately, of a "modular" nature, in the sense that certain structures and functions underlay others. Yet similar cells give rise to different types of tissues, similar tissues give rise to different types of organs, and so on. From this stems the notion that the functional and structural characteristics at any level in the hierarchy arise primarily from the interactions of the underlying structures. In summary, understanding biology at the system level is not tantamount to identify the characteristics of isolated parts of a cell or organism.

Originally launched in the 1960's these ideas are currently gaining momentum among hard core biologists, as information becoming available from genome sequencing allows one to get a comprehensive view on system level performance. It is realized that diseases are often multifunctional phenomena that cannot be ascribed to mutations of a single gene or to a single environmental factor. A global view integrating structure, dynamics, control and design becomes necessary to cope with them and with similar problems arising in a host of other situations. As a result, complexity is becoming a central issue in biology. In this section we outline the main ideas that are being developed in this context of a "bottom up" approach to biological complexity and comment on open questions and perspectives.

When addressing life processes in the perspective of complexity research one should refrain from transposing straightforwardly concepts and tools developed in the context of physical science. True, key aspects of biological order some of which are considered below are manifestations of the emergence of non-trivial behaviors as we know them from the study of physico-chemical systems. After all biological systems are open systems subjected to nonequilibrium conditions, and much of their activity is ensured by transport processes and by chemical reactions displaying a markedly nonlinear character. But on the other hand there are specificities differentiating life radically from phenomena occurring in ordinary matter. Most prominent amongst them is a *double causality*, to use Ernst Mayr's elegant expression: life processes are controlled by natural laws but also by genetic programs. To

Selected Topics

be sure, to the extent that the latter are the result of a long sequence of evolution and selection processes causality as we know it from physical science is eventually restored. Nevertheless, in present day life evolution, selection and information are encapsulated in the structure of biomolecules, which act at the right moment as prime sources of causality skipping the long series of physico-chemical events that led to their formation. Complexity research provides tools and proposes archetypes for doing justice to the specificity of biological systems while integrating them at the same time in the general problematics of complex systems encountered in nature.

6.6.1 Nonlinear dynamics and self-organization at the biochemical, cellular and organismic level

A familiar manifestation of biological complexity is the occurrence of unexpected dynamical behaviors arising from nonlinearity and encompassing most of the phenomenology outlined in Chapter 2.

One of the typical manifestations of such nonlinear behavior is self-sustained oscillations. They are observed at all levels of biological organization, from the molecular to the supercellular, or even to the social (population) one, with periods ranging from seconds to days to years. Among these the best understood are biochemical oscillations at the subcellular level. Their most characteristic feature is that they involve enzyme regulation at a certain stage of the reaction sequence. The enzymes responsible for regulation are usually not simple enzymes but, rather, *cooperative* (allosteric) ones in the sense that their conformation is affected by the fixation of certain metabolites and subsequently influences the catalytic activity of the enzyme. This cooperativity introduces, precisely, the nonlinearity necessary for complex behavior. Typically it manifests itself by the presence, in the rate equations, of terms in the form of a fraction whose numerator is a polynomial in the corresponding concentrations of degree higher than one and the denominator is the sum of a similar polynomial and a constant known as the *allosteric constant*.

Another common manifestation of nonlinearity in biology is the coexistence of multiple steady states. Two very interesting contexts in which this behavior is manifested are the functions of the nervous and the immune systems, where it is thought to provide a prototype of the phenomenon of *memory*. A less spectacular, but well-documented example is the ability of microorganisms to switch between different pathways of enzyme synthesis when subjected to a change of medium in which they are embedded. A source of nonlinearity common to all these phenomena is the almost stepwise response of the biomolecules to various effectors or even to their own or to the other unit's activity.

A most appealing example of symmetry breaking in living systems is morphogenesis. In the course of embryonic development one witnesses a sequence of events leading from a unique cell, the fertilized egg, to a multicellular organism involving specialized cells organized in an enormous variety of shapes and forms. Significantly, in developing tissues one frequently observes gradients of a variety of substances such as ions or relatively small metabolites. It has been conjectured that such gradients provide the tissue with a kind of "coordinate system" that conveys *positional information* to the individual cells, by means of which they can recognize their position with respect to their partners and differentiate accordingly. Symmetry-breaking bifurcations (cf. Sec. 2.6) provide a prototype for understanding these processes. Further examples of space-dependent nonlinear behavior in biology include the propagation of the nerve impulse, the calcium-induced propagating waves on cell membranes and the peculiar aggregates formed by unicellular organisms like amoebae or bacteria.

For a long time it was thought that chaos in biology is tantamount to nuisance. A logical consequence of this attitude was to view it as the natural reference for understanding certain forms of biological disorder. This idea has been implemented on a number of convincing examples related, for instance, to respiratory diseases or to arrhythmias of the cardiac muscle. Today it is realized that beyond this aspect chaos is likely to play a constructive role of the utmost importance in the highest levels of biological organization, particularly brain activity, by combining order and reliability on the one side, and variability and flexibility on the other. This may provide a mechanism of almost unlimited capacity for information storage and transmission, and for the recognition of environmental stimuli.

Population biology provides another example in which chaos may well be present. Experimental data are not clearcut, mainly because of the large environmental noise that is inevitably superimposed on the dynamical behavior. Still, taken in conjunction with mathematical models, they do suggest that irregular variations in time and space are ubiquitous. Such variations are expected to have an important role in problems involving competition and selection. The origin of chaos in population biology is to be sought in the fact that the rate of change of a population at a given time depends on the state of the previous generation, which is separated from the present state by a finite time lag. In many instances this dependence can be modeled successfully by the logistic map, eq. (2.18), or by delay equations such as eq. (2.28).

6.6.2 Biological superstructures

Less well-known and not sufficiently stressed compared to the foregoing is the ability of biological systems to transcend the level of the single organism and generate large scale structures and functions, many of them for the purposes of coping efficiently with their environment. An example of such biological "superstructures" is provided by the pheromone trails formed by social insects in view of solving, in a collective fashion, the problem of food recruitment and of the exploitation of environmental resources (Sec. 1.4.3). We hereafter survey briefly some further characteristic manifestations of the emergence of complex self-organizing patterns in social insects.

Collective construction. Bees, ants and termites build elaborate structures such as nests and underground galleries (Figs 6.13). In the latter case the behavior is guided by pheromone trails (as in Sec. 1.4.3) and stimulated by the privileged role of the excavation front. In the case of the nest the building activity is stimulated by the deposition of building material: information from the local environment and work in progress guides further activity, as opposed to direct communication among nest mates. Related to the above is also the formation of "cemeteries" in certain ant species, in the sense of the transport and deposition of corpses by carrier individuals. Experiment and theoretical modeling lead to the surprising result that the clusters so deposited are not distributed randomly: their distance is a (statistically) well defined, intrinsic property of the system. We are dealing here with a spatial structure of broken symmetry arising beyond the loss of stability of the spatially uniform state descriptive of random deposition, according to the scenario described in Secs 1.4.1 and 2.6. The study of the mechanisms intervening in pattern formation in social insects constitutes a privileged ground of experimentation and validation of mechanisms of morphogenesis in biology. In the case of ants the individual scale is in the millimeter range; in this scale the observer may thus intervene directly and straightforwardly. In this way the experimental study of the elementary mechanisms at work can be carried out quite naturally by deliberately decoupling the different phenomena from each other (measurements on an isolated individual in the experimental apparatus, response of an individual in the presence of other individuals, etc.). These "microscopic" characteristics can be confronted, by taking long time averages, to the experimental observations made on the system as a whole. As on the other hand the dimensions of the apparatus are typically less than a meter it is possible to act on the environment in a reliable and reproducible way, for instance by modifying the geometry of the apparatus.

Aggregation and self assembly. Bees and ants form aggregates involving

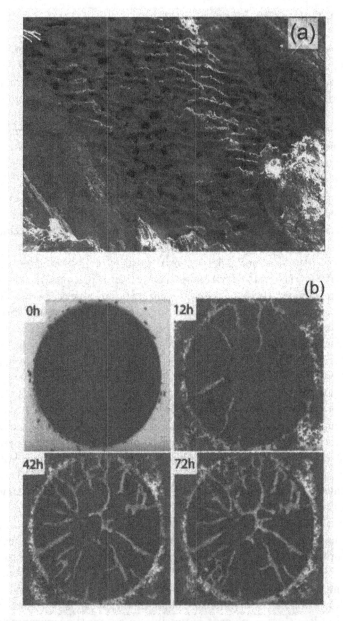

Fig. 6.13. Nest (a) (see Color Plates) and underground gallery network (b) realized by the ant species *Lasius Pallitarsis* and *Messor Sancta* (courtesy of Jerôme Buhl).

large numbers of individuals. This process usually involves physical contacts and leads to "living" bridges, chains, etc. allowing the species to achieve goals that are inaccessible to the single individual. Cooperativity is here related to the fact that the attachment and detachment probabilities of an individual to and from an aggregate depend not only on the mean density, but also on the size of this aggregate.

Social differentiation. Division of labor and hierarchical dominance relations are ubiquitous in ants and wasps. For instance, starting with undifferentiated wasps, ritualized pairwise contest eventually lead to a physically dominant female which usually becomes also a sexually dominant one, remaining in the nest and laying eggs, while the other females perform different tasks mainly outside the nest. Here self-organization can be linked to a special kind of cooperativity, whereby an individual that wins or loses a contest is more likely to win or lose subsequent ones.

There is a host of similar phenomena found in higher group-living animals as well. Penguins are known to cluster in order to create a warmer local environment; fish schools or bird flocks exhibit collective motion encompassing hundreds, thousands, or even millions of individuals thanks to which their endurance to achieve the desired performance or their defense against predation are dramatically enhanced; and so forth. In each case cooperativity - induced self-organization associated, for instance, with imitation seems to be at work. In addition to providing the key to the fundamental understanding of these very diverse phenomena such unifying principles and mechanisms may also prove useful in the management of man-made exploitation systems in connection, for instance, with farming.

These phenomena can be modeled at different levels. In a deterministic, mean-field approach one typically appeals to equations like (2.25) to (2.28). But owing to the pronounced variability of the constituent units a probabilistic approach often proves to be necessary. A favorite tool in practicing this approach is the Monte Carlo simulation (Sec. 3.6.1). Its merit is to address directly the processes of interest rather than solve evolution equations that might provide an ill-founded representation due to incomplete knowledge of the underlying evolution laws.

6.6.3 Biological networks

In their vast majority biological phenomena result from multi-step processes involving several intermediate elements. The large number of simultaneously ongoing activities makes the generation of each of these intermediate elements from scratch an impossible task. It is therefore essential to branch off elements participating in one process and utilize them in one or several other processes. As a result many essential expressions of the activity of living systems are to be viewed as the collective response of several coupled elements organized in the form of *networks*, from the regulation of the cell metabolism and the regulation of the genetic expression to the immune response and brain activity.

The metabolic level

Metabolic networks aim at producing, converting and storing energy in the form of energy-rich chemical bonds. They are based on the principle of enzyme catalysis: as most biochemical reactions would not proceed under physiological conditions at a measurable rate, synthesis is achieved by using highly specific catalysts, the enzymes, which speed up the required reaction by several orders of magnitude without being consumed. The vast majority of enzymes are proteins. In a few cases protein-RNA complexes or even RNA molecules alone can act as enzymes; furthermore, enzymes often combine to organelle-like chemical modules, known as multi-enzyme complexes. Many important steps in cell metabolism are combined into reaction cycles, e.g. the Krebs cycle. On the other hand metabolic compounds are required to be present in appropriate concentrations at the appropriate time. For this reason, the enzyme activity is modulated through positive or negative feedbacks by non-reacting "effectors" produced somewhere else in the chemical pathway. A universal mechanism of such regulation is the allosteric effect, where the effector induces a conformational change of the enzyme to a more or to a less active form. As mentioned in Sec. 6.6.1, this cooperativity is one of the principal sources of nonlinearity in biology. But, for the purposes of the present subsection, the point is that positive and negative feedback loops introduce more branches in an already quite intricate network. Traditional biochemists concentrate on the mechanistic analysis of single steps. The compilation of whole reaction cycles or of multi-step pathways is in many cases rewarding, as it leads to new properties resulting from the intertwining of different steps. An example is provided by the glycolytic pathway where for each glucose molecule degraded two molecules of ATP are produced: for appropriate rates of initial reactant injection sustained oscillations of all glycolytic intermediates are observed both *in vivo* and *in vitro*, with periods of the order of the minute. More complex dynamical behaviors have also been detected in *in vitro* experiments involving cell extracts. Such behaviors are believed to contribute to the flexibility of the cell's response to the external environment.

The genetic level

Let us turn next to the genetic level. We have repeatedly referred to the DNA molecule as an information carrier. Actually, this information is not flowing freely within living cells but remains instead jealously stored and "archived" within the double-stranded DNA helix, presumably in order to minimize errors. A first series of processes are therefore concerned with the expression of the information stored as DNA in the synthesis of proteins - the molecules responsible for the bulk of the structural and functional properties underlying life. They can be decomposed into two parts:

- *Transcription.* First an RNA copy referred as messenger RNA (mRNA), which is complementary to one of the strands of the relevant part of DNA is being made. This process is not spontaneous. It is catalyzed by appropriate "transcription factors" most important of which is the enzyme RNA polymerase, which are capable of recognizing and fixing themselves on specific DNA sites known as promoters lying just in front of the different coding parts. Once stimulated by such factors transcription begins at well-defined initiation sites near the promoter and ends at specific termination sites.

- *Translation.* The mRNA moves (which the DNA is unable to do) to specific nucleo-protein complexes within the cytoplasm known as ribosomes, where it acts like a template for the protein synthesis. The raw material for this synthesis are the aminoacids. They are present in the cell in a free state and carried in an activated form to the ribosome by yet another RNA species, the transfer RNA (tRNA).

It should be noticed that in higher organisms what is being expressed here concerns only a (minute) part of DNA, referred as *coding* DNA, in the form of a succession of genes whose structure contains the information for the protein synthesis, interrupted by fragments of *non-coding* DNA.

A second series of processes relates to *regulation.* Enzymes (most of which are proteins) are usually rather stable molecules (life times of the order of several minutes), whereas catalysis is a very fast process. Thus, it is not unusual to have a situation where the protein level in the cell tends to be too high. In response to this, the organism uses some substances to repress the synthesis of the macromolecules. In other circumstances it may happen that the organism is subjected to a sudden change of the environment. New types of proteins are needed to catalyze the reactions involving the new substrates available. In both cases the regulation of the level of gene expression is ensured by proteins called repressors or activators, acting at specific genetic sites. As it happens also in metabolic reactions, these regulatory proteins are often allosteric. A trick employed by living cells is that a metabolic substance can control the synthesis of the desired macromolecules by affecting the activity of an appropriate regulatory protein. The set of these elements and their connections through the above described interactions constitutes a genetic network. Two extreme examples of such networks encountered, respectively, in lower organisms and in higher ones, are the *lac operon* controlling the catabolism of a sugar, lactose, by the bacterium *Escherichia*

Coli; and the *p53 network*, controlling the responses to DNA damage and other forms of stresses affecting genetic expression. Regulation is reflected at the network level by the appearance of circuits in the form of positive or negative feedback loops. These are at the origin of complex behaviors, from multistability to oscillations and chaos, interfering in a decisive way with such important processes as cell division, cell differentiation and pattern formation in embryogenesis.

Large scale networks

We next consider large scale phenomena involving cell-to-cell interactions. A first instance is the immune system, protecting living organisms from invasion by pathogenic agents usually referred to as *antigens* (foreign cells or molecules, among others). The basic defense tools for achieving this are specialized cells known as B and T lymphocytes. B cells respond to antigens by producing antibody molecules, which have the ability to combine with the antigens and neutralize them. To this end they dispose of specific binding sites -the paratope- recognizing specific types of antigens. At the same time, they also possess sites presenting antigenic properties -the idiotopes- and are thus susceptible of binding with the paratopes of other antibodies. This dual role of the antibodies is one of the basic elements at the origin of a network structure of the immune system. A second basic element arises from the presence of T lymphocytes, two species of which, the helper (T_H) and the suppressor (T_S) cells exert a series of regulatory actions in the form of both negative and positive feedback loops, such as the maturation of B lymphocytes and the subsequent enhancement of antibody synthesis. The originality of the immune network compared to the metabolic and genetic networks is the presence of an ongoing process of evolution and selection. The diversity of antibodies capable of responding to a wide spectrum of aggressors is so high, that an organism cannot possibly keep in stock all antibodies in the required concentrations. To cope with this, organisms are endowed with a self-organizing process known as clonal selection. Specifically, recognition of the antibody by an antigen triggers growth and division of the corresponding B cells as a consequence of which the antibodies are in turn synthesized in large quantities. Mutations further increase the diversity of antibodies: B cells replicating their DNA and dividing make replication errors; if the altered antibody sequence produced by such cells has an increased binding strength to antigen the reproduction of the cell producing it is favored: an interesting version of the Darwinian mutation - selection idea , see also Sec. 6.6.5 below. The conjunction of all these elements -idiotypic network, regulation by B and T cell interactions and evolution- is at the origin of the rich variety of dynamical behaviors characterizing the immune system: the specificity of the immune response, the immune memory, as well as subtler phenomena such as unresponsiveness, tolerance and paralysis. These behaviors are related in one way or the other to the coexistence of several stable steady states, excitability and oscillations.

Neuronal systems constitute yet another striking realization -perhaps the most fascinating of all- of a network structure underlying biological function. Individual neurons can already perform complex tasks such as the propagation of electrical impulses, owing to the nonlinear dependence of the electrical current on the membrane potential and on the concentration of the ions involved. Undoubtedly, however, it is because of the intricate connections between neurons in the cortical tissue that the brain can sustain higher functions such as memory, recognition and information generation and processing. Contrary to the previously described three other networks, connections here imply "physical" wiring. But rather than being fixed, the effective cortical connectivity is highly dynamic, in the sense that the probability of transmission across an inter-neural connection depends on the previous patterns of activity. Connections are fairly dense within groups of neighboring neurons ensuring a particular function: a neuron may process as much as 10^5 signals coming from other neurons in the set. But long range connections between different cortical areas are comparatively quite sparse. Either way, connections emanating from a neuron can exert an activatory or inhibitory action on the neurons connected to it. Owing to the above mentioned highly nonlinear current-voltage relation characterizing single neurons the response tends to be almost discontinuous, in the sense that it attains a measurable value only when the stimulus exceeds some threshold. A variety of models integrating these characteristics have been proposed and analyzed in detail. They give rise to complex dynamical behaviors, from multistability and spatio-temporal chaos to more coordinated forms of activity in the form of clustering or synchronization (cf. Sec. 2.6). These are reminiscent of the behavior of some measurable macroscopic indicators of brain activity such as the electroencephalogram. Neurobiology led in the 1970's to an interesting spin-off, where networks of binary switches (referred as "neural nets") are being built and used as computational tools for relating input to output, or in connection with interpolation and optimization problems. Network parameters are usually determined by a "training set" which consists of problems similar to those one wants to solve, but for which the answers are known. Application of neural networks has turned to be particularly useful in cases where the information on complex problems is incomplete. Prediction of protein secondary structures is a typical example. Another fruitful application is pattern recognition. In this case the training set consists of several patterns. The network is typically able to reconstruct a pattern from an input that contains only a fragment of the original pattern. A major advantage is the ability to work with fuzzy input data and to provide approximate but essentially correct answers even in cases where the questions are not sharply posed.

Common to all major network types reviewed above is the presence of feedback circuits, ensuring a proper regulation of the system's operation (concentration levels of crucial chemical species, metabolic flows, etc.) as illustrated in Fig. 6.14a. By definition, a feedback circuit of order k ($k \geq 1$) involves steps starting from a certain element say x_1 and ending at this very element after having involved $k-1$ other intermediate elements: $x_1 \to \cdots x_j \cdots \to x_1$, $1 \leq j \leq k-1$. It is termed positive or negative according to whether each element globally exerts a positive action (activation) or a negative one (inhibition) on its own evolution.

Regulatory interactions encountered in biology are very nonlinear, a property reflected by a response curve to a stimulus in the form of a steep sigmoidal (almost step) function (Fig. 6.14b). This has prompted several authors and, in particular, Stuart Kauffman and René Thomas, to use logical methods to analyze biological regulatory networks. The main results of these investigations are that negative circuits generate homeostasis with or without oscillations, whereas positive ones generate multistability. What is more, a negative circuit turns out to be a *necessary* condition for stable periodicity and a positive one a *necessary* condition for multistability. These results suggest strongly that the dynamical behavior of a network is largely conditioned by the logical relations involved in its feedback circuits. More recently René Thomas pioneered a general approach applicable to a wide class of generic nonlinear dynamical systems, whereby circuits are identified in terms of the elements of the Jacobian matrix of the dynamical system (cf. eq. (2.10)). For instance, a circuit like $x_1 \xrightarrow{+} x_2 \xrightarrow{-} x_3 \xrightarrow{+} x_1$ is reflected by the matrix elements $J_{21} > 0$, $J_{32} < 0$ and $J_{13} > 0$. The interest of this correspondence is in the fact that the coefficients of the characteristic equation of the matrix, whose solutions determine the stability properties of a reference state, depend solely on the circuits present in the system. It should be pointed out that in a nonlinear system a circuit may be positive or negative, depending on the phase space region surrounding the reference state. For instance, if the system possesses more than one steady state the nature of each of these states will depend on the structure of the circuit at their location. Non-trivial behaviors, from multistability to chaos, require both appropriate circuits and appropriate nonlinearities. In particular, the nonlinearity has to be located on a "nucleus", i.e., a circuit involving the full set of the system's variables.

In the light of the above what is, then, the role of the global geometric structure of a network according to the classifications discussed in Sec. 6.5?

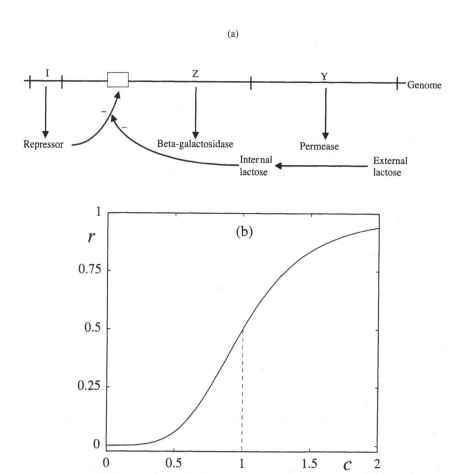

Fig. 6.14. (a) Example of genetic regulatory network underlying the functioning of the *lac operon*. Gene I is responsible for the synthesis of a repressor. The repressor binds to the genome and blocks the expression of genes Z and Y responsible, respectively, for the synthesis of the substances β-galactosidase and permease which facilitate the assimilation of the sugar lactose and its entry within the cell from the outside. This repressing action can be counteracted by the presence of internal lactose, which plays in this way a role of inducer of its own activity. (b) Typical response curve showing how the activity r of a cooperative module is regulated by the concentration of an effector C, $r = kC^n/(C_0^n + C^n)$, with a value of the *Hill coefficient* n equal to 2 and $k = 1, C_0 = 1$.

Several authors have concluded on the small world or the scale-free character of a large variety of biological networks. Qualitatively speaking, small-world like properties are likely to be relevant in networks in which signal transmission over large distances is crucial. An example might be the sparse long-range connectivity of the cortex. Although only a small fraction of the activity that occurs locally in the different modules can be reported to other areas by such an architecture, this may be just enough to ensure coordinated overall activity. Beyond this, whether the network complies or not with the full set of characteristics of small world networks is not especially crucial. As regards scale-free networks, we believe that part of the excitement about their ubiquity in biology is besides the point. First, as we saw above, it is the nature and in particular the logical structure of the circuits present in the network that matters the most for the overall behavior. This is very different from the full network topology, where less relevant structures are treated on equal footing with crucial ones. Second, part of the conclusions drawn depends on the particular representation adopted (this is the case for the exponent γ in the degree distribution, cf. Sec. 6.5.1) and on the choice of the elements to be included. Third, the argument about robustness and its concomitant evolutionary significance advanced in favor of the scale-free character of biological networks loses much of its relevance when one realizes that similar networks arise in many other areas, where the question of evolution does not arise. Conversely, networks having a definite significance in connection with evolution, such as the catalytic hypercycles discussed below, exhibit no scale-free structure. In the end, modularity and hierarchical structure is, perhaps, the single most important architectural feature underlying biological networks.

6.6.4 Complexity and the genome organization

The information content of present day DNA, to which we often referred in this book, rests on the sequence of the four bases A, G, C and T along the intertwined polynucleotide chains composing the double helix. In the language of Chapters 3 and 4 this makes DNA a text written in a four-letter alphabet, to be compared to the protein text which is written in the twenty-letter alphabet of the twenty naturally occurring aminoacids. The genetic code, prescribing how the information at the DNA level is conveyed at the level of protein synthesis, is a triplet code where three successive nucleotides code for each given aminoacid. It corresponds to the minimum possible degeneracy, in the sense that $4^3 = 64$ is the first power of 4 (representing the number of n-triplets that can be formed from 4 elementary units) to exceed 20. One refers to the successive nucleotide triplets as *codons*.

Selected Topics

In Sec. 6.6.3 we briefly described the transcription and translation steps through which the triplet code is materialized in the synthesis of proteins. We also alluded to the fact that in higher organisms the (protein) coding part of DNA, on which molecular biology has focussed for decades to develop its concepts and tools, is in fact only a minute portion (less than 2% in the case of the human genome) of the full DNA molecule. The remaining part appears in the form of "non-coding" sequences interrupting and separating the traditional protein-coding genes. Far from being "junk" as thought some time ago, it is now realized that it plays a vital, albeit still incompletely understood role. In particular, a number of these segments is transcribed into various active forms of RNA which fulfill some hitherto unsuspected functions controlling the development and the characteristics of all known organisms.

One view of the DNA molecule of higher organisms suggested from the foregoing is that of a symbolic sequence of two letters corresponding to two kinds of non-overlapping building blocks -the coding and non-coding regions- each of which is in turn a symbolic sequence of four letters corresponding to the four bases A, G, C, and T. Although perfectly reproducible from one generation to the next both of these intertwined sequences are manifestly aperiodic and basically unpredictable in the sense that their global structure cannot be inferred in a unique way from the knowledge of a part of it and, in the end, it is precisely this feature that allows one to speak of "information" at all. The question we would briefly like to discuss here is, to what extent the intricate hierarchical, aperiodic spatial structure described above can be characterized in a useful way using the concepts and tools of complex systems research. In addition to providing insights in view of a fundamental understanding of the physical mechanisms underlying biological information processing such a characterization could also be of practical value. In particular, the discovery of regularities of any kind beyond those provided by standard statistical approaches could help organizing the enormous amount of data available and extracting further relevant information.

In Chapter 5 we surveyed an array of methodologies for analyzing data sets generated by a complex system. The two most straightforward questions pertain to the nature of the probability distributions and of the correlations generated by the set. Both of them have received considerable attention in connection with DNA analysis and we briefly summarize hereafter the main results.

(i) At the global level of succession of coding and non-coding regions the statistics of the lengths of coding and of non-coding segments has been derived for several data sets pertaining to higher organisms, including human chromosomes. The principal conclusion is that the coding segment length distribution, as obtained by the best fit of the data, has an exponentially falling tail in the large length region whereas the non-coding segment length distribution falls off like a power law. Now, as seen in Chapter 3, short-tailed distributions possess moments of all orders and can be thought of as resulting from the random superposition of mutually independent variables having a common distribution with a finite mean and variance - a statement related closely to the central limit theorem. In contrast, long tailed distributions belong to an altogether different class and if, as it seems to be here the case, the characteristic exponent of the power law is sufficiently small, their variance is infinite (in the infinite length limit). The variable obeying to such a distribution can then in no way be viewed as a random superposition of mutually independent variables, a result suggesting the existence of correlations of some sort.

(ii) At the local level of succession of bases along a particular coding or non-coding region a variety of approaches for detecting deviations from a fully random succession has been proposed, from the computation of covariance matrices to wavelet analysis to the mapping into a problem of generalized random walk. The principal conclusion is that coding sequences are essentially random and can thus be viewed as resulting from a game of coin tossing, where "heads" and "tails" stand for a purine (A, G) or a pyrimidine (C, T), respectively. In contrast, non-coding ones exhibit long range correlations.

These thought provoking results are far from exhausting the fascinating question of the physics underlying genome organization. As seen in Chapter 5 such questions as sensitivity, recurrence and extreme value statistics are some further useful indicators of complexity of a data set and would undoubtedly deserve being considered in the present context. Even more obvious and relevant, though practically unexplored so far, would be the entropy analysis of biosequences along the lines of Secs 4.3 and 5.3.2, as it is specifically targeted to the theme of information.

Transposing the methods of complex systems research to unveil genome organization should be carried out with special care. Statistical analyses tend to mix up the different points of a data set, but this is not always legitimate when real world biosequences are concerned owing to the existence of grammatical rules (start and termination regions, sense of reading, etc.) that need to be respected. Furthermore, the data sets in the crucial asymptotic region of large sequence lengths may either not be sufficient to draw sharp conclusions, or present gaps owing to the occurrence of repeats featuring certain preferred length scales. All in all, despite the progress achieved it is our feeling that the analysis of biosequences in the perspective of complex systems reserves many surprises to come and is likely to raise new issues that will in turn stimulate progress at the level of fundamental theory.

Selected Topics   263

6.6.5   Molecular evolution

Genetic information and biological structures and functions as we know them today are the result of a long evolution. Evolutionary systems are particularly spectacular examples of complex self-organizing systems. Low probabilities of finding objects with desired properties by traditional methods of synthesis, owing to the combinatorial explosion discussed in Sec. 4.3, implies that central planning can neither generate nor maintain the organization of real world living organisms. Charles Darwin's principle of variation through replication-mutation processes followed by selection of the fittest by the environment provides inspiration in view of a solution to this major problem: replication produces a limited variation of the genetic information; the genetic information of each variant (mutant) is transferred to the descendants, and error propagation takes place during this stage; finally, the evolutionary success (the "fitness" or the "selective value") of the outcome is globally evaluated at the scale of the entire organism viewed as part of a population of competing entities, and this involves thousands of individual steps in the metabolism, reproduction and expression.

It is this latter step of the overall process that makes the search of the laws of evolution an exceedingly difficult task. Ordinarily, evolution at the organismic level is viewed as a historical process leading to a gradual complexification - from bacteria to man. Based on this idea phylogenetic kinships of different organisms are calculated, leading to plausible evolutionary trees. On the other hand, as Stephen Gould has pointed out, the events involved in evolution are so numerous and intertwined, so blurred by the presence of random and chaotic elements of all sorts, and are of such limited reproducibility, that standard methods of prediction do not apply. For instance, mass extinctions due to external causes perturb considerably the state that was supposed to be the starting point of the next step in the evolution. As for the preferential trend toward increasing complexity, he points out that to every more complex form of life corresponds a simpler equally well adapted one, for instance in the form of a parasite. These and similar observations call for a fresh look at the problem of evolution.

Evolution of species viewed as a historic process is bound to remain speculative as long as there is no way to confront its premises to basic physicochemical principles and to experiment. In this respect, there is a clear interest in formulating the problem of evolution at a more fundamental level: generation of primitive machineries for storing and decoding information; the development of primitive catalytic networks catalyzing the indispensable steps involved in this process. Below we outline some attempts in this direction and show how they are related to the concepts of complexity and self-organization.

Replication-mutation dynamics

Molecular evolution became a discipline in its own right following Manfred Eigen's seminal theoretical work on self-organization of biological macromolecules. The starting point is a formulation of replication-mutation dynamics in the language of chemical kinetics. Correct replication and mutation are parallel reaction channels for the synthesis of information carriers (I_j, I_ℓ, $j, \ell = 1, 2, 3, ...$) in which they play the role of "template" for their own reproduction or the production of mutant (error copy) forms:

$$A + I_j \overset{Q_{jj}k_j}{\rightarrow} 2I_j$$

$$A + I_j \overset{Q_{\ell j}k_j}{\rightarrow} I_\ell + I_j \qquad (6.33)$$

The symbol A stands for materials consumed in the replication process. The factor k_j in the rate constants for individual reaction channels measures the total number of copies of template I_j (correct ones as well as those containing errors) that are produced per time unit. The factor Q_{jj} is a measure of the accuracy of replication and counts the fraction of correct copies, and $Q_{\ell j}$ represents the frequency of the mutation $I_j \to I_\ell$. At high replication fidelity ($Q_{jj} = 1 - \delta$ where δ, the replication error frequency, is small compared to 1) populations converge to a stable stationary state with a deterministic distribution of mutants. The mutant distribution, also referred as *quasi-species*, is centered around a fittest type ("fittest" being tantamount to "having the largest replication rate constant"), called the *master sequence* (cf. also Sec. 3.7). This result depends crucially on the nonequilibrium constraints driving reactions (6.33). Without them detailed balance would compromise the systematic synthesis of information carriers, which would merely appear as the result of extremely improbable random fluctuations.

Geometrically, a quasi-species can be viewed as a configuration in an abstract space called *sequence space* (sequences are distributed in sequence space according to their distance which is expressed as the minimum number of mutations required to convert two sequences into each other). Its shape depends on the accuracy of replication and the distribution of fitness values. The smaller the error rate, or the larger the difference in fitness values, the narrower is the mutant distribution. The width of the quasi-species increases monotonously with increasing error rate. Error propagation through sequential generations becomes critical at some (minimal) replication fidelity Q_{min} which is determined by the fitness of the master sequence relative to the mean fitness of the remaining population ($Q_{min} = \sigma^{-1}$, where σ is the *superiority* of the master sequence). At the critical replication fidelity, a sharp transition is observed in population dynamics. Low fidelity in replication with error rates above the critical limit called the *error threshold* cannot sustain a stable stationary population. Instead, all information carriers and all molecular species have finite lifetimes, and the whole population migrates as a coherent structure in sequence space. At still higher error rates a second threshold (characteristically named *dispersion threshold*) is observed beyond which the migrating population ceases to stay coherent and becomes dispersed into isolated patches scattered in sequence space.

The occurrence of nonlinear spatial phenomena in replication-mutation systems in the form of traveling waves has also been established and verified experimentally, providing an excellent tool for the selection of faster growing variants. The existence of a potential function for replication-mutation systems allows one to view them as gradient systems, and thus neither oscillations, nor Turing patterns, nor any non-stationary spatio-temporal phenomena can occur.

For many model landscapes of fitness values (or replication rate constants in the simpler case), the existence of an error threshold has been proven and quantitative expressions for the critical error rates have been derived. Error thresholds in finite populations have been studied by stochastic models. Replication in small populations has to be more accurate than in large ones because the master sequence can be lost in future generations not only by erroneous replication but also by stochastic fluctuations. Replication dynamics in small populations on realistic fitness landscapes derived from the folding of RNA has been investigated by computer simulations. Again, error thresholds have been detected. Escape from local fitness optima may take hundreds of generations, so evolutionary optimization appears to proceed in steps. This provides a basis for the concept of "punctuated equilibria" pioneered by Gould and shows, clearly, the limitations of traditional evolution theory.

Replication-mutation dynamics is considered to be basic for evolutionary optimization and adaptation to the environment in asexually replicating populations. The concept of molecular quasi-species thus applies to RNA evolution *in vitro*, to viroids, RNA and DNA viruses as well as to other procaryotes. It has been verified directly in *vitro* by test tube experiments and *in vivo* in population of RNA viruses. Adaptations to changes in the environment have been detected in RNA synthesis, and analyzed through isolation and sequencing of RNA isolates. These studies provide interpretation of chemical or biochemical reactions as evolutionary phenomena on the molecular level and thus constitute the basis for a *molecular evolutionary biology* as a new area of research.

Catalyzed replication

In addition to the information carrier itself, which acts as a catalyst (template) of its own replication, an additional catalyst in the form of an enzyme known as replicase is usually needed. This leads to the following extension of scheme (6.33)

$$A + I_j + E_r \xrightarrow{Q_{jj}k_j^{(r)}} 2I_j + E_r$$

$$A + I_j + E_r \xrightarrow{Q_{lj}k_j^{(r)}} I_\ell + I_j + E_r \qquad (6.34)$$

Considering, in addition, each replicase E_r to be produced from the information carrier I_j, leads to the extra step

$$B + I_k \xrightarrow{b_{k_r}} I_k + E_r \qquad (6.35)$$

Dynamics of catalyzed replication without mutation has been studied extensively. Generally speaking concentrations of template and replicase can grow faster than exponential. This has indeed been verified *in vivo* by recording the concentrations of template and replicase in *Escherichia coli* cells infected by $Q\beta$ virus.

The simplest systems based on combinations of catalytic steps that give rise to a dynamics suppressing competition and guaranteeing survival of all participating members are called *hypercycles*. (Fig. 6.15) The principle of hypercycles is cyclic, mutual catalytic support: all components of the system can be arranged in a cycle such that every species catalyzes its precursor. For small numbers of components ($m \leq 4$) the system converges towards a stable stationary state, larger hypercycles ($m \geq 5$) show oscillations of concentrations. No deterministic chaos has been detected in the solutions of hypercycle equations. Small hypercycles ($m = 2, 3, 4$) are very useful models for symbiotic systems which integrate previous competitors into a larger functional unit.

If one generalizes specific catalytic action such that every component can catalyze in principle the production of any other component including its own synthesis, one obtains a system of differential equations of great flexibility. These so-called replicator equations exhibit for appropriate ranges of parameter values, the whole spectrum of complex dynamical behaviors known from nonlinear dynamics and surveyed in Chapter 2: multiple steady states, concentration oscillations, deterministic chaos, Turing patterns and propagating waves. These results can be extended to account for mutations in the limit of small mutation rates.

Molecular evolution experiments have been performed *in vitro*, and have largely validated the theoretical ideas summarized above. They concern RNA

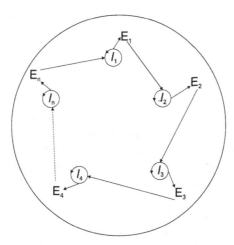

Fig. 6.15. A self-reproductive hypercycle. The RNA's I_i are capable of self-reproduction and provide information for the synthesis of enzymes E_i, which subsequently catalyze the production of RNA's I_{i+1} along the pathway.

replication, the specific amplification of a large variety of DNA sequences - optimally with chain lengths of some hundred nucleotides- as well as viral and microbial systems. These developments open the perspective of an *evolutionary biotechnology*: new biomolecules with predefined functions need not be fully designed but can be instead obtained by replication, mutation and selection.

In summary, molecular evolution leads to Darwinian scenarios close to those postulated in macroscopic biology. At the same time it clarifies the meaning of the "survival of the fittest" dogma, which in a way is tautological since the fittest is usually identified to the survivor! Indeed, fitness in molecular systems can be quantified and measured by techniques of physical chemistry independent of the outcome of a selection process. In this respect molecular evolution provides the integrating factor for a unification of molecular biology and molecular genetics on the one side, and conventional evolutionary biology on the other.

6.7 Equilibrium versus nonequilibrium in complexity and self-organization

In a liquid, molecules move in all possible directions and do not recognize each other over distances longer than a few-hundred-millionths of a centimeter. A liquid can therefore be regarded as a homogeneous material in which all points in space are equivalent. We now cool this systems slowly and uniformly. Below a characteristic temperature (0^0C for pure water at ordinary pressure) we obtain a crystal lattice, a new state of matter. Its properties like, for example, the density, are no longer identical as we move along a certain direction in space: the *translational symmetry* characterizing the liquid is broken. This property is concomitant with the appearance of new properties which prompt us to characterize the material as "ordered". For instance, the molecules perform small vibrations around regularly arranged spatial positions which, depending on the case, may lie on the vertices of a cube, the vertices of a regular hexagonal prism and the centers of its hexagonal bases, and so on.

What we are dealing with here is a case of equilibrium-mediated emergence of order belonging to the class of *phase transitions*, an important class of natural phenomena that are largely responsible for the polymorphism of matter as we encounter it in our experience. There are many other such cases giving rise to systems that one would definitely be tempted to qualify as "complex". For instance, under temperature, ionic composition, etc. conditions similar to those prevailing in a living cell, supramolecular architectures fulfilling well-defined functional properties such as membranes or organelles are spontaneously formed by self-assembly from their components.

Originally, equilibrium mediated ordered phenomena of the above kind were contrasted with the complex phenomena arising in systems undergoing nonlinear dynamics and subjected to nonequilibrium constraints - the kind of systems much of this book has been devoted to. The perception is, at present, more nuanced. In Sec. 3.7 we have already alluded to the possibility of subtle and constructive interplays between evolutionary processes out of equilibrium and processes occurring in an equilibrium environment. Below, we provide some prototypical illustrations of how such interferences can indeed take place.

6.7.1 Nucleation

One of the subtleties of phase transitions is that although different phases of matter coexist in thermodynamic equilibrium, the *transition* leading from one phase to the other does not occur spontaneously in equilibrium. Let us consider again the liquid to solid transition to which we referred earlier in this section. If we cool a water droplet down to 0^0 we will not see ice crystals appear spontaneously unless the liquid contains externally provided seeds in, e.g., the form of ice crystals. It is only when a mass of water will become sufficiently supercooled (at about -40^0C) that freezing will occur spontaneously.

Similarly, a protein crystal will form from a liquid solution containing protein molecules and salt ions only under conditions of supersaturation, where the content in solute is larger than what the solute-solvent equilibrium under the given ambient conditions can sustain. What is happening here is that under the aforementioned temperature and/or composition conditions the liquid and the solid are both stable states of matter and correspond, on these grounds, to minima of a thermodynamic potential like the free energy (Sec. 2.2). However, first, the free energy minimum corresponding to the supercooled liquid or to the supersaturated solution is more shallow that the one of the solid phase, a property that we express by referring to the former as "metastable" phases; and second, the transition between these two minima is bound to occur along a pathway involving higher free energy values, which can be "activated" by the system only because of the nonequilibrium locally created by supercooling or supersaturation. What is happening along such a pathway? First, a few randomly moving molecules of the liquid assemble by chance to a configuration closer to that of the crystal. Such group of molecules, which may also be viewed as a fluctuation, is called a *nucleus*. Small nuclei are unstable and disappear in time, due to the loss of energy that needs to be provided to create a surface of separation between the liquid and the solid (this would actually happen for nuclei of any size, if the liquid were stable rather than metastable). But nuclei of sufficient size will on the contrary be stable and will grow as they will be able to outweight the surface effects, thus becoming seeds from which the new phase will be formed. The minimum dimension that clusters created out of an initial phase by fluctuations must have in order that a new phase more stable than the initial one be formed defines the critical nucleus itself, in unstable equilibrium with the initial phase. The characteristic size of such a nucleus can be estimated by locally extremizing a free energy function involving a bulk term proportional to the volume (as expected from the property of extensivity) and a surface term proportional to the surface of separation, weighted by a factor called *surface tension* that plays a role analogous to the role of pressure in a bulk phase. The probability that such a nucleus be formed is, roughly, the inverse of transition time in eq. (3.17), where ΔU is usually identified with the *free energy barrier* and can be expressed in terms of the reference temperature and pressure, surface tension, and supercooling. Notice that once this barrier crossing has taken place subsequent growth occurs spontaneously, essentially though a diffusion like process.

In summary nucleation, the transition between two phases in equilibrium, is a dynamic process occurring under nonequilibrium conditions. To the extent that the new phase may correspond to very complex self-organized molecular arrangements, we are dealing here with a most interesting con-

270   Foundations of Complex Systems

structive interference between equilibrium and nonequilibrium.

Actually, the situation may be subtler and even more interesting than this. The classical picture of crystallization following the ideas pioneered by Josiah Willard Gibbs consists of the formation of critical nuclei that have already the crystalline structure. In this view the local density is thus the only variable -the only order parameter, in the terminology of Sec. 2.3- that matters, the crystalline cluster being (in general) denser than the fluid. Furthermore the critical nucleus is considered to be spherical, on the grounds of the isotropic character of the intermolecular interactions. In recent years both pictures have been called into question by theory, simulation and experiment starting with the important case of globular proteins. Specifically, in a rather wide range of external conditions (temperature, etc.) the favored nucleation path seems to be from the initial phase to high density disordered clusters and finally to clusters possessing crystalline order, over paths mo-

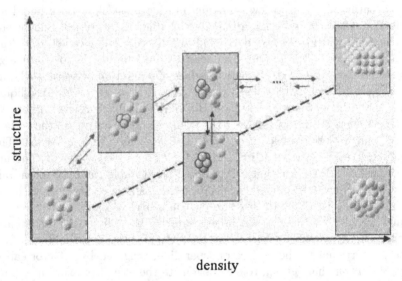

Fig. 6.16. Nucleation pathways in protein crystallization in the two-parameter space spanned by the density (horizontal) and crystallinity (vertical) axes. Classical nucleation as proposed by Gibbs proceeds along the diagonal and implies that the fluid densifies and gets ordered simultaneously. There is increasing evidence of non-classical nucleation in a variety of systems, where the initial fluid first forms preferentially a dense metastable fluid phase along the density axis, which subsequently constitutes the nucleus for the formation along the crystallinity axis, of the final crystal (see Color Plates).

ving directly to ordered clusters. Furthermore, the presence of the second, metastable disordered phase tends to enhance considerably the overall rate of nucleation. Put differently, two order parameters are here involved, associated with density and structure fluctuations (Fig. 6.16). This opens the way to a host of kinetic, nonequilibrium phenomena where different pathways compete to determine the final outcome of the process.

For instance, denoting by L_1 and L_2 the initial and metastable liquid phase and by S the solid phase, one has

$$L_1 \underset{k_{21}}{\overset{k_{12}}{\rightleftharpoons}} L_2, \quad L_1 \overset{k_{1s}}{\to} S, \quad L_2 \overset{k_{2s}}{\to} S \qquad (6.36)$$

where the rates k represent inverse mean transition times computed according to relations similar to eq. (3.17). It should be pointed out that because of the nonequilibrium character of the process and the presence of more than one variable, the free energy landscape is no longer the sole factor driving the evolution. The construction of "kinetic potentials" playing a role similar to free energy in the classical picture of nucleation is an interesting, partly open question.

The above described competition becomes even harsher if different types of solids can be produced with comparable likelihoods (e.g., orientationally disordered or ordered) and/or if more than one intermediate transient states can be present. This happens in the synthesis of a class of materials known as zeolites where the intermediate states exhibit different sorts of ordering. Here by favoring particular pathways among the many ones available one can tailormake a material possessing certain desired properties.

Let us come now to the symmetry properties of the critical nuclei. If, as it happens in globular proteins, the attractive part of the interactions is weak and short ranged (owing to the rather large effective radius of the protein molecules), then the first few particles involved in a cluster will tend to avoid being in the range of action of the other particle's repulsions as much as possible. This will result in a transient non spherically-symmetric structure which may simply not have the time necessary to reach the spherical form prior to the occurrence of the next kinetic step that may further exacerbate this initial asymmetry. Once the critical nucleus is formed this will mark a point of no return and the asymmetry will propagate thereafter to the macroscopic level, a property that has also been verified experimentally.

Interestingly, similar behavior is found for simple fluids below the triple point suggesting that crystallization involving passage through a metastable disordered phase may be a generic phenomenon. However, in that case, the metastable state is shorter-lived compared to the nucleation time.

6.7.2 Stabilization of nanoscale patterns

In any physical system the molecules interact, essentially by forces of electromagnetic origin. A typical example is the van der Waals forces prevailing in an electrically neutral fluid. These forces are short ranged: their intensity drops to very small values beyond an interparticle separation of the order of a few molecular diameters.

When the system is dilute or when the temperature is very high, intermolecular forces are not very efficient: kinetic energy dominates, and the material behaves in a disordered fashion. But when we lower the temperature (or compress the system), the roles of kinetic energy and of intermolecular forces tend to be reversed. Eventually the system is dominated by the interactions and adopts a configuration in which the corresponding potential energy is as small as possible. The spatial structure which results is precisely what we see in the liquid to solid phase transition. Being mediated by the intermolecular forces it displays a characteristic length -the lattice constant- which is microscopic, comparable to the range of these interactions. Moreover, the behavior of the material is time-independent if the environment itself is stationary. In this sense one can say that the order associated to a phase transition leads to "fossil" objects.

Both properties must be contrasted with those characterizing the transition phenomena in nonlinear dissipative systems driven far from equilibrium, see e.g. Sec. 1.4.1. Indeed, the space scales characterizing the states emerging beyond these latter transitions are macroscopic. Moreover, in addition to spatial patterns, a variety of time-dependent phenomena can arise and be sustained, whose characteristic scales are again macroscopic. The reason behind this basic difference is that a new organizing factor not amenable to intermolecular interactions is now at work: the nonequilibrium constraint which, by encompassing huge numbers of molecules, allows for new phenomena associated to states of matter enjoying long-range correlations. Whence the term "nonequilibrium phase transition" often used in connection with this class of transition phenomena.

Is anything interesting happening between these two extremes? As we saw in the preceding subsection, a metastable liquid phase will eventually be transformed into a stable solid one by a process of nucleation and subsequent diffusion-mediated growth. Suppose now that we suddenly bring, instead, the initial phase in a range of ambient parameters where it is unstable. In this "spinodal" region the phase will spontaneously decompose into clusters containing the stable state, skipping nucleation altogether and getting directly into the diffusion-mediated stage of growth. On the other hand, because of the phase instability thermodynamics imposes that the "'ordinary" diffusion coefficient should be negative: rather than smear out inhomogeneities an unstable phase tends, on the contrary, to generate them - at least in the initial stages of the process. What is one dealing with here is, therefore, an amended diffusion equation in which additional higher order derivative terms need to be included, accounting for the stabilizing effect of formation of interfaces separating the two phases.

As it turns out this extra term depends in a very sensitive way on the intermolecular interactions. Furthermore, in addition to spatial derivatives, both the ordinary and the higher order diffusion contain nonlinear contributions in the order parameter z involved in the process. But if we confine our interest to the initial stages of the phase transformation we may neglect nonlinear terms and obtain an equation of the form (to be compared with eq. (2.6)):

$$\frac{\partial z}{\partial t} = -D_0 \frac{\partial^2 z}{\partial r^2} - K \frac{\partial^4 z}{\partial r^4} \qquad (6.37)$$

where D_0 and K are positive.

Stability analysis along the lines of Sec. 2.3 shows that the uniform state is unstable toward inhomogeneous perturbations, thereby having the potential to generate spatial patterns of broken symmetry. But contrary to the Turing instability where growth near criticality is dominated by a privileged mode characterized by an intrinsic space scale there is here a continuum of unstable space scales involved, between a scale of the order of the inverse of $(D_0/K)^{1/2}$ and the system size. The fastest growing mode in this range corresponds to a scale of the order of the inverse of $(D_0/2K)^{1/2}$, which is typically in the nanometer range, i.e., intermediate between the microscopic and macroscopic scales referred in the beginning of this subsection. It will dominate the early stages of the process thereby creating a pattern of a new type, reflecting the coupling between macroscopic and microscopic-level processes. Sooner or later, however, this pattern will be corrupted by the other unstable modes and will gradually coarsen until a single stable phase invades the entire system.

Let now this phenomenon be coupled to a process subjected to nonequilibrium constraints. Under well-defined coupling conditions it can be shown that the modes with characteristic space scales larger than a certain value are now stabilized, leaving only a narrow band of unstable modes surrounding a privileged, fastest growing one. The conditions for a Turing like instability are thereby created, giving rise to a stable spatial pattern in the nanoscale range: we are witnessing the "freezing" of the phase transition by the action of the nonequilibrium constraint.

A variety of systems giving rise to freezing of this kind has been discovered. Of special interest is freezing of chemical origin in connection with catalytic processes going on on heterogeneous (solid) supports - the kind of processing utilized in the vast majority of chemical synthesis in the industrial scale. As an example gas phase oxygen and hydrogen molecules in contact with a surface of rhodium fix themselves on the surface with some probability, a process known as adsorption, and subsequently react. If the surface is pre-treated with potassium this species is acting as a promoter, by switching on a number of interactions and cross-diffusion effects that would otherwise not occur in the presence of oxygen and hydrogen alone. Eventually this leads to the nonequilibrium stabilization of nanoscale patterns, in a range in which the promoter-reactant mixture undergoes a phase instability from a well-mixed to an unmixed state. The nonequilibrium constraint is here ensured by the maintenance of fixed partial pressures of the reactants above the catalytic surface.

Based on similar mechanisms and principles, mechano-chemical couplings are expected to be at the origin of a new class of materials ("intelligent" gels, etc.) possessing potentially interesting physico-chemical properties.

6.7.3 Supramolecular chemistry

Nuclear matter displays new properties different from those of its elementary constituents, the quarks; atoms and molecules are capable of behaviors not accessible to nuclei and to electrons, not the least of which is the ability to form chemical bonds. How should one qualify such differences? How far can one go down to finer and finer scales and still speak meaningfully of complexity and self-organization?

The advent of quantum physics helped to clarify part of this question: the possibility to compute, ab initio, the properties of an atom or a molecule in terms of its constituents and their interactions explains chemical bonding and a host of other properties and shows that there are no new laws emerging from these configurations of matter. This is due largely to the fact that under the conditions encountered in standard situations ordinary atoms and molecules appear to be unique in their realization. True, there is always in the background an equilibrium between a molecule and its atoms in dissociated form, or between a left-handed (chiral) amino-acid and its right-handed homologue. However, the two members of these equilibria are separated by large energetic barriers rendering the conversion between them a very rare event on a relevant time scale. This reflects the huge success of the concept of covalent bonding, the dominant way to form stable molecules.

The situation becomes subtler if one looks at the question from an evolutionary perspective, as suggested already in Secs 3.7 and 6.6.5. According to the grand unified theories of the strong, electromagnetic and weak interactions of elementary particles, matter underwent a number of transitions at critical temperatures as it cooled from the initial exceedingly hot state prevailing after the Big Bang, owing to the expansion of the universe. Matter above the first transition was in a very symmetric state. Quarks, electrons, neutrinos and their antiparticles, all behaved identically. Below the transition point, however, differences were made manifest through symmetry breaking, of which the separation between matter and antimatter is the most striking manifestation. Eventually these differentiated particles became the raw material that made up atoms and molecules and thence stars, planets and living beings. In this sense, therefore, a transition mediated by the above three extremely short-ranged fundamental interactions reached a macroscopic level encompassing the present universe. This could not happen without expansion, which marked the passage from a state of thermodynamic equilibrium to a state in which the equilibrium between different constituents of matter as well as between matter and radiation was broken. In this respect, therefore, differentiated "complex" matter as we observe it today can be viewed as the "fossil" outcome of a global primordial nonequilibrium and of local short-ranged interactions.

Some time ago Jean-Marie Lehn launched the revolutionary idea that if evolution and self-organization can be reasonably regarded as terminated at the level of ordinary atoms and molecules, they can still go on -spontaneously or by man made design- at the supramolecular (but still well submacroscopic) level. The crux is to develop highly complex supramolecular entities from components interacting via non-covalent intermolecular forces. In view of the lability of these interactions ample space is left to dynamics. The reversibility of the associations which now manifests itself on relevant time scales allows for continuous changes in structure, which may be either internal (rearrangement of the components), or externally induced (exchange, incorporation, etc. of components). Furthermore, and much like in Sec. 6.7.1, it can generate intermediate kinetic products before reaching the final ones associated to the final state of chemical equilibrium.

A major achievement of this approach is the synthesis of products endowed with specific recognition properties similar to those of biopolymers. Specific binding requires a geometrically well-defined, relatively rigid structure with a surface complementarity to the substrate - the molecule that is being recognized. This is best achieved when the recognizing molecule has a "crypt" (cavity) that fits the substrate tightly. The receptor is then referred as *cryptand*, the resulting complex being a *cryptate*. This is what happens with the class of so-called macropolycyclic compounds which may define, depending on their three-dimensional structure, a 2-dimensional circular hole, or 3-dimensional spheroidal cavities where they are able to bury certain metal ions. The discriminating power, e.g. between Na^+ and K^+ is quite high: because the hydration water is displaced by the cryptand, the geometry of coordination of the metal ion must fit perfectly.

In addition to their recognition properties, many cryptands are endowed with the important function to get incorporated into membranes and other supramolecular architectures. The specific binding of a substrate unable to penetrate a membrane to a cryptand incorporated in the membrane opens a channel specific for that substance. Depending on the membrane architecture, the gate can even be unidirectional. Such ion gates are of primary importance in neurobiology, illustrating the potential of such devices. More generally supramolecular chemistry may lead, through a defined sequence of instructions and algorithms whereby the generation of a given intermediate depends on the previous one and conditions the appearance of the next one, to hierarchical self-organization along primary, secondary, tertiary, etc. structures. This opens the way to a new approach of synthesis of nanomaterials which in addition to the desired architecture may possess, thanks to their inherent dynamical dimension, unusual functional properties such as adaptation.

A fascinating perspective, in line with the discussion in Sec. 6.6.5, is that of an evolutionary supramolecular chemistry, where competition and selection in pools of dynamically interconverting supramolecular species could lead to optimization of certain functions under equilibrium conditions, without anticipating this function by external design. The imposition of a nonequilibrium constraint on this setting would introduce a direction in the evolution with the potentiality of novel forms and functions, in analogy to the results summarized in the previous subsection.

6.8 Epistemological insights from complex systems

Complexity's unique status comes from the fact that while it is rooted into fundamental physical and mathematical science, it has gradually given rise to a major scientific paradigm for addressing from a novel point of view a variety of problems of concern outside the realm of physics and mathematics. This has been possible thanks to the development of unifying and flexible methodologies and the realization that in large classes of phenomena arising in very diverse disciplines, beyond superficial resemblances common mechanisms are at work and underlie the observed behaviors. Inevitably, this unifying view interfered with practices and opinions prevailing in the disciplines in question. Beyond its technical level implications within each of these various fields, one of the outcomes of this confrontation has been to reassess the status of some general principles at the heart of science. In this section we outline some aspects of this reconceptualization. We emphasize that, much like in the post quantum mechanical era, these developments come essentially from "within" rather than from some abstract and heuristic considerations.

6.8.1 Complexity, causality and chance

In this book Complexity was addressed by means of a multi-level approach integrating structure and dynamics, macroscopic and microscopic descriptions, deterministic and probabilistic views. Now, determinism is closely related to causality and probability to chance. An integrated approach of this sort seems thus to connect two concepts that had been regarded as quite distinct throughout the history of science and of ideas in general. As we argue presently complexity research makes this connection fully legitimate while revealing, at the same time, some facets of causality and chance that would remain unsuspected in the traditional view.

Classical causality relates two qualitatively different kinds of events, the causes and the effects, on which it imposes a universal ordering in time. As Plato puts it, things come into being by necessity from an underlying cause (τό γιγνόμενον ὑπ'αἰτίου τινός ἐξ ἀνάγκης γίγνεσθαι, Timeus 28a). In fact, from the early greek philosophers to the founders of modern science causality has been regarded as a cornerstone, guaranteeing that nature is governed by objective laws and imposing severe constraints on the construction of theories aiming to explain natural phenomena.

Technically, causes may be associated to the initial conditions on the variables describing a system, or to the constraints (more generally, the control parameters) imposed on it. In a deterministic setting this fixes a particular trajectory (more generally, a particular behavior), and it is this unique "cause to effect" relationship that constitutes the expression of causality and is ordinarily interpreted as a dynamical law. But suppose that one is dealing with a complex system displaying sensitivity to the initial conditions as it occurs in deterministic chaos, or sensitivity to the parameters as it occurs in the vicinity of a bifurcation. Minute changes in the causes produce now effects that look completely different from a deterministic standpoint, thereby raising the question of predictability of the system at hand. Clearly, under these circumstances the causes acquire a new status. Without putting causality in question, one is led to recognize that its usefulness in view of making predictions needs to be reconsidered. It is here that statistical laws impose themselves as a natural alternative. While being formally related to the concept of chance, the point stressed throughout this book is that they need not require extra statistical hypotheses: when appropriate conditions on the dynamics are fulfilled statistical laws are *emergent properties*, that not only constitute an exact mapping of the underlying (deterministic) dynamics but also reveal key features of it that would be blurred in a traditional description in terms of trajectories. In a sense one is dealing here with a "deterministic randomness" of some sort. In fact, equations like the Liouville, master or Fokker-Planck equations (Sec. 3.2), governing the probability distributions associated to a complex system, are deterministic and causal as far as their mathematical structure is concerned: they connect an initial probability (the "cause") to a time-dependent or an asymptotic one (the "effect") in a unique manner, provided the system possesses sufficiently strong ergodic properties. By its inherent linearity and stability properties, stressed in Chapter 3, the probabilistic description re-establishes causality and allows one to still make predictions, albeit in a perspective that is radically different from the traditional one.

Another instance in which causality in its classical interpretation appears to be challenged is the presence of feedback loops, in connection with the issue of self-reference. If acting alone a feedback loop appears to entrain not only effects fed on the causes, but also causes fed on the effects. If that were indeed the case much of the field of complexity, where feedbacks are ubiquitous, would be subjected to either tautologies or logical paradoxes. In fact, feedback loops often act in conjunction with other processes rendering the system of interest open to its environment. In many real world situations this drives the system into a dynamically, or at least statistically, stable regime in which consistency and causality are restored.

In a different vein to the extent that initial conditions may be arbitrary, chance -here in its traditional interpretation- may be regarded as a probe of the different potentialities of a system: by switching on pathways that would remain dormant in the absence of variability (or fluctuations, to take up the setting of Chapter 3) chance is a prerequisite of diversification, evolution and complexity even though the latter may follow a deterministic course once the initial chance-driven impulse is provided.

Finally, modern societies have created a complex technological and economic tissue that contributes, in its turn, to redefine the relationship between causality and chance. Industrial and post-industrial products are often the result of design and control, whereby one interferes with natural evolution in order to quench undesired variability and achieve a certain predefined goal such as the synthesis of a material possessing certain desired properties. And financial markets often reflect the tendency of each agent to anticipate the behavior of his partners or the value that a certain product is likely to acquire in the near future. Clearly, in the first type of finality-driven process the role of causality is reinforced. But as seen in Sec. 1.4.4 the second type of process may be at the origin of complex behaviors associated with unexpected or even catastrophic events, in which the originally intended rationality is overwhelmed by a self-generated randomness through the rules of a game that the agents themselves had designed.

6.8.2 Complexity and historicity

It has been stressed throughout this book that the gap between the simple and the complex, between the microscopic level reversibility, permanence and determinism on the one side, and the macroscopic level irreversibility, evolution and randomness on the other, is much narrower than previously thought and can be bridged when certain conditions -satisfied by most natural systems- are met. These developments, which constitute the essence of the complexity paradigm, bring us closer to a "historical" representation of nature as we shall presently argue.

One may summarize our perception of historicity as a set of processes where the individual, with all is subjectivity, plays the primordial role; where the events, though unique, unfold in an irreversible and non-deterministic fashion; where everything remains open on a still undetermined future; where several outcomes are possible depending on the particular pathway that will eventually be adopted.

As we have seen, a most attractive formulation of the evolutionary and irreversible dimension of macroscopic phenomena is provided by thermodynamics. It is based on the introduction of entropy, a quantity determined uniquely by the instantaneous state. In an isolated system, starting with an arbitrary initial state one witnesses an evolution during which entropy will increase monotonously until it will attain its maximum that one associates to the state of thermodynamic equilibrium. This is the content of the second law of thermodynamics, which stipulates the existence of the time arrow underlying natural phenomena. This irreversible dimension does not convey automatically a historicity to the process concerned. Everybody would agree that the harmonic oscillator in the absence of friction is a prototype of ahistoricity. But in its amended version incorporating friction it is not history making either, since sooner or later it will settle in the position of equilibrium and nothing will happen thereafter. In addition to the arrow of time a minimal dynamical complexity is thus needed, before the historicity of a process acquires a meaning. The ideas outlined in this book lead us to the view that this decisive step is accomplished by the conjunction of the nonlinear structure and dissipative character of the evolution laws of the macroscopic variables, the nature of the initial state, and the constraints (such as the distance from thermodynamic equilibrium) inflicted on the system. This switches on a game of "challenge and response", singled out in a quite different context by Arnold Toynbee in his fascinating book "A study of history" as one of the regularities underlying the logic of History. Let us summarize some elements conferring credibility to such a view.

- A nonlinear system under constraint possesses, typically, several qualitatively different regimes to which correspond well-defined mathematical objects, the attractors. A given regime is reached from an ensemble of initial states that is specific to it and constitutes its basin of attraction. This evolution is irreversible (in a dissipative system) in the sense that, sooner or later, it leads inevitably to this regime. The evolutionary landscape -the phase space- is thus partitioned in cells inside which different destinies are realized. This is not in any sort of contradiction with the deterministic character of the evolution laws, since a given initial state can only evolve to only one of the attractors available. As seen in Chapter 2 the different cells are separated by frontiers constituted by unstable, and hence physically inaccessible, states. These frontiers are impermeable in the simplest cases but can become porous under certain conditions, adopting a highly fragmented, fractal structure. This is at the origin of an indeterminacy as to the attractor that will actually be reached when the system starts near this frontier, in the sense that minute differences in the initial state can lead to quite different realizations (cf. Sec. 2.1).

- The nature, the number and the accessibility of the attractors can be modified by varying the control parameters. As seen in Sec. 1.3 and throughout Chapter 2 these variations are marked by critical situations where the evolutionary landscape changes in a qualitative manner as new states are suddenly born beyond a threshold value of a parameter. This phenomenon of bifurcation confers to the system the possibility to choose and to keep the memory of past events, since different pathways can be followed under identical ambient conditions (cf. Fig. 6.17). As stressed repeatedly, all elements at our disposal indicate that the evolutionary scenarios that can be realized by a typical nonlinear system under constraint cannot be classified in an exhaustive manner: the evolution of complex systems is an open ended process, whose outcome cannot be anticipated. This open future, one of the leitmotivs of History, remains nevertheless compatible with the deterministic character of the underlying evolution laws! Thucydides, arguably the greatest historian of all times, attributes many key historical facts to human nature and suggests that since human nature remains the same over a very long time scale, under similar external conditions similar historical evolutions

Selected Topics

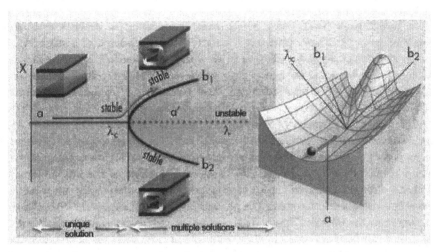

Fig. 6.17. The acquisition of historical dimension as a result of the phenomenon of bifurcation (Fig. 1.1), illustrated here on the example of the Rayleigh-Bénard convection (Fig. 1.3). Left panel: beyond the instability point the system must choose between the two new solutions b_1, b_2 that become available (two different directions of rotation in the case of a Rayleigh-Bénard convection cell). The yellow line depicts a particular evolution pathway. Right panel: mechanical analog of the process, where a ball rolling in the indicated landscape may end up in valley b_1 or valley b_2 beyond the bifurcation point λ_c (see Color Plates).

will be realized. This introduces to History a determinism of some sort. But while accepting the role of the individual Toynbee, who sees himself as a Thucydides disciple, expresses the conviction that events are actually never repeating each other in exactly the same way, thereby leaving to the future a margin of many possible realizations. The ideas summarized above, in conjunction with the concept of recurrence developed in Sec. 5.6, suggest that these two viewpoints may actually be part of the same overall logic. Let us also recall that a system functioning close to bifurcation presents an enhanced sensitivity, since minute differences in the choice of the initial state will generate evolutions towards different regimes. This sensitivity is manifested even if the frontiers between basins of attraction are impermeable.

- We have seen that among the regimes that can be realized through the phenomenon of bifurcation, the regime of deterministic chaos occupies a place of central importance on the grounds of the property of sensitivity to the initial conditions. This raises the fundamental question of predictability of certain phenomena, even if these are governed by deterministic laws. We have here yet another generic mechanism of evolution in which the future remains open, that one could again associate to the concept of historicity.

In the above setting evolution is viewed as an itinerary across different basins of attraction, as the initial state or the values of the control parameters are modified. It is the uncertainty on these quantities -a fundamental feature inherent in the very process of experimental measurement- that reveals the complexity of the underlying dynamics and its historical dimension. An uncertainty that also entails that a "state" actually perceived by an observer is in reality an ensemble of nearby values of the relevant variables. Since, in addition, in the most interesting cases each member of the ensemble will follow a different path, one is led to conclude that the only really meaningful questions are those concerning the probability that certain events be realized, rather than the detailed "pointwise" evolution of the individual trajectories. In other words the historical dimension of complex systems highlights the need for a probabilistic approach, thereby opening the door for the presence of elements of randomness in an initially deterministic setting.

The fundamental ingredients at the basis of a probabilistic description free of heuristic approximations have been discussed extensively throughout this book, see especially Secs 3.1 to 3.5 and Sec. 6.1, the main points being a complex dynamics -for instance in the form of deterministic chaos- and a large number of degrees of freedom. Once a self-consistent probabilistic description is available one may reassess the status of the evolution laws of the macroscopic variables. We have seen (eq. (3.7)) that variability is manifested in the form of stochastic perturbations superimposed on the macroscopic evolution laws. The evolution of the relevant variables takes then a form where the "deterministic" value associated with the macroscopic, mean field description is modulated by the fluctuations induced by these perturbations. Inasmuch as these fluctuations will have a marked random character, the path that will be followed by the system and the attractors that will be visited in the course of the evolution will acquire in this way a historicity even more marked than the one evoked above. This remarkable synergy between the individual level action (the fluctuations) and the collective level response is, once again, reminiscent of the fundamental role of the individual in our intuitive perception of historicity.

The use of probabilities also allows one to make predictions on system's future course, analogous to the ensemble forecasts evoked in Secs 5.5 and 6.3. Following some recent ideas by David Ruelle such "historical probabilities", should in principle allow one to assess, among others, the future occurrence of events conditioned by the particular situation prevailing at a given time. They may thus constitute a meaningful way to formulate questions like the probability of certain forms of life arising on primitive Earth or, on a shorter time scale, the probability of certain kinds of socio-economic events. While it seems unlikely that History will ever be scientificized, we may instead be witnessing the historicization of science.

6.8.3 Complexity and reductionism

One of the oldest debates in the history of ideas has been that opposing reductionism to holism.

Reductionism tends to derive higher level structures and functions of a system from the properties of the elementary building blocks that constitute it, the only ones to which it attributes an objective reality, thereby reducing all possible forms of organization to a physico-chemical origin. Today's reductionism is in many respects a modern version of materialism pioneered by the ionian philosophers and developed further by Democritus, who held the view that the diversity of forms encountered in everyday experience results from the transformation of a set of basic elements, the "atoms". A remarkably modern view indeed, of which the Newtonian paradigm evoked in the very first section of this book can be recognized as a distant heir, followed and reinforced some centuries later by Darwin's theory of evolution and the advent of molecular biology. Thanks to the immense success of these major enterprises and through the emphasis it places on the analytic approach and on the identification of elementary mechanisms, reductionism is traditionally regarded as the prototype for achieving the two major goals of science: unify and predict.

In contrast to the foregoing, holism insists on the specific character of a system as a whole. The strongest argument evoked in favor of such an attitude is drawn from biology and can be summarized by the (obvious) statement that a living being is a global structure, exhibiting structural and dynamical properties of its own. As such, holism appears as a more sophisticated version of vitalism and its corollaries of finality and teleology. In the early part of 5th century B.C. Parmenides challenged the materialistic view of ionian philosophers by providing powerful logical arguments in favor of the "One", his way to advocate a global view of the world. Plato and Aristotle devoted a great deal of ingenuity to reconcile the "One" with the "Multiple". Eventually, for almost 20 centuries, the balance settled in favor of the finalism and its corollaries. Challenged by the Newtonian revolution these views have nevertheless persisted albeit in more sophisticated and/or restricted forms, up to the present time. Let us quote two interesting historical examples. In the late 19th century the German biologist Hans Driesch was led to embrace vitalism following his own experiments on the early stages of embryonic development. He was struck by the fact that a same, normal individual can develop from different pathways starting with a complete ovum, each half of a divided ovum, or the fusion of two ova. This, he thought, contradicted the laws of physics and could only be accomplished by a vitalistic element governing the process and foreseeing the final goal. In a different context, the existence of variational principles in mechanics was interpreted by early investigators as the mathematical expression of finalism since, once again, they strongly emphasized the final, "desired" state which in many cases turned even out to correspond to an extremum ("optimum") of some sort.

Where does Complexity stand in this debate? The short answer is that rather than persisting in the opposition between reductionism and holism, we are moving in the light of complex systems research toward a postreductionist attitude where previously stressed polarized views become truisms as new bridges are being built and new syntheses are becoming possible. Let us start with the very concept of nonlinearity, at the heart of complexity. In a linear system the ultimate effect of the combined action of two different causes is merely the superposition of the effects of each cause taken individually. But in a nonlinear system the property of superposition breaks down: adding two elementary actions to one another can induce dramatic new effects reflecting the onset of cooperativity between the constituent elements. In particular, it can give rise to unexpected structures and events not foreseen in the properties of the constituent units. These statements, particularly the one on non-summability, sound definitely holistic. Yet they are consequences of the laws governing the evolution of the observables involved in the description, which fully agree with (and, in a sense specified in this book, derive from) the laws governing the structure and the interactions of the elementary building blocks of matter - a statement that sounds on the contrary as reductionist. In a similar vein, many aspects naively attributed to vitalism and finalism are just properties of dynamical systems. The fact that all initial conditions in the basin of attraction of an invariant set eventually tend to the same final state -the attractor- reflects the property of asymptotic stability rather than a finalism of any sort, unless one resorts to metaphors that can only confuse the issues. Similar comments can be made in connection with the ubiquity of feedback loops (Sec. 6.6.3).

It is the concept of emergence that crystallizes the best, both the distinction and the possible links between reductionism and holism. For a hard core reductionist, emergence is an expression of ignorance. For a hard core holist, it is a new characteristic accompanying each level of organization, an irreducible property of the elements participating in this level compared to the properties of the elements at a lower (more "elementary") level. Now, as we saw in this book (Sec. 2.3) complex systems research allows one to formulate emergence in a quantitative fashion, through the concepts of order parameters and normal forms. Going through the machinery that leads to this result one realizes that it derives, in principle, from the properties of the constituents at the lower level; but at the same time it is not reducible to them, in the sense that it cannot be predicted without accounting for the change of scale imposed by the appearance of a new level of description. All in all, analysis should not be confused with reduction. Beyond sterile oppositions between "top down" and "bottom up" views, what counts is the fantastic diversity of forms and functions starting with the same underlying rules, without any need of extra ad hoc hypotheses. But in the course of this unfolding of complexity new concepts and new regularities impose themselves, as new levels of description are revealed. Rather than try naively to reduce them to elementary physical events in the name of unification, it is operationally -and, in the end, ontologically as well- much more meaningful to regard them as "laws" governing these particular levels and interpret, predict or control on this basis the system at hand. This remains fully compatible with the search of universalities -a seemingly reductionist attitude- but, as stressed repeatedly, the universalities in question are of a new kind, better adapted to the complex realities encountered in our experience.

6.8.4 Facts, analogies and metaphors

We have seen that complexity research is offering solutions to a variety of problems of concern, many of them beyond the realm of physical sciences. In this process of "transfer of knowledge" analogy-based approaches are often necessary, since one may be lacking quantitative information on the underlying laws. They should be undertaken with care, particularly when political or ethical issues are involved, as it often happens when managing complex realities. The general idea is that, in many instances, feedback loops and networks are present, inducing a phenomenology reminiscent of that observed in natural sciences. Methods from complex systems research may then play a constructive role in analyzing the system at hand and predicting possible future trends. What is more, one may interfere with the system by introducing appropriate interactions to initiate an evolutionary process towards a desired solution. Consider, for instance, management. The hierarchy in traditional management is often leading to inertia. A bottom-up approach where small self-regulating units are combined into a large network offers a valuable alternative. Rapid communication within each unit and interaction between units leads to a higher order pattern that is not centrally managed and will be able to adapt promptly and flexibly.

Closely related to the foregoing is the issue of decision making. Decision making implies knowledge of the present and a capability to predict the future. But when the object of decision making is a complex system both of these conditions need to be redefined. As stressed in Secs 5.5, 6.3 and 6.4 in connection, in particular, with weather prediction and climatic change, the kind of information considered to be relevant for an ordinary system may well be incomplete or even irrelevant when dealing with a complex system. Furthermore, owing to the multiplicity of the states available and to the sensitivity in the initial conditions and in the parameters, several outcomes are possible, some of them with comparable likelihoods. Faced with such at first sight unexpected realities the process of decision making needs to be reinvented. First, by realizing that the horizon over which the impact of the decision can be assessed may well be quite limited; second, by exploring different evolution scenarios in view of the uncertainties inherent in the knowledge of the present and in the rules governing the future course of events, much like the weather forecaster resorts nowadays to ensemble predictions before issuing a reliable forecast. There is a tendency to bypass this admittedly uncomfortable way to proceed by appealing more and more to simulation of the situation of interest, whereby all variables and parameters that may enter in the description are incorporated off hand and evolved under plausible sets of rules. The availability of massive and highly performant computational facilities makes this option seem attractive and practicable. But it is important to realize that such an approach is also subjected to exactly the same uncertainties pointed out before: a complex system does not all of a sudden become simple if we augment the number of its variables and parameters to match the capacities of the most powerful existing computer! Furthermore, even if one had access to the perfect model and to the full initial data, lack of qualitative understanding would compromize the possibility to manage the results adequately and to have an a priori idea on what is really happening. The opposite attitude, namely, declare that decision making is futile in view of the uncertainties present is, naturally, equally unacceptable and contrary to the spirit in which complexity is to be understood.

Complexity is a natural phenomenon subjected to laws valid when certain premises are fulfilled. There is an inherent danger to transpose its principles and to exploit them in the wrong context. Science is not fully protected against aberrations. Fortunately it has developed defense mechanisms of its own, allowing it to react successfully to all sorts of confrontations and to re-emerge in a renovated and ever more vigorous form.

Color plates

Fig. 1.3.

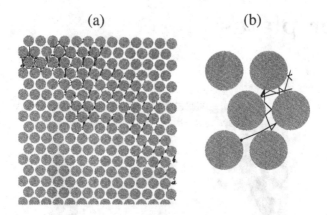

Fig. 6.2.

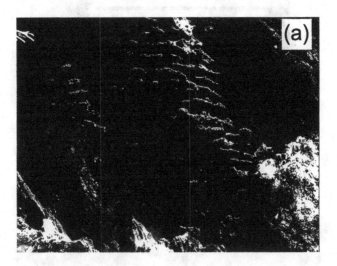

Fig. 6.13a.

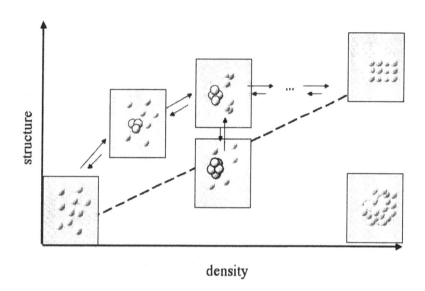

Fig. 6.16.

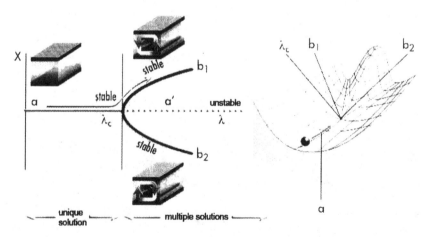

Fig. 6.17.

Suggestions for further reading

Textbooks and surveys on complexity, self-organization and emergence

General formulation and physical aspects

- R. Badii and A. Politi, *Complexity*, Cambridge University Press, Cambridge (1997).
- *Complex Systems*, Nature insight **410**, 241-284 (2001).
- K. Kaneko and I. Tsuda, *Complex systems : Chaos and beyond*, Springer, Berlin (2001).
- *La Complexité*, Pour la Science $N°$ 314, Paris (2003).
- G. Nicolis and I. Prigogine, *Exploring Complexity*, Freeman, New York (1989).
- S. Wolfram, *A new kind of science*, Wolfram Media Inc., Champaign, Illinois (2002).

Life sciences

- B. Goodwin, *How the leopard changed its spots : the evolution of complexity*, Touchstone, New York (1994).
- S. Kauffman, *The origins of order*, Oxford University Press, New York (1993).
- S. Kauffman, *At home in the universe*, Penguin, London (1995).

Multi-agent systems, sociology and finance

- P.W. Anderson, K.J. Arrow and D. Pines (eds), *The economy as an evolving complex system*, Addison-Wesley, Redwood City, CA (1988).
- B. Arthur, S. Durlaug and D. Lane (eds), *The economy as an evolving complex system II*, Addison-Wesley, New York (1997).
- L. Kocarev and G. Vattay (eds), *Complex dynamics in communication networks*, Springer, Berlin (2005).

- F. Schweitzer (ed.), *Self-organization of complex structures*, Gordon and Breach, New York (1997).

Self-organization, emergence

- C. Biebricher, G. Nicolis and P. Schuster, *Self-organization in the physico-chemical and life sciences*, EUR 16546 Report, European Commission (1995).
- S. Camazine, J.-L. Deneubourg, N. Franks, J. Sneyd, G. Theraulaz and E. Bonabeau, *Self-organization in biological systems*, Princeton University Press, Princeton (2001).
- H. Haken, *Synergetics*, Springer, Berlin (1977).
- J.H. Holland, *Emergence : from chaos to order*, Helix Books, New York (1998).
- *L'énigme de l'émergence*, Science et Avenir Hors-Série, $N°$ 143, Paris (2005).
- G. Nicolis and I. Prigogine, *Self-organization in nonequilibrium systems*, Wiley, New York (1977).
- P. Ortoleva, *Geochemical self-organization*, Oxford University Press, New York (1994).

Chapter 1

Textbooks and surveys on nonlinear science, stability, bifurcations and chaos

- P. Bergé, Y. Pomeau and C. Vidal, *L'ordre dans le chaos*, Herman, Paris (1984).
- J. Guchenheimer and Ph. Holmes, *Nonlinear oscillations, dynamical systems and bifurcations of vector fields*, Springer, Berlin (1983).
- G. Nicolis, *Introduction to nonlinear science*, Cambridge University Press, Cambridge (1995).
- H.G. Schuster, *Deterministic chaos*, VCH Verlag, Weinheim (1988).
- A. Scott (ed.), *Encyclopedia of nonlinear science*, Taylor and Francis, New York (2005).
- T. Tel and M. Gruiz, *Chaotic dynamics*, Cambridge University Press, Cambridge (2006).

Rayleigh-Bénard convection, chemical instabilities and related phenomena

- S. Chandrasekhar, *Hydrodynamic and hydromagnetic stability*, Oxford University Press, Oxford (1961).

- I. Epstein and J. Pojman, *An introduction to nonlinear chemical dynamics*, Oxford University Press, Oxford (1998).
- P. Mandel (ed.), *Quantum optics*, Physics Reports **219**, 77-350 (1992).
- P. Manneville, *Dissipative structures and weak turbulence*, Academic, Boston (1990).
- A. Nepomnyashchy, M.G. Velarde and P. Colinet, Interfacial phenomena and convection, Chapman and Hall/ CRC, Boca Raton (2002).
- C. Normand, Y. Pomeau and M.G. Velarde, Convective instability : A physicist's approach, Rev. Mod. Phys. **49**, 581-624 (1977).
- S. Scott, *Oscillations, waves, and chaos in chemical kinetics*, Oxford University Press, Oxford (1994).
- H. Swinney and J. Gollub, *Hydrodynamic instabilities*, Springer, Berlin (1984).
- D. Walgraef and N. Ghoniem (eds), *Patterns, defects and materials instabilities*, Kluwer Academic, Dordrecht (1990).

Atmospheric and climatic variability

- M. Ghil, R. Benzi and G. Parisi (eds), *Turbulence and predictability in geophysical fluid dynamics and climate dynamics*, North Holland, Amsterdam (1985).
- E.N. Lorenz, Irregularity: a fundamental property of the atmosphere, Tellus **36**, 98-110 (1984).
- C. Nicolis, Le climat peut-il basculer? La Recherche **232**, 584-587 (1991).
- C. Nicolis and G. Nicolis (eds), *Irreversible phenomena and dynamical systems analysis in geosciences*, Reidel, Dordrecht (1987).
- C. Nicolis and G. Nicolis, Chaos in dissipative systems : Understanding atmospheric physics, Adv. Chem. Phys. **91**, 511-570 (1995).

Food recruitment in ants and related phenomena

- R. Beckers, J.-L. Deneubourg and S. Goss, Trail laying behavior during food recruitment in the ant *Lasius niger*, Insectes sociaux **39**, 59-72 (1992).
- R. Beckers, J.-L. Deneubourg and S. Goss, Trails and U-turns in the selection of a path by the ant *Lasius niger*, J. Theor. Biol. **159**, 397-415 (1992).
- B. Hölldobler and E.O. Wilson, *The ants*, Springer, Berlin (1991).
- S.C. Nicolis and J.-L. Deneubourg, Emerging patterns and food recruitment in ants : an analytical study, J. Theor. Biol. **198**, 575-592 (1999).
- G.F. Oster and E.O. Wilson, *Caste and ecology in the social insects*, Princeton University Press, Princeton (1978).

Human systems

- P.M. Allen, Self-organization and evolution in urban systems, in *Cities and regions as nonlinear decision systems*, AAAS selected Symposia, **77**, Westview Press, Boulder (1984).
- L. Bachelier, *Théorie de la spéculation*, J. Gabay, Paris (1995).
- M. Batty, *Understanding cities with cellular automata, agent-based models, and fractals*, MIT Press, Cambridge, MA (2005).
- J.-P. Bouchard and M. Potters, *Theory of financial risks and derivative pricing*, Cambridge University Press, Cambridge (2000).
- W. Brock, D. Hsieh and B. Le Baron, *Nonlinear dynamics, chaos, and instability: Statistical theory and economic evidence*, MIT Press, Cambridge MA (1991).
- *Dynamic modelling in economics and finance*, Chaos, Solitons and Fractals **29**, 515-802 (2006).
- D. Helbing, Traffic and related self-driven many-particle systems, Rev. Mod. Phys. **73**, 1067-1141 (2001).
- H. Hurst, Long-term storage capacity of reservoirs, Trans. Am. Soc. Civ. Eng. **116**, 770-808 (1951).
- B. Mandelbrot, *Fractales, hasard et finance*, Flammarion, Paris (1997).
- E. Montroll and W. Badger, *Introduction to quantitative aspects of social phenomena*, Gordon and Breach, New York (1974).
- A. Orléan, Les désordres boursiers, La Recherche **232**, 668-672 (1991).
- P.Y. Oudeyer, *Self-organization in the evolution of speech*, Oxford University Press, New York (2006).
- R. Shiller, Do stock prices move too much to be justified by subsequent changes in dividends? Amer. Econ. Rev. **71**, 421-436 (1981).
- D. Sornette, *Why stock markets crash: Critical events in complex financial systems*, Princeton University Press, Princeton (2003).
- G. Zipf, *Human behavior and the principle of least effort*, Addison-Wesley, Cambridge MA (1949).

Chapter 2

Stable and unstable manifolds, Lyapunov exponents, Smale horseshoe, fractals

- J. Feder, *Fractals*, Plenum, New York (1988).
- C. Grebogi, E. Ott and J. Yorke, Chaos, strange attractors and fractal basin boundaries in nonlinear dynamics, Science **238**, 632-638 (1987).

- B. Mandelbrot, *Fractals : form, chance, and dimension*, Freeman, San Francisco (1977).
- E. Ott, *Chaos in dynamical systems*, Cambridge University Press, Cambridge (1993).
- S. Smale, *The mathematics of time*, Springer, Berlin (1980).

Non-integrable conservative systems

- V. Arnol'd, *Les méthodes mathématiques de la mécanique classique*, Mir, Moscow (1976).
- A. Lichtenberg and M. Lieberman, *Regular and stochastic motion*, Springer, New York (1983).
- J. Moser, *Stable and random motions in dynamical systems*, Princeton University Press, Princeton (1973).
- H. Poincaré, *Les méthodes nouvelles de la mécanique céleste*, Gauthier-Villars, Paris (1899). Reprinted in Vol. I to III by Dover Publ., New York (1957).

Dissipative systems, hydrodynamic modes, irreversible thermodynamics

- S. De Groot and P. Mazur, *Nonequilibrium thermodynamics*, North Holland, Amsterdam (1961).
- P. Glansdorff and I. Prigogine, *Thermodynamic theory of structure, stability and fluctuations*, Wiley, London (1971).
- L. Landau and E. Lifshitz, *Fluid mechanics*, Pergamon, Oxford (1959).
- G. Nicolis, Dissipative systems, Rep. Prog. Phys. **49**, 873-949 (1986).
- H. Öttinger, Beyond equilibrium thermodynamics, Wiley, Hoboken, N.J. (2005).

Qualitative analysis, bifurcations, normal forms, structural stability

- R. Abraham and C. Shaw, *The geometry of behavior*, Parts 1 to 4, Aerial Press, Santa Cruz (1983).
- A. Andronov, A. Vit and C. Khaikin, *Theory of oscillators*, Pergamon, Oxford (1966).
- V. Arnol'd, *Geometrical methods in the theory of ordinary differential equation*, Springer, Berlin (1983).
- J. Carr, *Applications of center manifold theory*, Springer, New York (1981).

- S. Chow, C. Li and D. Wang, *Normal forms and bifurcation of planar vector fields* Cambridge University Press, Cambridge (1994).
- O. Gurel and O. Rössler (eds), *Bifurcation theory and applications in scientific disciplines*, Ann. New York Acad. Sci. **316**, 1-708 (1979).
- Y. Ilyashenko and C. Rousseau (eds), *Normal forms, bifurcations and finiteness problems in differential equations*, Springer, Berlin (2004).
- Yu. A. Kuznetsov, *Elements of applied bifurcation theory*, Springer, Berlin (2004).
- N. Minorsky, *Nonlinear Oscillations*, Van Nostrand, Princeton (1962).
- L. Shilnikov, A. Shilnikov, D. Turaev and L. Chua, *Methods of qualitative theory in nonlinear dynamics*, World Scientific, Singapore (1998).
- R. Thom, *Stabilité structurelle et morphogénèse*, Benjamin, Reading MA (1972).

Asymptotic methods

- L. Cesari, *Asymptotic behavior and stability problems in ordinary differential equations*, Springer, Berlin (1963).
- W. Wasow, *Asymptotic expansions for ordinary differential equations*, Interscience, New York (1965).

Routes to chaos, strange non-chaotic attractors

- P. Collet and J.P. Eckmann, *Iterated maps in the interval as dynamical systems*, Birkhäuser, Basel (1980).
- M. Feigenbaum, Quantitative universality for a class of nonlinear transformations, J. Stat. Phys. **19**, 25-52 (1978).
- C. Grebogi, F. Ott, S. Pelikan and J. Yorke, Strange attractors that are not chaotic, Physica **13D**, 261-268 (1984).
- I. Gumowski and C. Mira, *Recurrences and discrete dynamical systems*, Springer, Berlin (1980).
- P. Manneville and Y. Pomeau, Different ways to turbulence in dissipative systems, Physica **1D**, 219-226 (1980).
- D. Ruelle and F. Takens, On the nature of turbulence, Comm. Math. Phys. **20**, 167-192 (1971).

Coupling-induced complexity

General

- V. Anishchenko, V. Astakhov, A. Neiman, T. Vadivasova and L. Schimansky-Geier, *Nonlinear dynamics of chaotic and stochastic systems*, Springer, Berlin (2002).

- K. Kaneko, Pattern dynamics in spatio temporal chaos, Physica **D34**, 1-41 (1989).
- S. Manrubia, A. Mikhailov and D. Zanette, *Emergence of dynamical order*, World Scientific, Singapore (2004).
- A. Pikovsky, M. Rosenblum and J. Kurths, *Synchronization: a universal concept in nonlinear science*, Cambridge University Press, Cambridge (2001).

Reaction-diffusion systems, Turing instability, propagating waves

- Y. Kuramoto, *Chemical oscillations, waves, and turbulence*, Springer, Berlin (1984).
- J. Murray, *Mathematical biology*, Springer, Berlin (1989).
- G. Nicolis, Nonlinear kinetics : at the crossroads of chemistry, physics and life sciences, Faraday Discuss. **120**, 1-10 (2001)).
- A. Turing, The chemical basis for morphogenesis, Phil. Trans. R. Soc. **B237**, 37-72 (1952).

Scale-free behavior in turbulence

- T. Bohr, M. Jensen, G. Paladin and A. Vulpiani, *Dynamical systems approach to turbulence*, Cambridge University Press, Cambridge (1998).
- U. Frisch, *Turbulence*, Cambridge University Press, Cambridge (1995).
- W. McComb, *The physics of fluids turbulence*, Clarendon Press, Oxford (1990).
- A.S. Monin and A. M. Yaglom, *Statistical fluid mechanics*, Vol. 1 and 2, MIT Press, Cambridge MA (1971 and 1975).
- D. Schertzer et al., Multifractal cascade dynamics and turbulent intermittency, Fractals **5**, 427-471 (1997).

Competition, growth, choice between options

- R. Beckers, J.-L. Deneubourg, S. Goss and J. Pasteels, Collective decision making through food recruitment, Insectes sociaux **37**, 258-267 (1990).
- J.R. Krebs and N.B. Davies (eds), *Behavioral ecology : an evolutionary approach*, Blackwell, Oxford (1984).
- A. Lotka, *Elements of physical biology*, Williams and Wilkins, Baltimore (1925).
- R. May, *Stability and complexity in model ecosystems*, Princeton University Press, Princeton (1973).
- P.F. Verhulst, Recherches mathématiques sur la loi d'accroissement de la population, Mem. Acad. Roy. Belg. **18**, 1-38 (1845).

- V. Volterra, *Leçons sur la théorie mathématique de la lutte pour la vie*, Gauthier-Villars, Paris (1931).

Modeling with delays

-T. Erneux, L. Larger, K. Green and D. Roose, Modeling nonlinear optics phenomena using delay differential equations, in *EQUADIFF 2003*, World Scientific, Singapore (2006).
- L. Glass , A. Beuter and D. Laroque, Time delays, oscillations and chaos in physiological control systems, Math. Biosc. **90**, 111-125 (1988).
- K. Gu, V. Kharitonov and J. Chen, *Stability of time delay systems*, Birkhäuser, Boston (2003).
- J. Hale and S. Lunel, *Introduction to functional differential equations*, Springer, New York (1993).

Chapter 3

Fluctuations

- S. Chandrasekhar, Stochastic problems in physics and astronomy, Rev. Mod. Phys. **15**, 1-87 (1943).
- L. Landau and E. Lifshitz, *Statistical physics*, Pergamon, Oxford (1958).

Probability theory, Markov processes

- W. Feller, *An Introduction to probability theory and its applications*, Wiley, New York, Vol. I (1968) and Vol. II (1971).
- C. Gardiner, *Handbook of stochastic methods*, Springer, Berlin (1983).
- I. Gihman and A.Skorohod, *The theory of stochastic processes* Vol.II, Springer, Berlin (1975).
- F. Moss et al (eds), *Noise and chaos in nonlinear dynamical systems*, Cambridge University Press, Cambridge (1990).
- T. Soong, *Random differential equations in science and engineering*, Academic, New York (1973).
- N. Van Kampen, *Stochastic processes in physics and chemistry*, North Holland, Amsterdam (1981).

Liouville and Frobenius-Perron equations, ergodicity and mixing, coarse-graining, symbolic dynamics

- T. Bedford, M. Keane and C. Series (eds), *Ergodic theory, symbolic dynamics, and hyperbolic spaces*, Oxford University Press, Oxford (1991).
- P. Grassberger and H. Kantz, Generating partitions for the dissipative Henon map, Phys. Letters **A113**, 235-238 (1985).
- B. L. Hao, *Elementary symbolic dynamics and chaos in dissipative systems*, World Scientific, Singapore (1989).
- C.S. Hsu and M.C. Kim, Method of constructing generating partitions for entropy evaluation, Phys. Rev. **A30**, 3351-3354 (1984).
- A. Lasota and M. Mackey, *Probabilistic properties of deterministic systems*, Cambridge University Press, Cambridge (1985).
- D. Mc Kernan and G. Nicolis, Generalized Markov coarse graining and spectral decompositions of chaotic piecewise linear maps, Phys. Rev. **E50**, 988-999 (1994).
- G. Nicolis and P. Gaspard, Towards a probabilistic approach to complex systems, Chaos, Solitons and Fractals **4**, 41-57 (1994).
- G. Nicolis and C. Nicolis, Master equation approach to deterministic chaos, Phys. Rev. **A38**, 427-433 (1988).
- Ya. Sinai, Markov partitions and C-diffeomorphisms, Funct. Anal. Appl. **2**, 61-82 (1968).
- Ya. Sinai, *Introduction to ergodic theory*, Princeton University Press, Princeton (1977).

Stochastic potential, variational principles

- R. Graham, Noise and extremum principles in macroscopic nonequilibrium systems, in *Synergetics and dynamical instabilities*. G. Caglioti, H. Haken and L. Lugiato (eds), North Holland, Amsterdam (1988).
- G. Hu, Lyapunov function and stationary probability distributions, Z. Phys. **B65**, 103-106 (1986).
- R. Kubo, K. Matsuo and K. Kitahara, Fluctuation and relaxation of macrovariables, J. Stat. Phys. **9**, 51-96 (1973).
- H. Lemarchand, Asymptotic solution of the master equation near a nonequilibrium transition : the stationary solutions, Physica **101A**, 518-534 (1980).
- H. Lemarchand and G. Nicolis, Stochastic analysis of symmetry-breaking bifurcations : master equation approach, J. Stat. Phys. **37**, 609-629 (1984).
- G. Nicolis and R. Lefever, Comment on the kinetic potential and the Maxwell construction in nonequilibrium chemical phase transitions, Phys. Letters **62A**, 469-471 (1977).

- G. Nicolis, Thermodynamics today, Physica **A213**, 1-7 (1995).
- M. Suzuki, Statistical mechanics of nonequilbrium systems II - extensivity, fluctuation and relaxation of Markovian macrovariables, Progr. Theor. Phys. **55**, 383-399 (1976).
- M. Suzuki, Passage from an initial unstable state to a final stable state, Adv. Chem. Phys. **46**, 195-278 (1981).

Beyond the mean-field description

Long range correlations, nonequilibrium phase transitions

- C. Gardiner, K. Mc Neil, D. Walls and I. Matheson, Correlations in stochastic theories of chemical reactions, J. Stat. Phys. **14**, 307-331 (1976).
- P. Gaspard, The correlation time of mesoscopic chemical clocks, J. Chem. Phys. **117**, 8905-8916 (2002).
- H. Lemarchand and G. Nicolis, Long range correlations and the onset of chemical instabilities, Physica **82A**, 521-542 (1976).
- M. Malek-Mansour, J.W. Turner and A. Garcia, Correlation functions for simple fluids in a finite system under nonequilibrium constraints, J. Stat. Phys. **48**, 1157-1186 (1987).
- M. Malek-Mansour, C. Van den Broeck, G. Nicolis and J.W. Turner, Asymptotic properties of Markovian master equations, Ann. Phys. **131**, 283-313 (1981).
- G. Nicolis and M. Malek-Mansour, Nonequilibrium phase transitions and chemical reactions, Progr. Theor. Phys. Suppl. **64**, 249-268 (1978).
- G. Nicolis and M. Malek-Mansour, Onset of spatial correlations in nonequilibrium systems : a master equation description, Phys. Rev. **A29**, 2845-2853 (1984).
- H. Spohn, Long range correlations for stochastic lattice gases in a nonequilibrium steady state, J. Phys. **A16**, 4275-4291 (1983).
- D. Walgraef, G. Dewel and P. Borckmans, Nonequilibrium phase transitions and chemical instabilities, Adv. Chem. Phys. **49**, 311-355 (1982).
- V. Zaitsev and M. Shliomis, Hydrodynamic fluctuations in the vicinity of a convection threshold, Sov. Phys. JETP **32**, 866-869 (1971).

Closing the hierarchy of moment equations

- G. Eyink and J. Restrepo, Most probable histories for nonlinear dynamics: tracking climatic transitions, J. Stat. Phys. **101**, 459-472 (2000).
- K. Kürten and G. Nicolis, Moment equations and closure schemes in chaotic dynamics, J. Phys. **A31**, 7331-7340 (1998).

- C. Nicolis and G. Nicolis, Closing the hierarchy of moment equations in nonlinear dynamical systems, Phys. Rev. **E58**, 4391-4400 (1998).

Anomalous kinetics in low-dimensional and disordered systems

For a general presentation and a review of literature see :
- D. ben-Avraham and S. Havlin, *Diffusion and reactions in fractals and disordered systems*, Cambridge University Press, Cambridge (2000).
- A. Mikhailov and A. Loskutov, *Foundations of synergetics II* , Springer, Berlin (1996).

Complex kinetics :

- S. Prakash and G. Nicolis, Dynamics of the Schlögl models on lattices of low spatial dimension, J. Stat. Phys. **86**, 1289-1311 (1997).
- A. Provata, G. Nicolis and F. Baras, Oscillatory dynamics in low-dimensional supports: a lattice Lotka-Volterra model, J. Chem. Phys. **110**, 8361-8368 (1999).

Transitions between states, stochastic resonance

- R. Benzi, G. Parisi, A. Sutera and A. Vulpiani, A theory of stochastic resonance in climatic change, Tellus **34**, 10-16 (1982).
- L. Gammaitoni, P. Hänggi, P. Jung and F. Marchesoni, Stochastic resonance, Rev. Mod. Phys. **70**, 223-287 (1998).
- P. Hänggi and P. Talkner, Reaction rate theory : fifty years after Kramers, Rev. Mod. Phys. **62**, 251-341 (1990).
- B. Matkowsky and Z. Schuss, The exit problem for randomly perturbed dynamical systems, SIAM J. Appl. Math. **33**, 365-382 (1977).
- F. Moss, A. Bulsara and M. Shlesinger (eds), *Stochastic resonance in physics and biology*, J. Stat. Phys. **70**, 1-512 (1993).
- C. Nicolis, Stochastic aspects of climatic transitions - Response to a periodic forcing, Tellus **34**, 1-9 (1982).

Simulating complex systems

Stochastic and Monte Carlo simulations

- N. Bouleau and D. Lépingle, *Numerical methods for stochastic processes*, Wiley, New York (1994).
- D. Gillespie, *Markov Processes*, Academic, New York (1992).
- M. Newman and G. Barkema, *Monte Carlo methods in statistical physics*,

Clarendon Press, Oxford (1999).
- R. Ziff, E. Gulari and Y. Barshad, Kinetic phase transitions in an irreversible surface-reaction model, Phys. Rev. Letters **56**, 2553-2556 (1986).

Microscopic simulations

- F. Baras and M. Malek-Mansour, Microscopic simulations of chemical instabilities, Adv. Chem. Phys. **100**, 393-474 (1997).
- W. Hoover, *Computational statistical mechanics*, Elsevier, Amsterdam (1991).
- M. Mareschal, Microscopic simulations of complex flows, Adv. Chem. Phys. **100**, 317-392 (1997).

Cellular automata, multi-agent simulations, games

- H. Bierman and L. Fernandez, *Game theory with economic applications*, Addison-Wesley, New York (1998).
- E. Bonabeau, M. Dorigo and G. Theraulaz, Swarn intelligence, Oxford University Press, Oxford (1999).
- J. Maynard Smith, *Evolution and the theory of games*, Cambridge University Press, Cambridge (1982).
- T. Toffoli and N. Margolus, *Cellular automata machines* M.I.T. Press, Cambridge MA (1987).
- S. Wolfram, *Cellular automata and complexity*, Reading MA (1994).

Disordered systems and complexity

- K. Binder, Spin glasses: experimental facts, theoretical concepts and open questions, Rev. Mod. Phys. **58**, 801-976 (1986).
- W. Fontana, W. Schnabl and P. Schuster, Physical aspects of evolutionary optimization and adaptation, Phys. Rev. **A40**, 3301-3321 (1989).
- A. Hartmann and H. Rieger (eds), *New optimization algorithms in physics*, Wiley, New York (2004).
- S. Kauffman and S. Levin, Towards a general theory of adaptive walks on rugged landscapes, J. Theor. Biol. **128**, 11-45 (1987).
- G. Parisi, *Statistical field theory*, Addison-Wesley, Redwood City CA (1988).

Suggestions for Further Reading

Chapter 4

Fundamentals of information theory

- A. Khinchine, *Mathematical foundations of information theory*, Dover, New York (1957).
- R. Landauer, Information is physical, Physics today, May issue, p. 23-29 (1991).
- J.S. Nicolis, *Dynamics of hierarchical systems*, Springer, Berlin (1986).
- J.R. Pierce, *An introduction to information theory : symbols, signals, and noise*, Dover, New York (1957).
- C. Shannon and W. Weaver, *The mathematical theory of communication*, University of Illinois Press, Urbana (1949).

Information entropy across bifurcation, information entropy balance, information versus thermodynamic entropy

- D. Daems and G. Nicolis, Entropy production and phase space volume contraction, Phys. Rev. **E59**, 4000-4006 (1999).
- H. Haken, Information and information gain close to nonequilibrium phase transitions. Numerical results, Z. Phys. **B62**, 255-259 (1986).
- J.L. Luo, C. Van den Broeck and G. Nicolis, Stability criteria and fluctuations around nonequilibrium states, Z. Phys. **B56**, 165-170 (1984).
- G. Nicolis and V. Altares, Physics of nonequilibrium systems, in *Synergetics and dynamical instabilities*, G. Caglioti, H. Haken and L. Lugiato (eds), North Holland, Amsterdam (1988).
- G. Nicolis and D. Daems, Nonequilibrium thermodynamics of dynamical systems, J. Phys. Chem. **100**, 19187-19191 (1996).
- G. Nicolis and C. Nicolis, What can we learn from thermodynamics on stochastic and chaotic dynamics ? in *Stochastic and chaotic dynamics in the lakes*, D. Broomhead, E. Luchinskaya, P. Mc Clintock and I. Mullin (eds), American Institute of Physics (2000).

Block entropies, scaling rules, selection

- W. Ebeling and G. Nicolis, Word frequency and entropy of symbolic sequences: a dynamical perspective, Chaos, Solitons and Fractals **2**, 635-650 (1992).
- W. Ebeling, T. Pöschel and K. Albrecht, Entropy, transinformation and word distribution of information - carrying sequences, Bifurcation and Chaos

5, 51-61 (1995).
- P. Gaspard and X.J. Wang, Sporadicity : between periodic and chaotic dynamical behaviors, Proc. Nat. Acad. Sci. U.S.A. **85**, 4591-4595 (1988).
- P. Gaspard and X.J. Wang, Noise, chaos, and (ϵ, τ) entropy per unit time, Phys. Reports **235**, 291-345 (1993).
- S. Narasimhan, J. Nathan and K. Murthy, Can coarse-graining induce long-range correlations in a symbolic sequence? Europhys. Letters **69**, 22-28 (2005).
- G. Nicolis, C. Nicolis and J.S. Nicolis, Chaotic dynamics, Markov partitions, and Zipf's law, J. Stat. Phys. **54**, 915-924 (1989).
- G. Nicolis, G. Subba Rao, J. Subba Rao and C. Nicolis, Generation of spatially asymmetric, information rich structures in far from equilibrium systems, in *Structure, coherence and chaos in dynamical systems*, L. Christiansen and R. Parmentier (eds), Manchester University Press, Manchester (1989).
- J.S. Nicolis, *Chaos and information processing*, World Scientific, Singapore (1991).
- J.S. Nicolis, Superselection rules modulating complexity: an overview, Chaos, Solitons and Fractals **24**, 59-63 (2005).
- O. Rössler, Continuous chaos - four prototype equations, Ann. N.Y. Acad. Sci. **316**, 376-392 (1979).
- P. Schuster, Taming combinatorial explosions, Proc. Nat. Acad. Sci. U.S.A. **97**, 7678-7680 (2000).

Large deviations, fluctuation theorems

- D. Andrieux and P. Gaspard, Fluctuation theorem for transport in mesoscopic systems, J. Stat. Mech. P01011, 1-23 (2006).
- P. Gaspard, Time-reversed dynamical entropy and irreversibility in Markovian random processes, J. Stat. Phys. **117**, 599-615 (2004).
- P. Gaspard, Fluctuation theorem for nonequilibrium reactions, J. Chem. Phys. **120**, 8898-8905 (2004).
- J. Kurchan, Fluctuation theorem for stochastic dynamics, J. Phys. **A31**, 3719-3730 (1998).
- J. Lebowitz and H. Spohn, A Gallavotti-Cohen type symmetry in the large deviation functional for stochastic dynamics, J. Stat. Phys. **95**, 333-365 (1999).
- C. Maes, The fluctuation theorem as a Gibbs property, J. Stat. Phys. **95** 367-392 (1999).
- C. Maes and K. Netocny, Time-reversal and entropy, J. Stat. Phys. **110**, 269-310 (2003).

Non-Shannonian information measures, thermodynamic formalism, dimensions and entropies of chaotic systems

- C. Beck and F. Schlögl, *Thermodynamics of chaotic systems*, Cambridge University Press, Cambridge (1993).
- T. Bohr and T. Tel, The thermodynamics of fractals, in *Directions in chaos 2*, B.L. Hao (ed.), World Scientific, Singapore (1988).
- B. Eckardt and D. Yao, Local Lyapunov exponents in chaotic systems, Physica **D65**, 100-108 (1993).
- M. Gell-Mann and C. Tsallis (eds), *Nonextensive entropy - interdisciplinary applications*, Oxford University Press, New York (2004).
- P. Grassberger, Temporal scaling at Feigenbaum points and non-extensive thermodynamics, Phys. Rev. Letters **95**, 140601 1-4 (2005).
- P. Grassberger, R. Badii and A. Politi, Scaling laws for invariant measures on hyperbolic and non-hyperbolic attractors, J. Stat. Phys. **51**, 135-178 (1988).
- J.S. Nicolis, G. Mayer-Kress and G. Haubs, Non-uniform chaotic dynamics with implications to information processing, Z. Naturf. **38a**, 1157-1169 (1983).
- D. Ruelle, *Thermodynamic formalism*, Addison-Wesley, Reading MA (1978).

Complexity versus dissipation

- C. Nicolis, Entropy production and dynamical complexity in a low-order atmospheric model, Q.J.R. Meteorol. Soc. **125**, 1859-1878 (1999).
- G. Nicolis and C. Nicolis, Thermodynamic dissipation versus dynamical complexity, J. Chem. Phys. **110**, 8889-8898 (1999).

Algorithmic complexity, theory of computation

- C. Bennett, The thermodynamics of computation - a review, Int. J. Theor. Phys. **21**, 905-940 (1982).
- C. Bennett, Dissipation, information, computational complexity and the definition of organization, in *Emerging Syntheses in Science*, D. Pines (ed.), Addison-Wesley, Reading MA (1988).
- G. Chaitin, A theory of program size formally identical to information theory, J. Assoc. Comput. Mach. **22**, 329-340 (1975).
- G. Chaitin, *Algorithmic information theory*, Cambridge University Press, Cambridge (1987).
- G. Chaitin, MetaMath? The quest for Omega, http://arxiv.org/abs/math.-HO/0404335.

- R. Gandy and C. Yates (eds), *The collected works of A.M. Turing : mathematical logic*, North Holland, Amsterdam (2001).
- P. Grassberger, Toward a quantitative theory of self-generated complexity, Int. J. Theor. Phys. **25**, 907-938 (1986).
- A. Kolmogorov, Logical basis for information theory and probability theory, IEEE Transactions on information theory IT-14, 662-664 (1968).
- R. Landauer, Dissipation and noise immunity in computation and communication, Nature **335**, 779-784 (1988).
- M. Li and P. Vitanyi, *An introduction to Kolmogorov complexity and its applications*, Springer, New York (1997).
- S. Lloyd and H. Pagels, Complexity as thermodynamic depth, Ann. Phys. **188**, 186-213 (1988).
- R. Solomonov, A formal theory of inductive inference 1 and 2, Information and Control **7**, 1-22 and 224-254 (1964).
- W. Zurek (ed.), *Complexity, entropy and the physics of information*, Addison-Wesley, New York (1989).

Chapter 5

Classical data analysis

- R.L. Bras and I. Rodrigues-Iturbe, *Random functions and hydrology*, Dover, New York (1985).
- A. Papoulis, *Probability, random variables, and stochastic processes*, Mc Graw-Hill, New York (1984).
- M. Priestley, *Spectral analysis and time series I and II*, Academic, London (1981).
- D.S. Wilks, *Statistical methods in the atmospheric sciences*, Elsevier, Amsterdam (2006).

Nonlinear data analysis

- N. Abraham et al (eds), *Quantitative measures of dynamical complexity in nonlinear systems*, Plenum, New York (1989).
- D. Broomhead and G. King, Extracting qualitative dynamics from experimental data, Physica **D20**, 217-236 (1986).
- J. Farmer and J. Sidorowich, Predicting chaotic time series, Phys. Rev. Letters **59**, 845-848 (1987).
- P. Grassberger and I. Procaccia, Characterization of strange attractors, Phys. Rev. Letters **50**, 346-349 (1983).

- H. Kantz and T. Schreiber, *Nonlinear time series analysis*, Cambridge University Press, Cambridge (1997).
- G. Mayer-Kress (ed.), *Dimensions and entropies in chaotic systems*, Springer, Berlin (1986).
- C. Nicolis, W. Ebeling and C. Baraldi, Markov processes, dynamical entropies and the statistical prediction of mesoscale weather regimes, Tellus **49A**, 108-118 (1997).
- G. Sugihara and R. May, Nonlinear forecasting as a way of distinguishing chaos from measurement error in time series, Nature **344**, 734-741 (1990).
- F. Takens, Detecting strange attractors in turbulence, in Lecture Notes in Math. Vol. 898, Springer, New York (1981).
- A.A. Tsonis and J. B. Elsner, Nonlinear prediction as a way of distinguishing chaos from random fractal sequences, Nature **358**, 217-220 (1992).

Monitoring, data assimilation

- E. Kalnay, *Atmospheric Modeling, data assimilation and predictability*, Cambridge University Press, Cambridge (2003).
- E.N. Lorenz and K. Emanuel, Optimal sites for supplementary weather observations : Simulation with a small model, J. Atmos. Sci. **55**, 399-414 (1998).
- S. Lovejoy, D. Schertzer and P. Ladoy, Fractal characterization of inhomogeneous geophysical measuring networks, Nature **319**, 43-44 (1986).
- C. Nicolis, Optimizing the global observational network : a dynamical approach, J. Appl. Meteorol. **32**, 1751-1759 (1993).
- R. Vose and M. Menne, A method to determine station density requirements for climate observing networks, J. Climate **17**, 2961-2971 (2004).

Growth of initial errors

- J. Lacarra and O. Talagrand, Short range evolution of small perturbations in a barotropic model, Tellus **40A**, 81-95 (1988).
- E.N. Lorenz, Atmospheric predictability as revealed by naturally occurring analogs, J. Atmos. Sci. **26**, 636-646 (1969).
- C. Nicolis, Probabilistic aspects of error growth in atmospheric dynamics, Q.J.R. Meteorol. Soc. **118**, 553-568 (1992).
- C. Nicolis and G. Nicolis, Dynamics of error growth in unstable systems, Phys. Rev. **A43**, 5720-5723 (1991).
- C. Nicolis and G. Nicolis, Finite time behavior of small errors in deterministic chaos and Lyapunov exponents, Bifurcation and Chaos **3**, 1339-1342 (1993).

- C. Nicolis, G. Nicolis and Q. Wang, Sensitivity to initial conditions in spatially distributed systems, Bifurcation and Chaos **2**, 263-269 (1992).
- C. Nicolis, S. Vannitsem and J.F. Royer, Short range predictability of the atmosphere : Mechanisms for superexponential error growth, Q.J.R. Meteorol. Soc. **121**, 705-722 (1995).
- *Predictability* (Vol. I), European Centre for Medium-Range Weather Forecasts, Reading U.K. (1995).
- A. Trevisan, Impact of transient error growth on global average predictability measures, J. Atmos. Sci. **50**, 1016-1028 (1993).

Model errors

- C. Nicolis, Dynamics of model error: some generic features, J. Atmos. Sci. **60**, 2208-2218 (2003).
- C. Nicolis, Dynamics of model error: the role of unresolved scales revisited, J. Atmos. Sci. **61**, 1740-1753 (2004).
- C. Nicolis, Can error source terms in forecasting models be represented as Gaussian Markov noises ? Q.J.R. Meteorol. Soc. **131**, 2151-2170 (2005).
- C. Nicolis, Dynamics of model error: the role of boundary conditions, J. Atmos. Sci. **64**, 204-215 (2007).
- T.N. Palmer, A nonlinear dynamical perspective on model error: A proposal for non-local stochastic-dynamic parametrization in weather and climate prediction models, Q.J.R. Meteorol. Soc. **127**, 279-304 (2001).
- H. Rabitz, M. Kramer and D. Dacol, Sensitivity analysis in chemical kinetics, Ann. Rev. Phys. Chem. **34**, 419-461 (1983).
- R. Tomovic and M. Vukovratovic, *General sensitivity theory*, Elsevier, New York (1972).
- J.J. Tribbia and D.P. Baumhefner, The reliability of improvements in deterministic short-range forecasts in the presence of initial state and modeling deficiencies, Mon. Weather Rev. **116**, 2276-2288 (1988).

Recurrence

- V. Balakrishnan, G. Nicolis and C. Nicolis, Recurrence time statistics in chaotic dynamics. I. Discrete time maps, J. Stat. Phys. **86**, 191-212 (1997).
- V. Balakrishnan and M. Theunissen, Power-law tails and limit laws for recurrence time distributions, Stoch. Dyn. **1**, 339-343 (2001).
- V. Balakrishnan, C. Nicolis and G. Nicolis, Recurrence time statistics in chaotic dynamics : multiple recurrences in intermittent chaos, Stoch. Dyn. **1**, 345-359 (2001).
- W. Feller, Fluctuation theory of recurrent events, Trans. Bull. Amer.

Math. Soc. **67**, 98-119 (1949).
- M. Hirata, Poisson law for axiom A diffeomorphisms, Ergod. Th. Dynam. Sys. **13**, 533-556 (1993).
- M. Kac, *Probability and related topics in physical science*, Interscience, New York (1959).
- Y.C. Lai, Y. Nagai and C. Grebogi, Characterization of the natural measure by unstable periodic orbits in chaotic attractors, Phys. Rev. Letters **79**, 649-652 (1997).
- D. Mayer, On the distribution of recurrence times in nonlinear systems, Letters Math. Phys. **16**, 139-143 (1988).
- C. Nicolis, Atmospheric analogs and recurrence time statistics : toward a dynamical formulation, J. Atmos. Sci. **55**, 465-475 (1998).
- G. Nicolis and C. Nicolis, Recurrence, in A. Scott (ed.), *Encyclopedia of nonlinear science*, Taylor and Francis, New York (2005).
- H. Poincaré, Sur le problème des trois corps et les équations de la dynamique, Acta Matematica **13**, 1-270 (1890).
- N. Slater, Gaps and steps for the sequence $n\theta$ mod 1, Proc. Camb. Phil. Soc. **63**, 1115-1123 (1967).
- M. Theunissen, C. Nicolis and G. Nicolis, Recurrence times in quasi-periodic motion: statistical properties, role of cell size, parameter dependence, J. Stat. Phys. **94**, 437-467 (1999).

Extreme events

- V. Balakrishnan, C. Nicolis and G. Nicolis, Extreme value distributions in chaotic dynamics, J. Stat. Phys. **80**, 307-336 (1995).
- P. Embrechts, C. Klüppelberg and Th. Mikosch, *Modeling extreme events*, Springer, Berlin (1999).
- W. Feller, The asymptotic distribution of the range of sums of independent random variables, Ann. Math. Statist. **22**, 427-432 (1951).
- E. Gumbel, *Statistics of extremes*, Columbia University Press, New York (1958).
- V. Klemes, The Hurst phenomenon: a puzzle?, Water Resour. Res. **10**, 675-688 (1974).
- M. Leadbetter and H. Rootzen, Extremal theory for stochastic processes, Ann. Prob. **16**, 431-478 (1988).
- B. Mandelbrot, Noah, Joseph, and operational hydrology, Water Resour. Res. **4**, 909-918 (1968).
- C. Nicolis, V. Balakrishnan and G. Nicolis, Extreme events in deterministic dynamical systems, Phys. Rev. Letters **97**, 210602 1-4 (2006).
- T. Yalcinkaya and Y.C. Lai, Phase characterization of chaos, Phys. Rev.

Letters **79**, 3885-3888 (1997).
- C. Yeung, M. Rao and R. Desai, Temporal fluctuations and Hurst exponents of Goldstone and massive modes, Phys. Rev. Letters **73**, 1813-1816 (1994).

Chapter 6

Boltzmann equation, paradoxes and their resolution

- S. Brush, *Kinetic theory*, Vols I and II, Pergamon, London (1966).
- M. Kac, *Probability and related topics in physical science*, Interscience, New York (1959).
- G. Nicolis, Nonequilibrium statistical mechanics, in A. Scott (ed.), in *Encyclopedia of nonlinear science*, Taylor and Francis, New York (2005).
- G. Uhlenbeck and G. Ford, *Lectures in statistical mechanics*, American Mathematical Society, Providence (1963).

Generalized kinetic theories

- R. Balescu, *Equilibrium and nonequilibrium statistical mechanics*, Wiley, New York (1975).
- N. Bogolubov, Problems of a dynamical theory in statistical physics, in *Studies in statistical mechanics* Vol. I, J. De Boer and G. Uhlenbeck (eds), North Holland, Amsterdam (1962).
- I. Prigogine, *Nonequilibrium statistical mechanics*, Interscience, New York (1962).
- L. Van Hove, Quantum mechanical perturbations giving rise to a statistical transport equation, Physica **21**, 517-540 (1955).

Microscopic chaos and nonequilibrium statistical mechanics

- J.R. Dorfman, *An introduction to chaos in nonequilibrium statistical mechanics*, Cambridge University Press, Cambridge (1999).
- P. Gaspard, *Chaos, scattering and statistical mechanics*, Cambridge University Press, Cambridge (1998).
- W. Hoover, *Time reversibility, computer simulation, and chaos*, World Scientific, Singapore (1999).
- N. Krylov, *Works on the foundations of statistical physics*, Princeton University Press, Princeton (1979).
- O. Penrose, *Foundations of statistical mechanics*, Pergamon, Oxford (1970).

- I. Prigogine, *From being to becoming*, Freeman, San Francisco (1980).
- I. Prigogine and S.A. Rice (eds), *Resonances, instability and irreversibility*, Adv. Chem. Phys. **99**, 1-456 (1997).
- D. Ruelle, Smooth dynamics and new theoretical ideas in nonequilibrium statistical mechanics, J. Stat. Phys. **95**, 393-468 (1999).

Statistical mechanics of dynamical systems

- V. Arnold and A. Avez, *Ergodic problems of classical mechanics*, Benjamin, New York (1968).
- R. Bowen, Equilibrium states and the ergodic theory of Anosov diffeomorphisms, in *Lecture notes in mathematics* Vol. 470, Springer, Berlin (1975).
- R. Bowen, Invariant measures for Markov maps in the interval, Commun. Math. Phys. **69**, 1-17 (1979).
- M. Courbage and G. Nicolis, Markov evolution and H-theorem under finite coarse-graining in conservative dynamical systems, Europhys. Letters **11**, 1-6 (1990).
- J.P. Eckmann and D. Ruelle, Ergodic theory of chaos and strange attractors, Rev. Mod. Phys. **57**, 617-656 (1985).
- P. Gaspard, r-adic one-dimensional maps and the Euler summation formula, J. Phys. **A25**, L483-L485 (1992).
- P. Gaspard, Diffusion in uniformly hyperbolic one-dimensional maps and Appell polynomials, Phys. Letters **A168**, 13-17 (1992).
- P. Gaspard, G. Nicolis, A. Provata and S. Tasaki, Spectral signature of the pitchfork bifurcation : Liouville equation approach, Phys. Rev. **E51**, 74-94 (1995).
- P. Gaspard and S. Tasaki, Liouvillian dynamics of the Hopf bifurcation, Phys. Rev. **E64**, 056232 1-17 (2001).
- H. Hasegawa and W. Saphir, Unitarity and irreversibility in chaotic systems, Phys. Rev. **A46**, 7401-7423 (1992).
- G. Nicolis, S. Martinez and E. Tirapegui, Finite coarse-graining and Chapman-Kolmogorov equation in conservative dynamical systems, Chaos, Solitons and Fractals **1**, 25-37 (1991).
- D. Ruelle, Resonances of chaotic dynamical systems, Phys. Rev. Letters **56**, 405-407 (1986).
- D. Ruelle, *Chaotic evolution and strange attractors*, Cambridge University Press (1989).
- Ya. Sinai (ed.), *Dynamical systems, ergodic theory and applications*, Springer, Berlin (2000).
- T. Tel, P. Gaspard and G. Nicolis (eds), *Chaos and irreversibility* Focus issue, Chaos **8**, 309-461 (1998).

Small systems

- C. Bustamante, J. Liphardt and F. Ritort, The nonequilibrium thermodynamics of small systems, Physics today, July issue 43-48 (2005).

Fluctuation theorems

- B. Cleuren, C. Van den Broeck and R. Kawai, Fluctuation theorem for the effusion of an ideal gas, Phys. Rev. **E74**, 021117 1-10 (2006).
- D. Evans, E. Cohen and G. Morriss, Probability of second law violations in shearing steady flows, Phys. Rev. Letters **71**, 2401-2404 (1993).
- G. Gallavotti and E. Cohen, Dynamical ensembles in nonequilibrium statistical mechanics, Phys. Rev. Letters **74**, 2694-2697 (1995).
- O. Narajan and A. Dhar, Reexamination of experimental tests of the fluctuation theorem, J. Phys. **A37**, 63-76 (2004).

Statistical properties of irreversible work

- G. Crooks, Nonequilibrium measurements of free energy differences for microscopically reversible Markovian systems, J. Stat. Phys. **90**, 1481-1487 (1998).
- T. Hatano, Jarzynski equality for the transitions between nonequilibrium steady states, Phys. Rev. **E60**, R5017-R5020 (1999).
- C. Jarzynski, Nonequilibrium equality for free energy differences, Phys. Rev. Letters **78**, 2690-2693 (1997).
- J. Liphardt et al., Equilibrium information from nonequilibrium measurements in an experimental test of Jarzynski's equality, Science **296**, 1832-1835 (2002).
- T. Speck and U. Seifert, Distribution of work in isothermal nonequilibrium processes, Phys. Rev. **E70**, 066112 1-4 (2004).

Transport in a fluctuating environment

- R.D. Astumian, Thermodynamics and kinetics of a Brownian motor, Science **276**, 917-922 (1997).
- F. Jülicher, A. Ajdari and J. Prost, Modeling molecular motors, Rev. Mod. Phys. **69**, 1269-1281 (1997).
- H. Qian, Entropy production and excess entropy in a nonequilibrium steady state of single macromolecules, Phys. Rev. **E65**, 021111 1-6 (2002).
- P. Reimann, Brownian motors : noisy transport far from equilibrium, Phys. Rep. **361**, 57-265 (2002).

- C. Van den Broeck, R. Kawai and P. Meurs, From Maxwell demon to Brownian refrigerator, Phys. Rev. Letters **93**, 090601 1-4 (2004).

Atmospheric dynamics

- R. Barry and A. Perry, *Synoptic climatology: methods and applications*, Methuen, London (1973).
- J.R. Holton, *An introduction to dynamic meteorology*, Academic, New York (1972).
- P. Thompson, Uncertainty of initial state as a factor in the predictability of large-scale atmospheric flow patterns, Tellus **9**, 275-295 (1957).

Low-order models

- J. Charney, Planetary fluid dynamics, in *Dynamic Meteorology*, p. 97-351, E. Morel (ed.), Reidel, Dordrecht (1973).
- J. Charney and J. De Vore, Multiple flow equilibria in the atmosphere and blocking, J. Atmos. Sci. **36**, 1205-1216 (1979).
- J. Egger, Stochastically driven large scale circulations with multiple equilibria, J. Atmos. Sci. **38**, 2609-2618 (1981).
- E.N. Lorenz, Deterministic non-periodic flow, J. Atmos. Sci. **20**, 130-141 (1963).
- E.N. Lorenz, Attractor sets and quasi-geostrophic equilibrium, J. Atmos. Sci. **37**, 1685-1699 (1980).
- E.N. Lorenz, Low-order models of atmospheric circulation, J. Meteor. Japan **60**, 255-267 (1982).
- C. Nicolis, Irreversible thermodynamics of a simple atmospheric flow model, Bifurcation and Chaos **12**, 2557-2566 (2002).
- A. Shil'nikov, G. Nicolis and C. Nicolis, Bifurcation and predictability analysis of a low-order atmospheric circulation model, Bifurcation and Chaos **5**, 1701-1711 (1995).
- C. Sparrow, *The Lorenz equations*, Springer, Berlin (1982).
- S. Vannitsem and C. Nicolis, Predictability experiments on a simplified thermal convection model : the role of spatial scales, J. Geophys. Res. **99**, 10377-10385 (1994).

More detailed models

- M. Ehrendorfer, Predicting the uncertainty of numerical weather forecasts: a review, Meteorol. Z., N.F. **6**, 147-183 (1997).

- E. Kalnay, *Atmospheric modeling, data assimilation and predictability*, Cambridge University Press, Cambridge (2003).
- J. Marshall and F. Molteni, Toward a dynamical understanding of planetary-scale flow regimes, J. Atmos. Sci. **50**, 1792-1818 (1993).
- A.J. Simmons and A. Hollingsworth, Some aspects of the improvement in skill of numerical weather prediction, Q.J.R. Meteor., **128**, 647-677 (2002).
- S. Vannitsem and C. Nicolis, Lyapunov vectors and error growth patterns in a T21L3 quasi-geostrophic model, J. Atmos. Sci. **54**, 347-361 (1997).

Data analysis, modeling and predicting with probabilities

- R. Buizza, P.L. Houtekamer, G. Pellerin, Z. Toth and M. Wei, A comparison of the ECMWF, MSC and NCEP Global Ensemble Prediction, Mon. Wea. Rev., **133**, 1076-1097(2005).
- G. Demarée and C. Nicolis, Onset of Sahelian drought viewed as a fluctuation-induced transition, Q.J.R. Meteorol. Soc. **116**, 221-238 (1990).
- C. Keppenne and C. Nicolis, Global properties and local structure of the weather attractor over Western Europe, J. Atmos. Sci. **46**, 2356-2370 (1989).
- R. Mureau, F. Molteni and T. Palmer, Ensemble prediction using dynamically constrained perturbations, Q.J.R. Meteorol. Soc. **119**, 299-323 (1993).
- A.A. Tsonis and J.B. Elsner, Chaos, strange attractors, and weather, Bull. Amer. Meteor. Soc. **70**, 14-23 (1989).

Climate dynamics

Low-order models

- R. Benzi, G. Parisi, A. Sutera and A. Vulpiani, Stochastic resonance in climatic change, Tellus **34**, 10-16 (1982).
- C. Crafoord and E. Kallén, A note on the condition for existence of more than one steady-state solution in Budyko-Sellers type models, J. Atmos. Sci. **35**, 1123-1125 (1978).
- C. Nicolis, Stochastic aspects of climatic transitions-response to a periodic forcing, Tellus **34**, 1-9 (1982).
- C. Nicolis, Self-oscillations and predictability in climate dynamics - periodic forcing and phase locking, Tellus **36A**, 217-227 (1984).
- C. Nicolis, Long term climatic variability and chaotic dynamics, Tellus **39A**, 1-9 (1987).
- C. Nicolis and G. Nicolis, Stochastic aspects of climatic transitions - additive fluctuations, Tellus **33**, 225-234 (1981).

- B. Saltzman, A. Sutera and A. Hansen, A possible marine mechanism for internally-generated long-period climate cycle, J. Atmos. Sci. **39**, 2634-2637 (1982).

Dynamics of averaged observables

- K. Fraedrich, Estimating weather and climate predictability on attractors, J. Atmos. Sci. **44**, 722-728 (1987).
- C. Nicolis and G. Nicolis, Is there a climatic attractor? Nature **311**, 529-532 (1984).
- C. Nicolis and G. Nicolis, From short scale atmospheric variability to global climate dynamics: toward a systematic theory of averaging, J. Atmos. Sci. **52**, 1903-1913 (1995).
- B. Saltzman, Climatic system analysis, Adv. Geophys. **25**, 173-233 (1983).
- S. Vannitsem and C. Nicolis, Dynamics of fine scale variables versus averaged observables in a simplified thermal convection model, J. Geophys. Res. **100**, 16367-16375 (1995).

Climatic change in the perspective of complex systems

- A. Berger (ed.), Climatic variations and variability: facts and theories, Reidel, Dordrecht (1981).
- H. Lamb, *Climate: Present, past and future*, Pergamon Press, Oxford (1977).
- C. Nicolis, Transient climatic response to increasing CO_2 concentration: some dynamical scenarios, Tellus **40A**, 50-60 (1988).
- C. Nicolis and G. Nicolis, Passage through a barrier with a slowly increasing control parameter, Phys. Rev. **E62**, 197-203 (2000).
- C. Nicolis and G. Nicolis, Noisy limit point bifurcation with a slowly increasing parameter, Europhys. Letters **66**, 185-191 (2004).

Networks

Geometric and statistical properties

- R. Albert and A.L. Barabasi, Statistical mechanics of complex networks, Rev. Mod. Phys. **74**, 47-97 (2002).
- R. Pastor-Satorras, M. Rubi and A. Diaz-Guilerra (eds), *Statistical mechanics of complex networks*, Springer, Berlin (2003).
- C. Song, S. Havlin and H. Makse, Self-similarity of complex networks, Nature **433**, 392-395 (2005).
- G. Uhlenbeck and G. Ford, The theory of linear graphs with applications

to the theory of the virial development of the properties of gases, in *Studies in statistical mechanics* Vol. I., J. De Boer and G. Uhlenbeck (eds), North Holland, Amsterdam (1962).
- D. Watts, *Small worlds : The dynamics of networks between order and randomness*, Princeton University Press, Princeton (1999).

Dynamics on networks, evolutionary aspects

- L. Gallos and P. Argyrakis, Absence of kinetic effects in reaction-diffusion processes in scale-free networks, Phys. Rev. Letters **92**, 138301 1-4 (2004).
- A. Garcia Cantu and G. Nicolis, Toward a thermodynamic characterization of chemical reaction networks, J. Nonequil. Thermodyn. **31**, 23-46 (2006).
- A. Mikhailov and V. Calenbuhr, *From cells to societies*, Springer, Berlin (2002).
- J.D. Noh and H. Rieger, Random walks on complex networks, Phys. Rev. Letters **92**, 118701 1-4 (2004).
- G. Nicolis, A. Garcia Cantu and C. Nicolis, Dynamical aspects of interaction networks, Bifurcation and Chaos **15**, 3467-3480 (2005).
- R. Thomas (ed.), *Kinetic logic*, Springer, Berlin (1979).

Biological complexity

- R. Barbauet et al (eds), Integrative biology and complexity in natural systems, International Union of Biological sciences (2003).
- G. Nicolis and I. Prigogine, *Self-organization in nonequilibrium systems*, Wiley, New York.
- E. Schrödinger, *What is life?*, Cambridge University Press, Cambridge (1944).
- O. Solbrig and G. Nicolis (eds), Perspectives on biological complexity, International Union of Biological sciences (1991).
- *The puzzle of complex diseases*, Science **296**, 685-703.
- L. von Bertalanffy, *General systems theory*, George Braziller, New York (1969).
- C. Waddington (ed.), *Towards a theoretical biology*, Vol. I, Edinburgh University Press, Edinburgh (1968).
- *Whole-istic Biology*, Science **295**, 1661-1682 (2002).

Nonlinear dynamics, chaos and structure formation in biology

- A. Beuter, L. Glass, M. Mackey and M. Titcombe (eds), *Nonlinear dynamics in physiology and medicine*, Springer, New York (2003).

- S. Camazine, J.-L. Deneubourg, N. Franks, J. Sneyd, G. Theraulaz and E. Bonabeau, *Self-organization in biological systems*, Princeton University Press, Princeton (2001).
- L. Glass and M. Mackey, *From clocks to chaos : the rhythms of life*, Princeton University Press, Princeton (1988).
- A. Goldbeter, *Biochemical oscillations and cellular rhythms*, Cambridge University Press, Cambridge (1996).
- W. Freeman, *Neurodynamics*, Springer, Berlin (2000).
- K. Lehnertz, C. Elger, J. Arnhold and P. Grassberger (eds), *Chaos in brain?* World Scientific, Singapore (2000).
- R. May, Chaos and the dynamics of biological populations, Proc. R. Soc. Lond. **A413**, 27-44 (1987).
- J.S. Nicolis, *Chaos and information processing*, World Scientific, Singapore (1991).
- L. Segel, *Modeling dynamical phenomena in molecular and cellular biology*, Cambridge University Press, Cambridge (1984).
- R. Thomas and R. D'Ari, *Biological feedback*, CRC Press, Boca Raton FL (1990).
- A. Winfree, *The geometry of biological time*, Springer, New York (1980).
- L. Wolpert, Positional information and the spatial pattern of cellular differentiation, J. Theor. Biol. **25**, 1-47 (1969).

Networks and modularity in biology

- P. Bowers, S. Colens, D. Eisenberg and T. Yeates, Use of logic relationships to decipher protein network organization, Science **306**, 2246-2249 (2004).
- J. Han et al, Evidence for dynamically organized modularity in the yeast protein-protein interaction network, Nature **430**, 88-93 (2004).
- N. Jerne, Towards a network theory of the immune system, Ann. Immunol. Inst. Pasteur **125C**, 373-389 (1974).
- M. Kaufman, J. Urbain and R. Thomas, Towards a logical analysis of the immune response, J. Theor. Biol. **114**, 527-561 (1985).
- S. Kauffman, Metabolic stability and epigenesis in randomly constructed genetic nets, J. Theor. Biol. **22**, 437-467 (1969).
- *Networks in biology*, Science **301**, 1863-1877 (2003).
- R. Thomas, Boolean formalization of genetic control circuits, J. Theor. Biol. **42**, 565-583 (1973).
- R. Thomas, Regulatory networks seen as asynchronous automata : a logical description, J. Theor. Biol. **153**, 1-23 (1991).
- B. Vogelstein, D. Lane and A. Levine, Surfing the p53 network, Nature **408**, 307-310 (2000).

- A. Wagner, *Robustness and evolvability in living systems*, Princeton University Press, Princeton (2005).

Biosequence analysis

- Y. Almirantis and A. Provata, An evolutionary model for the origin of non-randomness, long-range order and fractality in the genome, Bio Essays **23**, 647-656 (2001).
- A. Arneodo, E. Bacry, P. Graves and J. Muzy, Characterizing long-range correlations in DNA sequences from wavelet analysis, Phys. Rev. Letters **74**, 3293-3296 (1995).
- N. Dokholyan, V. Sergey, S. Buldyrev, S. Havlin and E. Stanley, Distribution of base pair repeats in coding and noncoding DNA sequences, Phys. Rev. Letters **79**, 5182-5185 (1997).
- L. Gatlin, *Information theory and the living system*, Columbia University Press, New York (1972).
- H. Herzel and I. Grosse, Measuring correlations in symbol sequences, Physica **A216**, 518-542 (1995).
- H. Herzel, E. Trifonov, O. Weiss and I. Grosse, Interpreting correlations in biosequences, Physica **A249**, 449-459 (1998).
- W. Li and K. Kaneko, Long-range correlations and partial $1/f^\alpha$ spectrum in a noncoding DNA sequence, Europhys. Letters **17**, 655-660 (1992).
- C.K. Peng, S. Buldyrev, A. Goldberger, S. Havlin, S. Sciortino, M. Simons and E. Stanley, Long range correlations in nucleotide sequences, Nature **356**, 168-170 (1992).
- A. Provata and Y. Almirantis, Statistical dynamics of clustering in the genome structure, J. Stat. Phys. **106**, 23-56 (2002).
- A.A. Tsonis, J.B. Elsner and P.A. Tsonis, On the existence of scaling in DNA sequences, Biochem. Biophys. Res. Commun. **197**, 1288-1295 (1993).

Evolution

- W. Ebeling and R. Feistel, Theory of self-organization and evolution : the role of entropy, value and information, J. Nonequil. Thermod. **17**, 303-332 (1992).
- M. Eigen and P. Schuster, *The Hypercycle*, Springer, Berlin (1979).
- M. Eigen and W. Gardiner, Evolutionary molecular engineering based on RNA replication, Pure and Appl. Chem. **56**, 967-978 (1984).
- S. Gould, *The structure of evolutionary theory*, Harvard University Press, Cambridge MA (2002).
- S. Gould, *Wonderful life*, Norton, New York (1989).

- G. Joyce, Directed molecular evolution, Sci. Am. **267**, 48-55 (1992).
- S. Kauffman, Applied molecular evolution, J. Theor. Biol. **157**, 1-7 (1992).
- E. Mayr, *What makes biology unique?*, Cambridge University Press, Cambridge (2004).
- P. Schuster, Dynamics of molecular evolution, Physica **D22**, 27-41 (1986).
- P. Schuster, How does complexity arise in evolution?, Complexity **2**, 22-30 (1996).
- P. Stadler and P. Schuster, Mutation in autocatalytic reaction networks. An analysis based on perturbation theory, J. Math. Biol. **30**, 597-632 (1992).

Equilibrium versus nonequilibrium

- L. Landau and E. Lifshitz, *Statistical physics*, Pergamon, Oxford (1959).
- L. Landau and E. Lifshitz, *Physical kinetics*, Pergamon, Oxford (1981).

Nucleation, from classical to unconventional

- O. Galkin and P. Vekilov, Nucleation of protein crystals : critical nuclei, phase behavior, and control pathways, J. Crystal Growth **232**, 63-76 (2001).
- C. Kirschhock et al, Design and synthesis of hierarchical materials from ordered zeolitic binding units, Chem. Eur. J. **11**, 4306-4313 (2005).
- J. Lutsko and G. Nicolis, Theoretical evidence for a dense fluid precursor to crystallization, Phys. Rev. Letters **96**, 046102 1-4 (2006).
- J. Mullin, *Crystallization*, Butterworth, Oxford (1997).
- G. Nicolis, V. Basios and C. Nicolis, Pattern formation and fluctuation-induced transitions in protein crystallization, J. Chem. Phys. **120**, 7708-7719 (2004).
- G. Nicolis and C. Nicolis, Enhancement of the nucleation of protein crystals by the presence of an intermediate phase : a kinetic model, Physica **A323**, 139-154 (2003).
- P. ten Wolde and D. Frenkel, Enhancement of protein crystal nucleation by critical density fluctuations, Science **277**, 1975-1978 (1997).
- P. Vekilov, Dense liquid precursor for the nucleation of ordered solid phases from solution, Crystal growth and design **4**, 671-685 (2004).

Nanoscale patterns

- D. Carati and R. Lefever, Chemical freezing of phase separation in immiscible binary mixtures, Phys. Rev. **E56**, 3127-3136 (1997).
- J. Chanu and R. Lefever (eds), Inhomogeneous phases and pattern formation, Physica **A213**, 1-276 (1995).

- Y. De Decker, H. Marbech, M. Hinz, S. Günther, M. Kiskinova, A. Mikhailov and R. Imbihl, Promoter-induced reactive phase separation in surface reactions, Phys. Rev. Letters **92**, 198305 1-4 (2004).
- B. Huberman, Striations in chemical reactions, J. Chem. Phys. **65**, 2013-2019 (1976).
- J. Langer, Statistical theory of the decay of metastable states, Ann. Phys. **54**, 258-275 (1969).
- J. Langer, Theory of spinodal decomposition in alloys, Ann. Phys. **65**, 53-86 (1971).
- Q. Tran-Cong, J. Kawai, Y. Nishikawa and H. Jinnai, Phase separation with multiple length scales in polymer mixtures induced by autocatalytic reactions, Phys. Rev. **E60**, R1150-R1153 (1999).
- R. Yoshida, T. Takahashi, T. Yamaguchi and H. Ichijo, Self-oscillating gels, Adv. Mater. **9**, 175-178 (1997).

Self-assembly, supramolecular chemistry

- J. Atwood, J. Davies, D. Mac Nicol, F. Vögtle and J.-M. Lehn (eds), *Comprehensive supramolecular chemistry*, Pergamon, Oxford (1996).
- J.M. Lehn, *Supramolecular chemistry : concepts and perspectives*, VCH Verlag, New York (1995).
- J.M. Lehn, Toward complex matter: Supramolecular chemistry and self-organization, Proc. Nat. Acad. Sci. U.S.A **99**, 4763-4768 (2002).
- L. Lindoy and I. Atkinson, *Self-assembly in supramolecular systems*, Roy. Soc. Chemistry, Cambridge (2000).

Epistemological aspects

- A. Beckermann, H. Flohr and J. Kim (eds), *Emergence or reduction : Essays on the prospects of non-reductive physicalism*, De Gruyter, New York (1992).
- H. Driesch, *Philosophie des Organischen*, Quelle und Meyer, Leipzig (1899).
- J. Kim, *Mind in a physical world*, MIT Press, Cambridge MA (1998).
- A. Koestler and J. Smythies (eds), *Beyond reductionism*, Hutchinson, London (1969).
- *L'univers est-il sans histoire?* Sciences et Avenir Hors-Série $N°$ 146, Paris (2006).
- I. Prigogine, *The end of certainty*, The Free Press, New York (1997).
- L. Sève, *Emergence, complexité et dialectique*, Odile Jacob, Paris (2005).
- A. Toynbee, *A study of history*, Oxford University Press, Oxford (1972).

Index

adaptive behavior 3, 16, 95, 100, 195, 247, 265, 276
additivity 102, 120, 123
adenosine diphosphate (ADP) 217
    triphosphate (ATP) 216, 217, 254
agents 95
algorithmic complexity 124–128
aminoacid sequence 98, 99
Anderson, Paul viii
ants (see social insects)
arrow of time 34, 37, 195, 279
atmospheric dynamics 217
    detailed models of 222–223
    low order models of 218–221
atmospheric variability 11
attraction basin 28
attractor 28, 34, 38, 49, 72, 74, 81, 88, 121, 127, 132, 160, 234, 284
    chaotic 33, 108, 111, 114, 172, 179, 224, 233
    fixed point 28, 31
    periodic (limit cycle) 31, 57, 90, 230
    quasi-periodic (torus) 31
    reconstruction from time series 139–143
    strange non-chaotic 33, 108
autonomous system 27
auto-regressive models 135
averaging 65, 160, 170, 182, 222, 230–233

balance equation 37, 39, 53, 116, 201, 219, 221
Barabasi, Albert-Laszlo 239
Bernoulli processor 106
bifurcation 5–7, 9, 10, 17, 19, 20, 41–44, 48, 54, 59, 64, 86, 104, 126, 166, 250, 277, 280, 281
    and long range correlations 78
    cascade 44, 51, 220
    global 48, 49
    limit point 46
    pitchfork 44
    probabilistic analog of 79–81
    sensitivity of 6
    transcritical 46
biology 55, 62, 83, 95, 98, 148, 247, 248
    large scale structures in 251–253
    networks in 253–260
    nonlinear behavior in 249–250
bistability 85
Bogolubov, Nikolai 202
Boltzmann equation 198
Boltzmann, Ludwig 196
bottom-up mechanism 11, 17, 20, 248
Bowen, Rufus 208
Brownian motion, fractional 192, 194

Cantor set 32, 33
carrying capacity 60

catastrophe theory 47, 98
causality 277
cellular automata 94
central limit theorem 79, 80, 88, 262
Chaitin, Gregory 125
chance 277
chaos 5, 7, 49, 54, 106
   fully developed 108, 178, 187, 240
   in continuous time systems 111, 219, 220
   intermittent 110, 178, 187, 241
   microscopic 204
Charney, Julius 221
chemical kinetics 37, 54
choice 6, 17–20, 60, 61, 96, 101, 120
climatic change 233
   variability 13
climate dynamics 226
   low order models of 227–230
clustering 63, 236
coarse graining 75, 103, 105, 108, 110, 173, 178, 204
coding sequences 255, 260
Cohen, Eddie 210
collective behavior 15–17, 95, 208, 251, 254, 282
   properties 36
   variables 59
collision operator 198, 203
complexity measures 24, 103, 127, 128
computation 95, 124
conditional probability 105
conservative systems 27
constraint 4, 8, 117, 207, 209–211, 215, 219, 264, 268, 273-277, 280
control parameter 4–6, 8, 25, 27, 33, 36, 42–45, 47, 48, 59, 84, 88, 111, 212, 214, 278, 280, 282
critical value of 5, 42
convection, thermal 9
convexity 102, 201
cooperativity 4, 17, 19, 26, 54, 61, 95, 249, 253, 254, 284
correlation function 78, 133
correlations 22, 52, 77, 78, 107, 108, 110–113, 120, 122, 144, 193, 194, 202, 203, 262
coupling 54–57, 81
   global 62, 63, 244
   nearest neighbor 56, 63, 244
covariance matrix 133
crystallization 270
cusp map 190, 241

Darwin, Charles 263
data analysis 223
   linear 132–135
   nonlinear 139–143
data assimilation 14, 157
decision making 95, 285
delay embedding theorems 142
delays 61
deoxyribonucleic acid (DNA) 77, 99, 100, 247, 255, 261
detailed balance 116, 209, 264
determinism 23, 277, 279, 281
deterministic description 25
diffusion 72, 196, 199, 207
   generalized 272, 273
dimension
   Euclidean 33
   fractal 51, 60, 121
   from time series data 139-143
   generalized 33
   Renyi 121, 122
Diophantine inequality 176
disorder 4, 96

Index

dissipative structure 9
dissipative systems 27
  and irreversibility 37
  attractors in 28
distribution
  binomial 79, 200, 237
  Fréchet 183, 184
  Gaussian 79, 200
  Gumbel 183, 184
  Pareto 22, 34, 184
  Poisson 79, 117, 183, 237
  Weibull 184
Dow Jones index 20, 30
Driesch, Hans 283
dynamical systems 25

Ehrenfest, Paul and Tatiana 200
Eigen, Manfred 264
eigenvalue 31, 81, 207, 237, 238
embryonic development 56, 250
emergence 3, 41, 44, 284
energy balance 128, 219, 227, 228
ensemble 224, 280, 282
  prediction 160, 224, 282, 286
entropy 39
  and second law of thermodynamics 40
  balance 40
  block 105
  flux 40
  information (Shannon) 102
  Kolmogorov-Sinai 105
  production 40, 207, 210
  Renyi 120–123, 129
enzyme 217, 254, 255, 266
  allosteric 61, 249
  catalysis 254
equilibrium 4, 8, 39, 40, 47, 73, 78, 81, 97, 116, 119, 120, 129, 199, 201, 211, 214–216, 268, 275, 280

Erdös, Paul, 237
ergodicity 73
ergodic states 74
error growth 14, 160, 164, 170
escape rate 205, 207
evolution 4–8, 39, 40, 48, 52, 104, 125, 279
  biological 16, 24, 60, 99, 100
  criteria 35, 128
  law 5, 25, 28, 61, 70, 72, 126, 127, 148, 160, 167, 280, 282
  molecular 263–267
excitability 54, 257
exit time 144-147
extreme events 180
  and the Hurst phenomenon 191–194
  signatures of deterministic dynamics in 185–191
  statistical approach to 182–185

Farey tree 177
feedback 4, 227, 238, 240, 246, 254
  loop (circuit) 254, 256, 258, 278, 284, 285
Feller, William 191
fitness 263, 264, 267
  function 97, 99
  landscape 98, 265
fixed point (see attractor)
fluctuations 5, 41, 65, 68, 69, 83, 84, 104
  critical behavior of 78, 80
fluctuation theorems 119, 210
fluctuation-dissipation theorem 214
flux 53
  entropy 40, 116
  probability 71
Fokker-Planck equation 72, 76, 82, 103, 117, 170, 191, 215, 225, 226, 278

food searching 15, 16
forecasting 62, 136, 147, 158–160,
   218, 222, 227, 233
   ensemble 160, 225
   statistical 135, 146, 226
fractal 32, 33
   law 34
Frobenius-Perron equation 66, 70,
   122, 173
Frobenius theorem 74
frustration 97
fundamental matrix 30, 167

Gallavotti, Giovanni 210
games 95
Gaussian process 135, 136
generalized potential 75, 85
genetic code 260
genome organization 260–262
genotype 99
geopotential height 135, 151
Gibbs, Josiah Willard 270
glaciation 13, 227, 229
global warming 234
Gödel, Kurt 125
Gould, Stephen 265
graphs 236, 244
   random 237, 238

halting problem 125
Hamilton's equations 34, 93
heat conduction 8
historicity 6, 279
holism 283
horseshoe 32, 34
human genome 261
   systems 2, 17, 19–23, 59, 171,
   209, 280
Hurst, Harold 15
Hurst exponent 5, 22, 191–194
Huygens, Christiaan 62

H-theorem 199–204, 207
hydrodynamic modes 36
hydrodynamics 37, 199
hypercycles 266
hypersymbols 111

immune system 28
information 19, 62, 77, 95, 99, 101,
   107, 120, 127, 249, 255, 260
   and chaotic dynamics 51, 143
   and selection 106
   and symbolic dynamics 103, 111
   as an emergent property 104
   entropies 102, 105, 201
   entropy balance 116
initial conditions 25, 218, 277
instability 5, 14, 29, 41, 42, 52, 55,
   56, 59, 64, 72, 74, 76, 85,
   108, 122, 128, 129, 153, 161,
   179, 212, 214, 233, 246, 272–
   274, 281
   Rayleigh-Bénard 17, 54, 93, 125,
   219
integrability conditions 48
integrable systems 35
intermittency 110
invariant sets 27, 28
irreversibility (see also arrow of time)
   37, 195, 196, 200, 207, 208,
   211, 213, 279
   Markovian processes and 74, 200
   microscopic chaos and 204, 207
irreversible work 213
isolated systems 35, 39, 40, 108,
   119, 120, 279

Jacobian matrix 42, 167, 258
Jarzynski, Christopher 213

Kalman filter 137
Kauffman, Stuart 258
kinesin 216

kinetic potential 46–48, 57, 129, 271
  theory 196, 202, 203
Kolmogorov-Chaitin complexity (see algorithmic complexity)
Kolmogorov-Sinai entropy (see entropy)

lac operon 255, 258
Landau, Lev 44
Langevin equation 71, 214, 215, 225
large deviations 117, 209
law of large numbers 79
Lehn, Jean-Marie 275
limit cycle (see attractor)
linear regression 136
linear stability analysis 30, 41–43
Liouville equation 66, 70, 75, 122, 173, 202–205
logistic map 51, 61, 63, 89, 150, 162, 187, 240, 250
  like curve 160
Lorentz gas 205
Lorenz models 219, 220
Lyapunov exponents 31, 51, 60, 108, 114, 117, 122, 149, 161, 162, 167–169, 206, 207, 220, 233, 243
  local 122, 123, 129, 162–164, 242, 249

macroscopic description 34, 36–38, 41, 45, 46, 53, 67, 71, 78, 82
Mandelbrot, Benoît 22
manifolds, invariant 27, 28, 31, 48, 51, 139, 174, 176, 179, 284
  stable and unstable 28–30, 204
Markov process 70, 200–202, 204, 226
  partitions 76, 108, 119, 173, 204
master equation 70–72, 74–76, 91, 103, 106, 116, 117, 200–202

Maxwell-Boltzmann distribution 199
Maxwell, James Clerk 196
Mayr, Ernst 248
mean-field description 38, 82–84, 91
memory 249
mesoscopic description 67, 70, 71, 74, 77, 82, 88, 91, 103, 106, 108, 115, 119, 199, 214
metastability 270
microscopic description 34, 35, 93, 119, 204, 206, 207, 211, 213
mixing 74, 81, 161
molecular biology 261, 267, 283
  dynamics simulation 93
motors 216
moments 78, 88, 89, 123, 170, 231, 262
  and the closure problem 82–84
monitoring 3, 122, 131, 138, 151, 153, 223, 235
Monte Carlo simulation 91, 92, 94, 147, 156, 226, 253
morphogenesis 250
multifractal 34, 51, 123, 157
mutation (see also replication-mutation dynamics) 60, 67, 99, 246, 248, 256, 264, 266

nanoscale systems 65, 83, 209, 276
  patterns 272–274
Navier-Stokes equations 37
networks 235
  and logical analysis 258
  dynamical origin of 239
  dynamics on 244
  genetic 255
  geometrical and statistical properties of 236
  immune 256
  metabolic 254
  neural 257

Newton's equations 34, 93
Mewtonian paradigm 1–3, 35, 248, 283
Newton, Isaac 1
noise 72, 88, 89, 98, 99, 101, 104, 127, 131, 135, 137, 139, 142, 150, 151, 170, 191, 192, 214, 215, 228, 250
non-coding sequences 255, 261
nonequilibrium states (systems) 39–41, 52, 78, 107, 108, 119, 127, 207, 208–211, 215, 216, 248, 264, 268–274, 276
  phase transitions 81, 83
non-integrability 204
nonlinearity 4, 5, 35, 46, 72, 128, 284
  in biology 249, 254, 258
nonlinear prediction 148–151
normal form 41, 44–46, 48, 56, 59, 60, 75, 85, 129, 143, 285
nucleation 85, 268, 270
nucleic acids 99

Onsager matrix 39
open systems 40, 41, 248
order 4, 10, 20, 40, 41, 51, 52, 268, 270, 272
order parameter 43, 44, 46–48, 56, 59, 60, 75, 85, 129, 270, 271, 273, 285
Ornstein-Uhlenbeck process 192
oscillations 54, 249, 257, 258, 266

partition (see coarse graining, symbolic dynamics)
Pearson correlation 134, 150
period doubling 51
perturbation 5, 27–31, 41–43, 85, 133, 164, 166, 203, 225, 246, 247, 273, 282

p53 network 256
phase dynamics 43, 56, 57, 63, 169, 194, 203, 220, 230
phase space 26–29, 33, 38, 48, 65, 73, 81, 117, 140–142, 148, 202–204, 207, 208, 210, 218, 220, 224, 280
phase transitions 4, 44, 46, 51, 53, 81, 208, 268, 272
  freezing of by nonequilibrium constraints 273
phenomenological coefficients 39
  relations 40
phenotype 99
Plato 277
Poincaré, Henri 49, 172
Poincaré map 49
  recurrence time 173, 200
Poisson process 138
polynucleotide chain 260
potential barrier 87, 229
power spectrum 57, 133, 139, 233
predictability 159, 222, 230
prediction (see also forecasting) 4, 8, 11, 14, 20, 61, 128, 136, 164, 170, 171, 180, 184, 263, 277, 278, 282
Prigogine, Ilya 41, 202, 204
principal component analysis 134
probabilistic description 23, 51, 64, 65, 68, 72, 77, 82, 88, 172, 278, 282
probability distribution (see also distribution) 22, 65, 68, 78, 120, 132, 164, 196, 201, 278
  cumulative 186
probability flux 71
proteins 98, 247, 254, 255, 261, 270, 271

quantum description 35, 67

Index   327

quasi-periodicity 74, 141, 176, 191
quasi steady-state approximation 55

random force 71, 72, 216, 218
   variable 68, 80, 91, 175, 178, 183, 191
   walk 52
randomness 3, 9, 23, 51, 68, 122, 129, 279, 282
   deterministic 278
range 15, 191
ratchets 215, 216
rate limiting step 55
Rayleigh-Bénard convection 8, 10, 21, 44, 125
reaction-diffusion equations 37
recruitment 15–17. 59, 94, 251
recurrence 171–175, 201
   and dynamical complexity 176–180
   (Zermelo) paradox 199
reductionism 283
reference state 28, 41
regulation 249, 254–256, 258
renewal equation 175
renormalization group 51, 81
Rényi, Alfréd 237
replication 108, 217, 256, 263, 265–267
   -mutation dynamics 263–267
residence time 226
resonance 82, 203
reversibility, microscopic 201, 215, 279
   (Loschmidt) paradox 199
ribonucleic acid (RNA) 77, 99, 255, 265
Rössler attractor 111
Ruelle, David 208, 282

sampling (see also monitoring) 138

scale free 15, 58
scaling 33, 34, 46, 51, 52, 57–59, 109, 143, 145, 150, 174, 181, 187
Schrödinger, Erwin 41, 100
selection 106–115, 263
   natural 99
self-organization 4, 20, 94, 249, 253, 264, 267, 274–276
self-similarity 52, 99
sensitivity 88, 131, 231, 262, 281, 286
   to the initial conditions 7, 31, 50, 114, 140, 205, 221, 223, 277
   to the parameters 6, 14, 166, 277
sequence (see also time series) 77, 96, 98, 99, 103, 104, 106, 117, 124, 125, 144, 180, 260–262, 264, 267
   space 99, 264, 265
Shannon, Claude 102
Sinai, Yakov 208
small systems 65, 83, 208–217
social insects 15–19, 251–253
sociology 94–96, 236, 239
spin glasses 96
sporadic systems 110
SRB measures 208
stability 5, 6, 9, 28–31, 72, 74, 246, 258, 278
   asymptotic 29, 34, 37, 74, 127, 196, 284
   marginal 42
   structural 48, 166, 247
stationarity 13, 73
steady states 28
   multiple 54, 83, 85, 87, 221, 249, 257, 258, 266
stochastic processes 72, 108, 118,

119, 121, 137, 146, 175, 179, 233
differential equation 191, 228
matrix 70, 74, 201
resonance 88, 229
Stosszahlansatz 198, 199, 202
supramolecular chemistry 274
switching (see also transitions between states) 13, 49, 88, 100, 278
symbolic dynamics 76, 77, 89, 103, 108
and time series analysis 143–148
symmetry breaking 10, 44, 56, 81, 250, 275
synchronization 57, 63, 244, 246, 257

Takens, Floris 142
tangent space 30, 163
tent map 187, 240, 241
thermodynamic description 38–41
limit 83, 116, 208
potential 39
thermodynamics, second law of 40, 74, 199, 201, 202, 211–213, 279
thermostating 93, 117, 207, 210
Thomas, René 246, 258
time reversal 34, 195, 196, 200, 207, 210
time series (see also sequence) 5, 92, 135–139, 140–143, 151, 233
Toynbee, Arnold 280

trajectory (see phase space)
transcription 217, 255, 261
transition probability 70, 74
transitions between states 4, 70, 84–88, 103, 148, 180, 201, 202
translation 217, 255, 261
transport 2, 37, 53, 54–59, 61, 199, 204, 207, 236, 239, 248, 251
anomalous 211
in a fluctuating environment 214–217
turbulence 10, 57–59, 126
Turing, Alan 125
Turing instability 56

unfolding 47
universality 8, 35, 38, 39, 45–49, 52, 72–75, 128, 183, 203

Van Hove, Léon 202
variational principles 35, 46, 75
Verhulst, Pierre François 60

wave fronts 57
in atmospheric dynamics 11, 55, 220
in biology 250, 265, 266
weather regimes 143, 172
block entropy analysis of 145
exit time distributions of 144
Wiener process 191, 192
Wolfram, Stephen 95

Zipf, George 22
Zipf-Mandelbrot law 22, 23, 114, 115